Contents

Preface to the Third Edition

The dramatic advances in computing power made over the last twenty years have combined with theoretical progress to make modern time series analysis a very different subject (as symbolised by dropping the hyphen used in earlier editions). Thus, the third edition has turned out to be quite distinct from its predecessors, apart from the classical elements retained in the opening chapters. Nevertheless, the book's role as an applied companion volume to *Kendall's Advanced Theory Of Statistics* is preserved, and cross-references are given where necessary. The notation is generally similar, but some differences will be noticed, such as the change of signs on the moving average coefficients.

The central part of the book develops univariate models in the time domain and includes an extensive discussion on forecasting. Time spent on leave at the London School of Economics led to interesting discussions on structural modelling with Andrew Harvey and others. As a result, I find myself turning increasingly to the structural approach: a development of which I think MGK would have approved.

The chapter on the frequency domain has been extended and updated; this is self-contained and could be read straight after Chapter 5 or even omitted by those seeking only a discussion in the time domain. Later chapters cover the extension to models containing explanatory variables and the multivariate case. The book ends with a brief review of other topics.

Exercises have been added to all but the first and last chapters to make the book more useful as a course text, as well as to help the reader's understanding of the material. A data appendix is provided for those wishing to try model building for themselves, surely the ultimate path to understanding. Another appendix gives a brief guide to computer programs; the different formats of some of the diagrams reflect the use of these several packages in performing the analyses.

Acknowledgements

I should like to thank Ildiko Schall for typing and retyping the manuscript. Also, I should like to thank David Reilly for his permission to reproduce

Figure 7.7. Finally, I should like to thank the London School of Economics for their hospitality during my sabbatical leave and several colleagues there, especially Andrew Harvey, for interesting discussions on time series.

Keith Ord
State College, Pennsylvania, USA
New Year's Eve, 1989

Revisions to the third edition

I am grateful to a number of colleagues who pointed out errors in the original third edition, particularly Bruce Bowerman, Carla Inclan and Richard Martin. Doubtless there are still some mistakes, but the number has been considerably reduced thanks to their efforts

Keith Ord
November 17th, 1992

1

General ideas

1.1 Time is perhaps the most mysterious concept in a mysterious universe. But its esoteric nature, fortunately, will not concern us in this book. For us, time is what it was to Isaac Newton, a smoothly flowing stream bearing the phenomenal world along at a uniform pace. We can delimit points of time with ease and measure the intervals between them with great accuracy. In fact, although errors and random perturbations are frequent and important for many of the variables with which we shall be concerned, only exceptionally shall we have to consider errors of measurement in time itself. Such difficulties as arise in measuring time intervals are mostly man-made (e.g. the way in which the Christian world fixes Christmas Day but allows Easter Sunday to vary over wide limits).

1.2 From the earliest times man has measured the passage of time with candles, clepsydras or clocks, has constructed calendars, sometimes with remarkable accuracy, and has recorded the progress of his race in the form of annals. The study of time series as a science in itself, however, is of quite recent origin. Recording events on a chart whose horizontal axis is marked with equal intervals to represent equal spaces of time must have occurred a thousand years ago; for example, the early monkish chants recorded on the eleven-line musical stave are a form of time series. In this connection Fig. 1.1 (from Funkhauser, 1936) is of some interest as the earliest diagram known in the Western world containing the essential concepts of a time graph. It dates from the tenth, possibly eleventh century, and forms parts of a manuscript consisting of a commentary of Macrobius on Cicero's *In Somnium Scipionis*. The graph was apparently meant to represent a plot of the inclinations of the planetary orbits as a function of time. The zone of the zodiac is given on a plane with the horizontal (time) axis divided into thirty parts and the ordinate representing the width of the zodiacal belt. The precise astronomical significance need not detain us; the point is that, even at this early stage, the time abscissa and the variable ordinate were in use, albeit in a crude and limited way.

1.3 Notwithstanding the invention of coordinate geometry by Descartes, the pictorial representation of time series was a late development. As late as 1879,

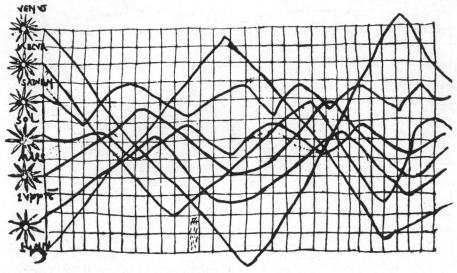

Fig. 1.1   Graph of an early time series

Stanley Jevons, whose book *The Principles of Science* was by no means intended for schoolboys, felt it necessary to devote some space to the use of graph paper. Possibly the first (and certainly one of the earliest) writers to display time charts in the modern way was William Playfair, one of his diagrams being reproduced in Fig. 1.2; the diagram was published in 1821. Playfair, the brother of the mathematician known to geometers as the author of Playfair's axiom on parallel lines, made a number of claims to priority in diagrammatic presentation which, whether justified or not, at least demonstrated the general lack of awareness of such procedures.

1.4   In the nineteenth century, theoretical statistics was not the unified subject it has since become. Work in the physical sciences was largely independent of work in economics or sociology, and at that time the ideas of physics were entirely deterministic; that is, a phenomenon tracked through time was imagined as behaving completely under deterministic laws. Any imperfection, any failure of theory to correspond with fact, was either dealt with by modifying the theory in a deterministic direction (as, for example, in the discovery of Neptune) or attributed to errors of observation. During the latter part of the nineteenth century, attempts were made to apply the methods which had been so successful in the physical sciences to the biological and behavioural sciences. The deterministic approach was adopted, together with the rest of the mathematical apparatus which had been developed. At that point the modern theory of statistics began with the realization that, although individuals might not behave deterministically, aggregates of individuals were themselves subject to laws which could often be summarized in fairly simple mathematical terms. For a historical account of these developments, see Stigler (1986).

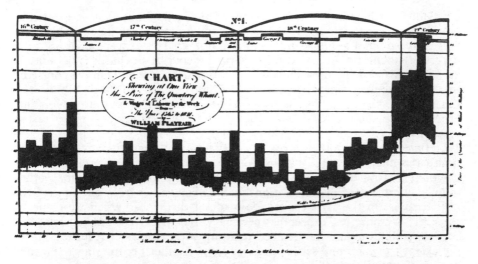

Fig. 1.2 Chart combining a graph and a histogram: from Playfair's *A letter on our Agricultural Distress (1821)*

1.5 Time series, however, resisted this change in viewpoint longer than any other branch of statistics. Until 1925 or thereabouts, a time series was regarded as being generated deterministically; the evident departures from trends, cycles or other systematic patterns of behaviour that were observed in Nature were regarded as 'errors' analogous to errors of observation. They occurred all too frequently in some fields, but were regarded in much the same way an engineer regards 'noise', as a fortuitous series of disturbances on a systematic pattern. In particular, fluctuating phenomena such as the trade 'cycles' of the nineteenth century were subject to Fourier analysis only if they were generated by a number of harmonic terms. The failure of such models to account for much of the observed variation in such things as trade, population and epidemics, though disappointing, did not deter the search for underlying cyclical movements of a strictly harmonic kind. Indeed, the belief in their existence is not yet dead.

1.6 In 1927, Udny Yule (see his *Statistical Papers*, 1971) broke new ground with a seminal idea that underlies much of the subsequent work in time series analysis. Working on sun-spot numbers, which obviously fluctuate in a manner that cannot be due entirely to chance, Yule was struck by the irregularities in the series, both in amplitude and in the distances between successive peaks and troughs. The illustration that he used to explain his fresh approach is classical: if we have a rigid pendulum swinging under gravity through a small arc, its motion is well known to be harmonic, that is to say, it can be represented by a sine or cosine wave, and the amplitudes are constant, as are the periods of the swing. But if a small boy now pelts the pendulum irregularly with peas, the motion is disturbed. The pendulum will still swing, but with irregular amplitudes and irregular intervals. Instead of leading to behaviour in which any difference between theory and observation is attributable to an evanescent error, the peas provide a series of shocks which *are incorporated into the future motion of the system*. This concept leads us to the theory of stochastic processes, of which the theory of stochastic time series is an important part. Its usefulness will be illustrated frequently in the course of this book. Stochastic processes also encompass the study of the distribution of times between events, such as the flow of customers joining a queue. This aspect of the subject will not be discussed further here, and the interested reader is referred to Cox and Miller (1968).

1.7 In contrast to other areas of statistics, the characteristic feature of time-series analysis is that the observations occur in temporal order, which is not quite as trite as it sounds. The implication is that we shall, among other things, be interested in the relationship between values from one time to the next; that is, in the *serial correlations* along the series. When we come to consider several series, it becomes necessary to consider not only correlations between series, but also the serial correlations within each series. Furthermore, the extent to which one series leads, or lags, another becomes an important part of statistical modelling.

1.8 The study of relative position in time leads naturally to the thought that processes may exhibit spatial, or both spatial and temporal, dependence. Such questions are beyond the scope of this book, but have been examined by several authors, notably Bennett (1979), Cliff and Ord (1981), and Ripley (1981).

Discrete time series

1.9 When a variable is defined at all points in time, we say that the time series is *continuous*. Examples include the temperature at a given location, the price of a commodity on an open market, or the position of a projectile. Other variables are created by aggregation over a period, such as rainfall, industrial production or total passenger miles recorded by an airline. Yet others are defined only at discrete points in time such as annual crop yields of harvest, monthly salaries or the majority of a political party at a general election. Sometimes we cannot choose the times at which data are recorded as, for example, with harvest yields. In other cases the choice may be limited or dictated by convention, as with the monthly publication of many government statistics. Under more controlled conditions, the recording times may be at choice; for example, regular surveys of political opinion may be carried out as often as funds allow, or medical staff may check a patients's pulse every hour. Finally, many continuous records, such as temperature and barometric pressure recorded on a rotating drum, or the alpha rhythm of the brain on an encephalograph, may be digitized; that is, the continuous record is converted into a time series reported at regular intervals. In all these cases, the set of time points is finite, and we speak of a *discrete* time series, whether the random variable being measured is continuous or discrete. All the statistical methods we shall consider relate to such discrete series; further, we shall usually assume that the data are recorded at regular points, or are aggregated over regular intervals of time. The analysis of irregularly recorded observations will be discussed briefly in Section 15.8. Nevertheless, it is useful to think of the time series as continuous in time when developing certain theoretical concepts.

1.10   In official statistics, the term 'continuity' is used in a different sense. Index numbers, such as the Index of Retail Prices, were historically based on a *fixed* set of weights, or 'basket of goods'. Over time, consumers' tastes change and new products become available so that the 'basket' must be changed, leading to a new set of weights. Each revision of the weights technically produces a discontinuity in the series, although the two segments can be spliced together at the point of change. Thus, continuity is taken to mean comparability over the period concerned. We shall not use continuity in this sense, although the notion of comparability finds more formal definition in the concept of *stationarity* introduced in Chapter 5.

1.11   There are two conflicting factors to be considered when the length of time between observations, or recording frequency, is at choice. First, for economic reasons, we do not want to take more observations than is necessary; yet, on the other hand, we do not want to miss important features of the phenomenon being studied. For example, if we are interested in seasonal variations, we should take several observations (quarterly, monthly) per year. However, if we wish to ignore seasonal variation it may be possible to take only one observation per year or to aggregate over shorter time periods. Likewise, examination of a patient's overall health may require measurement of the pulse rate once an hour, whereas the detection of possible heart irregularities will utilise the continuous scan of an electrocardiograph (ECG), perhaps digitized on intervals of one tenth of a second or less.

In time series analysis, as we shall often find, there are few rules of universal application; a great deal depends on the purpose of the study.

Calendar problems

1.12   Under experimental conditions, we can usually ensure that observations are regularly spaced; for social and economic data, however, problems may arise. To quote only the most obvious examples, the months are of different lengths and Nature failed to make the solar year an integral number of days. Further, a month may contain either four or five weekends. Movable feasts and public holidays contribute their own share of confusion, especially Easter, which may fall in either the first or second quarter of the year. Even series derived from experimental observation on the factory floor are not immune as interruptions in production may occur due to strikes, mechanical breakdowns, material shortages and even meal breaks.

1.13   A variety of methods for 'cleaning up' data are available, which we note briefly:

(a)   Many figures recorded for calendar months can be adjusted by scaling to a standard month of 30 days, e.g., multiplying the figure for February by 30/28, that for March by 30/31 and so on. The time periods for which such data are recorded remain unequal, but this is rarely a major problem. It should be noted that the total for twelve 'corrected' months will not be exactly equal to the annual figure, even if corrected for a year of 360 days.

(b) Adjustments for production and similar series may be made by using the number of working days per month.

(c) Short-term effects may sometimes be eliminated by aggregation. We may work with half-yearly periods rather than quarters to avoid the effects of a movable Easter or we may record weekly averages to avoid the effects of weekends; and so on.

(d) Data relating to *value* are problematical because of changes in the value of money. The best approach seems to be that of deflating the value series by a suitable price index.

1.14   As is evident from this discussion, such data-recording problems cannot be simply ignored. Yet much of the time series literature tends to assume these difficulties away. A notable exception is the US Bureau of the Census X-11 seasonal adjustment procedure, which we discuss in Section 4.9. The X-11 system provides a variety of adjustment procedures for trading day and other calendar effects. More recently, Bell and Hillmer (1983b) have developed a regression procedure to adjust for calendar day variations as part of the time series model building paradigm described in Chapters 5–7. This is clearly an area where further development would be desirable.

The length of a time series

1.15   When we refer to the 'length' of a time series, we tend to think of the elapsed time between the recorded start and finish. Indeed, this is appropriate when the phenomenon is recorded continuously. However, common usage in time series analysis decrees that a series is of length 60 when 60 observations have been recorded at regular intervals, whether elapsed time covers one minute, one hour or five years.

1.16   A more important point concerns the amount of information in the series as measured by the number of terms. In ordinary statistical work, we are accustomed to thinking of the amount of information in a random sample as being proportional to the size of the sample. Whether this is a correct usage of the word 'information' is arguable, but it is undoubtedly true that the variance of many of the estimates that we derive from random samples is inversely proportional to the sample size. This idea needs modification in time series analysis because successive values are not independent. A series of $2n$ values (even if it extends over twice the time) may not provide twice as much information as a series of n values. Further, if we sample a given time period more intensively by recording values at half the previous interval of observation, thereby doubling the number of observations, we do not add much to our knowledge if successive observations are highly positively correlated. The consequence is that n, the number of observations, is not a full measure of the information content. We shall see that the precision of estimates that we obtain from data involves the internal structure of the series as well as the sample size.

Some examples of time series

1.17 The statistical methods we shall develop will be illustrated by application to a variety of observed series. We now describe some of the principal series we shall examine.

Table 1.1 Annual yields per acre of barley in England and Wales from 1884 to 1939 (data from the *Agricultural Statistics*)

Year	Yield per acre (cwt)	Year	Yield per acre (cwt)	Year	Yield per acre (cwt)
1884	15.2	1903	15.1	1922	14.0
85	16.9	04	14.6	23	14.5
86	15.3	05	16.0	24	15.4
87	14.9	06	16.8	25	15.3
88	15.7	07	16.8	26	16.0
89	15.1	08	15.5	27	16.4
90	16.7	09	17.3	28	17.2
91	16.3	10	15.5	29	17.8
92	16.5	11	15.5	30	14.4
93	13.3	12	14.2	31	15.0
94	16.5	13	15.8	32	16.0
95	15.0	14	15.7	33	16.8
96	15.9	15	14.1	34	16.9
97	15.5	16	14.8	35	16.6
98	16.9	17	14.4	36	16.2
99	16.4	18	15.6	37	14.0
1900	14.9	19	13.9	38	18.1
01	14.5	20	14.7	39	17.5
02	16.6	21	14.3		

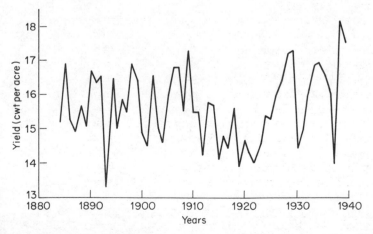

Fig. 1.3 Graph of the data of Table 1.1 (yields per acre of barley in England and Wales, 1884–1939, annual)

Table 1.2 Sheep population of England and Wales for each year from 1867 to 1939 (data from the *Agricultural Statistics*)

Year	Population (10 000)	Year	Population (10 000)	Year	Population (10 000)
1867	2 203	1892	2 119	1917	1 717
68	2 360	93	1 991	18	1 648
69	2 254	94	1 859	19	1 512
70	2 165	95	1 856	20	1 338
71	2 024	96	1 924	21	1 383
72	2 078	97	1 892	22	1 344
73	2 214	98	1 916	23	1 384
74	2 292	99	1 968	24	1 484
75	2 207	1900	1 928	25	1 597
76	2 119	01	1 898	26	1 686
77	2 119	02	1 850	27	1 707
78	2 137	03	1 841	28	1 640
79	2 132	04	1 824	29	1 611
80	1 955	05	1 823	30	1 632
81	1 785	06	1 843	31	1 775
82	1 747	07	1 880	32	1 850
83	1 818	08	1 968	33	1 809
84	1 909	09	2 029	34	1 653
85	1 958	10	1 996	35	1 648
86	1 892	11	1 933	36	1 665
87	1 919	12	1 805	37	1 627
88	1 853	13	1 713	38	1 791
89	1 868	14	1 726	39	1 797
90	1 991	15	1 752		
91	2 111	16	1 795		

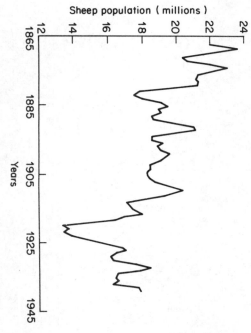

Fig. 1.4 Graph of data in Table 1.2 (sheep population)

Table 1.1 gives the yields of barley in England and Wales for the 56 years 1884–1939. Table 1.2 gives the sheep population of England and Wales for the 73 years 1867–1939. Table 1.3 gives the miles flown by British airlines for the 96 months January 1963 to December 1970. Table 1.4 gives the immigration into the United States for the 143 years 1820–1962. Table 1.5 gives the number of births of babies, according to the hour at which they were born, for certain US hospitals. Finally, Table 1.6 gives for 1960–71 the quarterly average index of share prices on the London exchange as compiled by the *Financial Times*. The six series are plotted in Figs. 1.3–1.8. The selected series are typical of those which arise in practice; other series are presented in Appendix A at the end of the book, so that the reader may attempt to analyse some different examples.

1.18 Barley yields, by definition, occur only once each year. The sheep population, although continuously in existence, is observed only once a year at

Table 1.3 UK airlines: aircraft miles flown, by month (thousands)

	1963	1964	1965	1966	1967	1968	1969	1970
Jan.	6 827	7 269	8 350	8 186	8 334	8 639	9 491	10 840
Feb.	6 178	6 775	7 829	7 444	7 899	8 772	8 919	10 436
Mar.	7 084	7 819	8 829	8 484	9 994	10 894	11 607	13 589
Apr.	8 162	8 371	9 948	9 864	10 078	10 455	8 852	13 402
May	8 462	9 069	10 638	10 252	10 801	11 179	12 537	13 103
June	9 644	10 248	11 253	12 282	12 950	10 588	14 759	14 933
July	10 466	11 030	11 424	11 637	12 222	10 794	13 667	14 147
Aug.	10 748	10 882	11 391	11 577	12 246	12 770	13 731	14 057
Sept.	9 963	10 333	10 665	12 417	13 281	13 812	15 110	16 234
Oct.	8 194	9 109	9 396	9 637	10 366	10 857	12 185	12 389
Nov.	6 848	7 685	7 775	8 094	8 730	9 290	10 645	11 595
Dec.	7 027	7 602	7 933	9 280	9 614	10 925	12 161	12 772

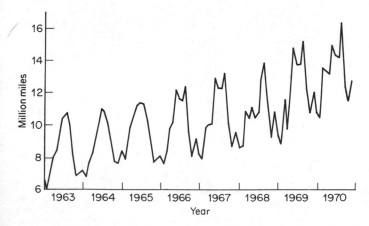

Fig. 1.5   Graph of the data of Table 1.3 (UK airlines: miles flown by month)

Table 1.4 Immigration into USA (Dewey 1963)

Year	Number	Year	Number	Year	Number
1820	8 385	1870	387 203	1920	430 001
1821	9 127	1871	321 350	1921	805 228
1822	6 911	1872	404 806	1922	309 556
1823	6 354	1873	459 803	1923	522 919
1824	7 912	1874	313 339	1924	706 896
1825	10 199	1875	227 498	1925	294 314
1826	10 837	1876	169 986	1926	304 488
1827	18 875	1877	141 857	1927	335 175
1828	27 382	1878	138 469	1928	307 255
1829	22 520	1879	177 826	1929	279 678
1830	23 322	1880	457 257	1930	241 700
1831	22 633	1881	669 431	1931	97 139
1832	48 386	1882	788 992	1932	35 576
1833	58 640	1883	603 322	1933	23 068
1834	65 365	1884	518 592	1934	29 470
1835	45 374	1885	395 346	1935	34 956
1836	76 242	1886	334 203	1936	36 329
1837	79 340	1887	490 109	1937	50 244
1838	38 914	1888	546 889	1938	67 895
1839	68 069	1889	444 427	1939	82 998
1840	84 066	1890	455 302	1940	70 756
1841	80 289	1891	560 319	1941	51 776
1842	104 565	1892	579 663	1942	28 781
1843	69 994	1893	439 730	1943	23 725
1844	78 615	1894	285 631	1944	28 551
1845	114 371	1895	258 536	1945	38 119
1846	154 416	1896	343 267	1946	108 721
1847	234 968	1897	230 832	1947	147 292
1848	226 527	1898	229 299	1948	170 570
1849	297 024	1899	311 715	1949	188 317
1850	295 984	1900	448 572	1950	249 187
1851	379 466	1901	487 918	1951	205 717
1852	371 603	1902	648 743	1952	265 520
1853	368 645	1903	857 046	1953	170 434
1854	427 833	1904	812 870	1954	208 177
1855	200 877	1905	1 026 499	1955	237 790
1856	200 436	1906	1 100 735	1956	321 625
1857	251 306	1907	1 285 349	1957	326 867
1858	123 126	1908	782 870	1958	253 265
1859	121 282	1909	751 786	1959	260 686
1860	153 640	1910	1 041 570	1960	265 398
1861	91 918	1911	878 587	1961	271 344
1862	91 985	1912	838 172	1962	283 763
1863	176 282	1913	1 197 892		
1864	193 418	1914	1 218 480		
1865	248 120	1915	326 700		
1866	318 568	1916	298 826		
1867	315 722	1917	295 403		
1868	277 680	1918	110 618		
1869	352 768	1919	141 132		

Note: The years are ended June 30, except for certain earlier years up to 1868, and occasional adjustments have been made to ensure comparability.

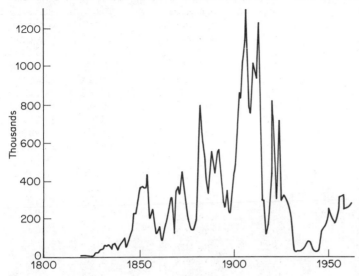

Fig. 1.6   Graph of the data of Table 1.4 (immigration into the USA, 1820–1962, annual)

a fixed date (June 4) so that seasonal movements are omitted. The airline data present a characteristic pattern of seasonal variation on a rising trend. Immigration, also on an annual basis, shows fluctuations, some of which can be identified with events such as war. The birth data are exceptional in that we have graphed the square root of the number rather than the number of births themselves (on the grounds that births probably follow a Poisson distribution and the square root transformation stabilises the variance), and in that form they reveal a remarkable cyclical pattern. The F.T. index-numbers are typical of fluctuations in the stock market over a period of time.

The objectives of time series analysis

1.19   It would be inappropriate to launch into a microscopic analysis of the various reasons for analysing time series, but a few general comments are in order because they often determine the methods of analysis to be used. Broadly speaking, we may identify five major types of investigation.

(a)   At the most superficial level, we take a particular series and construct a simple system, usually of a more or less mathematical kind, which *describes* its behaviour in a concise way.

(b)   Penetrating a little deeper, we may try to *explain* its behaviour in terms of other variables and to develop a structural model of behaviour. Stated another way, we set up the model as a hypothesis to account for the observations.

(c)   We may, from either (a) or (b), use the resulting model to *forecast* the behaviour of the series in the future. From (a) we work on the assumption

Table 1.5 Number of normal human births in each hour in four hospital series, transformed to y = square root of number of births (Bliss 1958; King 1956)

Hour starting	$\sqrt{\text{births}} = y$ in hospital				Total	Observed $\bar{y}$	Expected Y
	A	B	C	D			
Mt 12	13.56	19.24	20.52	21.14	74.46	18.6150	18.463
AM 1	14.39	18.68	20.37	21.14	74.58	18.6450	18.812
2	14.63	18.89	20.83	21.79	76.14	19.0350	19.129
3	14.97	20.27	21.14	22.54	78.92	19.7300	19.393
4	15.13	20.54	20.98	21.66	78.31	19.5775	19.587
5	14.25	21.38	21.77	22.32	79.72	19.9300	19.697
6	14.14	20.37	20.66	22.47	77.64	19.4100	19.716
7	13.71	19.95	21.17	20.88	75.71	18.9275	19.641
8	14.93	20.62	21.21	22.14	78.90	19.7250	19.479
9	14.21	20.86	21.68	21.86	78.61	19.6525	19.240
10	13.89	20.15	20.37	22.38	76.79	19.1975	18.941
11	13.60	19.54	20.49	20.71	74.34	18.5850	18.602
M 12	12.81	19.52	19.70	20.54	72.57	18.1425	18.246
PM 1	13.27	18.89	18.36	20.66	71.18	17.7950	17.897
2	13.15	18.41	18.87	20.32	70.75	17.6875	17.579
3	12.29	17.55	17.32	19.36	66.52	16.6300	17.315
4	12.92	18.84	18.79	20.02	70.57	17.6425	17.121
5	13.64	17.18	18.55	18.84	68.21	17.0525	17.011
6	13.04	17.20	18.19	20.40	68.83	17.2075	16.993
7	13.00	17.09	17.38	18.44	65.91	16.4775	17.067
8	12.77	18.19	18.41	20.83	70.20	17.5500	17.229
9	12.37	18.41	19.10	21.00	70.88	17.7200	17.468
10	13.45	17.58	19.49	19.57	70.09	17.5225	17.767
11	13.53	18.19	19.10	21.35	72.17	18.0425	18.106
Total	327.65	457.54	474.45	502.36	1762.00	18.3542	

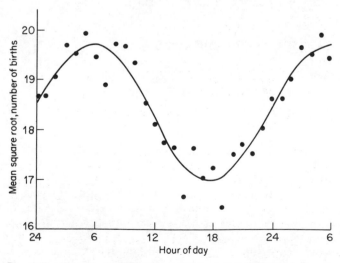

Fig. 1.7 Graph of the data of Table 1.5 (births by hour of day)

Table 1.6 Financial Times Index of leading equity prices: quarterly averages, 1960–71

Year and quarter		Index	Year and quarter		Index	Year and quarter		Index
1960	1	323.8	1964	1	335.1	1968	1	409.1
	2	314.1		2	344.4		2	461.1
	3	321.0		3	360.9		3	491.4
	4	312.9		4	346.5		4	490.5
1961	1	323.7	1965	1	340.6	1969	1	491.0
	2	349.3		2	340.3		2	433.0
	3	310.4		3	323.3		3	378.0
	4	295.8		4	345.6		4	382.6
1962	1	301.2	1966	1	349.3	1970	1	403.4
	2	285.8		2	359.7		2	354.7
	3	271.7		3	320.0		3	343.0
	4	283.6		4	299.9		4	345.4
1963	1	295.7	1967	1	318.5	1971	1	330.4
	2	309.3		2	343.1		2	372.8
	3	295.7		3	360.8		3	409.2
	4	342.0		4	397.8		4	427.6

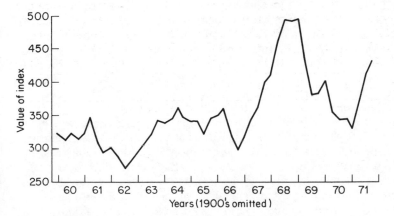

Fig. 1.8 Graph of the data of Table 1.6 (F.T. index quarterly averages)

that, even though we may be unaware of the basic mechanism which is generating the series, there is sufficient momentum in the system to ensure that future behaviour will be like the past. From (b) we have, we hope, more insight into the underlying causation and can make projections into the future more confidently.

(d) Using a structural model, as in (b), we may seek to *control* a system, either by generating warning signals of future untoward events or by examining what would happen if we alter either the inputs to the system or its parameters.

(e) More generally, we may wish to consider several jointly dependent variables, known as a vector process. In such cases we are approaching the more general subject area of statistical model-building as, for example, in the simultaneous equation systems developed in econometrics (cf. Johnston 1984). Indeed, in recent years the research in multiple time series and in econometrics has drawn much closer together (cf. Harvey 1981; Hendry and Richard 1983).

Decomposition

1.20 A survey of the examples of time series already given, and of the many others which are doubtless known to the reader, suggests that we may usefully consider the general series as a mixture of four components:

(a) a trend, or long-term movement;
(b) fluctuations about the trend of greater or less regularity;
(c) a seasonal component;
(d) a residual, irregular, or random effect.

It is convenient to represent the series as a sum of these four components, and one of the objectives may be to break the series down into its components for individual study. However, we must remember that, in so doing, we are imposing a model on the situation. It may be reasonable to suppose that trends are due to permanent forces operating uniformly in more or less the same direction, that short-term fluctuations about these long movements are due to a different set of causes, and that there is in both some disturbance attributable to random events, giving rise to the residual. But that this is so, and that the effects of the different causes are additive, are assumptions which we must always be ready to discard if our model fails to fit the data.

1.21 Perhaps the easiest components to understand are those which are undoubtedly due to physical factors, e.g. diurnal variations of temperature, the tidal movements associated with the lunar months, and seasonal variation itself. We must be careful not to confuse such effects with fluctuations of a pseudo-cyclical kind such as trade 'cycles', or with sunspot 'cycles' in which there is no known underlying astronomical phenomenon of a periodic kind.[*] The definition of seasonality, however, is by no means as easy as one might think. A glance at Fig. 1.5 will illustrate one of the problems. In this series of air-miles travelled, there are undoubtedly seasonal effects, a peak around Christmas, another at Easter, and one in the summer, all due to holiday travel. But the recurrence at Easter varies with Easter itself and therefore does not occur at the same date each year; and the pattern of the variation is altering from year to year, owing partly to the increased volume of traffic and partly to the spread of the period over which holidays are now taken. In short, our seasonal effect itself has a trend.

---

[*] Although it has been suggested that an apparent four-yearly swing in the British economy is due to the man-made fact that General Elections must be held at no greater than five-yearly intervals.

1.22 As we shall see when we come to a detailed study, it seems that trend and seasonality are essentially entangled, and we cannot isolate one without, at the same time, trying to isolate the other. Conceptually, however, they are distinct enough. Trend is generally thought of as a smooth broad movement, non-oscillatory in nature, extending over a considerable period of time. However, it is a relative term. What appears as a trend in climate to a drainage engineer may be nothing more than a temporary observation or a short-term swing to the geologist, whose time scale is very much longer.

1.23 If we can identify trend and seasonal components and then eliminate them from the data, we are left with a fluctuating series which may be, at one extreme, purely random, or at the other, a smooth oscillatory movement. Usually, we have something between these extremes; some irregularity, especially in imperfect data, but also systematic effects due to successive observations being dependent. We prefer to call this systematic effect an *oscillation* rather than a *cycle*, unless it can be shown to be genuinely cyclical in the pattern of recurrence, and in particular, that its peaks and troughs occur at equal intervals of time. Very few economic series are cyclical in this strict sense.

Notation

1.24 As noted earlier we shall usually assume that the series is observed at equal intervals of time, and as a rule no generality is lost if we take these intervals as units. Hence, we may denote a series by using subscripts, such as y_1, y_2, y_3, etc., the observation at time t being y_t. Here we suppose observations to begin at $t = 1$, but if necessary we can represent previously occurring values by y_0, y_{-1}, y_{-2}, etc.

Plan of the book

1.25 Having presented an overview of time series analysis, we now outline briefly how we shall chart our path through the subject.

1.26 Chapters 2–4 deal with what might be termed the 'classical' approach to time series analysis. In Chapter 2 we describe various tests of randomness since, if the series is devoid of structure, further analysis is pointless. Chapter 3 describes various methods of trend removal and Chapter 4 discusses seasonality and seasonal adjustment; these methods are still very relevant since they form the basis of nearly all procedures for the adjustment of official series.

1.27 Chapters 5–9 describe the analysis of a single series in the time domain. Chapter 5 contains a development of the widely used class of linear stochastic models, the autoregressive moving-average (ARMA) schemes, which form the basis for much of the theoretical and applied work in the area. Chapter 6 describes the sampling properties of the serial correlations which are used to assess the form of dependence in a series; these coefficients provide a natural basis for model specification, or *identification*.

The model-building paradigm we shall follow is summarised in Fig. 1.9 and derives from the basic work of Box and Jenkins (1976) in this area. Chapter 7 then goes on to describe estimation procedures for ARMA schemes and Chapter 8 discusses univariate or autoprojective forecasting procedures. Many of the forecasting procedures in current use were developed directly as multiple component forecasting models rather than as integrated models for the time series itself. This leads naturally to the use of the 'state-space' approach, outlined in Chapter 9, which has drawn increasing attention from statisticians in recent years.

1.28   In Chapter 10 and the first part of Chapter 11, we examine the behaviour of time series in the frequency domain, first for single series and then for several series. Frequency-domain analysis provides the natural vehicle for identifying strictly cyclical phenomena and is especially useful in engineering and the physical sciences. In other areas, its primary value is as a descriptive tool since exact cycles are rarely observed.

1.29   In Chapters 11–13 we examine structural models for a single dependent

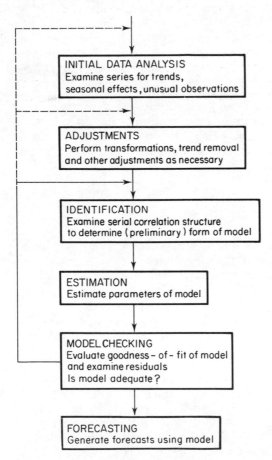

Fig. 1.9   A general paradigm for univariate time series modelling

series. The use of transfer functions and intervention analysis enables us to combine the benefits of structural modelling from the regression area with the explicit modelling of error processes underlying univariate time series analysis. Chapter 14 provides an overview of recent developments in vector time series models and Chapter 15 summarises other developments in the field which have not been considered explicitly in this book.

Bibliography

1.30 The emphasis in this volume is upon the methodological aspects of time series analysis rather than the formal development of theory. We shall often refer to Kendall *et al.* (1983), *The Advanced Theory of Statistics*, Volume 3, Chapters 45–51, as a source for such theoretical developments. Volumes 1 (Stuart and Ord, 1987) and 2 (Stuart and Ord, 1991) will also be used as standard references for theoretical results. To assist readers who may have access to different editions, we shall refer to sections in the *Advanced Theory* rather than pages. Other useful books on the theory of time series are T. W. Anderson (1971), Brillinger (1981), Fuller (1976), Hannan (1970) and Priestley (1981). Two useful volumes of papers describing recent research in the frequency domain and the time domain, respectively, are edited by Brillinger and Krishnaiah (1983) and Hannan, Krishnaiah, and Rao (1985). An overview of recent developments is given in the review papers by Newbold (1981, 1984), whereas current developments in forecasting are evaluated in a special issue of the *International Journal of Forecasting* (1988, number 4).

2

Tests of randomness

2.1 In the opening chapter, we considered the classical decomposition of a time series into trend, seasonal, oscillatory and irregular components. In many cases the presence of such components is very clear, as with the downward trend in the sheep series of Table 1.2 or the marked seasonal pattern in aircraft miles flown, shown in Table 1.3. However, there are cases, such as the barley data of Table 1.1, where the issue is not as clear-cut and a more accurate test is needed. Other instances may arise where some components have been removed, or filtered out, and we wish to test the residuals to see whether any structure remains. In this chapter we shall concentrate upon tests of randomness where the null hypothesis is that the observations are independent and identically distributed. That is, the observations are equally likely to have occurred in any order.

2.2 A considerable variety of possible tests of randomness is available, but certain selection criteria may be specified.

(a)   The test should not make any restrictive assumptions about the underlying distribution.

(b)   When new observations arise over time, it should be possible to update the test statistics without having to perform the calculations *ab initio*.

(c)   The choice of test should depend upon the alternative hypothesis we have in mind.

In the rest of this chapter, we consider some simple tests corresponding to intuitive ideas about departures from randomness. These tests use only the relative positions of the observations in the series and make no specific distributional assumptions, except that the random variables are continuous.

Turning points

2.3 A simple departure from a random series arises when the series exhibits well-defined turning points. A turning point is defined as either a 'peak' when a value is greater than its two neighbouring values, or a 'trough' when the value is less than its two neighbours. A simple test is given by counting the number of peaks and troughs, a particularly simple procedure when the series has been plotted. In order to carry out the test we must determine the distribution of the number of turning points in a random series.

2.4   Consider a time series consisting of n values $y_1, y_2, ..., y_n$. Three consecutive points are needed to define a turning point; thus

$$y_i > y_{i-1} \quad \text{and} \quad y_i > y_{i+1} \qquad \text{defines a peak at time } i,$$

whereas

$$y_i < y_{i-1} \quad \text{and} \quad y_i < y_{i+1} \qquad \text{defines a trough at } i.$$

Turning points are possible at the $(n-2)$ times $2, 3, ..., n-1$ since only one inequality can be checked for y_1 and y_n. If we define the indicator variables

$$U_i = 1, \text{ when there is a turning point at time } i$$
$$= 0, \text{ otherwise,}$$

then the number of turning points is given by

$$p = \sum_{i=2}^{n-1} U_i. \qquad (2.1)$$

Consider now the three value $\{y_{i-1}, y_i, y_{i+1}\}$. When the series is random, the six possible orders are equally likely:

$$123 \quad 132 \quad 213$$
$$231 \quad 312 \quad 321$$

where '3' denotes the largest value and '1' the smallest. Four of these six yield a turning point, so that the expected values of U_i is

$$E(U_i) = 1 \cdot (\tfrac{4}{6}) + 0 \cdot (\tfrac{2}{6}) = \tfrac{2}{3}$$

and

$$E(p) = \Sigma \, E(U_i) = \tfrac{2}{3}(n-2). \qquad (2.2)$$

By an extension of this argument (see Kendall *et al.* 1983, Section 45.18), we find that the variance of p is

$$\text{var}(p) = \frac{16n - 29}{90}.$$

Further, the distribution of p approaches normality rapidly as n increases. Therefore, we may carry out the test using

$$z = \frac{p - \tfrac{2}{3}(n-2)}{\{\text{var}(p)\}^{1/2}}. \qquad (2.3)$$

When the null hypothesis is true, the distribution of z is, approximately, the standard normal.

Example 2.1   In the barley data of Table 1.1 there are 56 values, but at two points (1906/7 and 1910/11) the values in successive years are equal. We shall consider each of these as a single point and reduce the number n to   There are 35 turning points in the series. The expected number is $\frac{2}{3}(52) = 34.67$. Agreement is so close that no test is necessary.

For the record, we note that

$$\text{var}(p) = \frac{16(54) - 29}{90} = 9.278$$

and

$$z = \frac{(35 - 34.67)}{3.04} = 0.11.$$

Phase-length

2.5   Another feature of interest in a series is the length of time between successive turning points; from a trough to the next peak is a run up, whereas peak to trough represents a run down. Thus, if y_i is a trough and y_{i+d} is the next peak, there is a run up, or *phase* of length d.

To define a phase, a run up say, of length d, we need the specific pattern

$$\longleftarrow \text{run up of d terms} \longrightarrow$$
$$y_{i-1} > y_i < y_{i+1} < y_{i+2} < \cdots < y_{i+d} > y_{i+d+1}.$$
$$\uparrow \qquad\qquad\qquad\qquad \uparrow$$
$$\text{trough} \qquad\qquad\qquad \text{peak}$$

Arguments similar to those for the turning points test (Kendall *et al.* 1983, Section 45.19) show that the number of phases of length d, N_d, has expected value

$$E(N_d) = 2(n - d - 2)(d^2 + 3d + 1)/(d + 3)! \tag{2.4}$$

for $1 \leqslant d \leqslant n - 3$. The total number of phases, N, has approximate expected value

$$E(N) \doteq \tfrac{1}{3}(2n - 7). \tag{2.5}$$

The distribution of N tends to normality for large n, see Levene (1952). Gleissberg (1945) tabulated the actual distribution for $n \leqslant 25$.

2.6   The distributions of the ratios N_d/N do not tend to normality. However, Wallis and Moore (1941) showed that when observed and expected numbers are compared for phases of length $d = 1, 2$ and $\geqslant 3$, the usual χ^2 statistic may be used with the following approximate percentage points for the upper tail:

α	0.10	0.05	0.01
value	5.38	6.94	10.28

Example 2.2   In the barley data of Table 1.1, there are 34 phases. As before, we take $n = 54$; only complete phases are counted, starting at the peak of 1885 and ending at the peak of 1938. Their actual lengths and the theoretical values given by (2.4) are as shown in Table 2.1. The values are so close that a test is hardly necessary; the χ^2 statistic has the value 0.83, clearly not significant.

Table 2.1

Phase-length	No. of phases observed	Theoretical
1	23	21.25
2	7	9.17
$\geqslant 3$	4	3.25
Total	34	33.67

Tests for trend

2.7 The phase-length and turning point tests could be used to look for trends, but their primary value is in detecting cyclical effects in a series. More direct tests for trend are obtained by comparing successive terms and examining them for decreases or increases. The simplest such test is the difference-sign test which counts the number of points of increase in the series. That is, we define the indicator variable

$$U_i = 1 \qquad \text{if } y_{i+1} > y_i$$
$$= 0 \qquad \text{if } y_i < y_{i+1}.$$

Then the number of points of increase,

$$c = \sum_{i=1}^{n-1} U_i \tag{2.6}$$

has $E(c) = \frac{1}{2}(n-1)$ and $\text{var}(c) = \frac{1}{12}(n+1)$. The distribution was tabulated by Moore and Wallis (1943), although the approach to normality is again fairly rapid.

2.8 The difference-sign test is clearly useless for detecting oscillatory behaviour as the number of points of increase would be approximately $\frac{1}{2}n$. Conversely, a test based on turning points will perform poorly as a test for trend, because random fluctuations imposed on a mild trend will have much the same set of turning points as if the trend were absent. A more appropriate test for trend is to regress y on t and then test the significance of the regression coefficient. Such a test is powerful when the suspected trend is close to linear. Another alternative, which may be used to detect any monotone trend, is a test based on the relative ordering of all pairs of observations.

2.9 Given a time series $y_1, y_2, ..., y_n$, let

$$q_{ij} = 1 \qquad \text{if } y_i > y_j \text{ when } j > i$$
$$= 0 \qquad \text{otherwise.}$$

Then let

$$Q = \sum_{i<j} \sum q_{ij}. \tag{2.7}$$

Under the null hypothesis that the series is random

$$P(q_{ij} = 1) = P(q_{ij} = 0) = \frac{1}{2}. \tag{2.8}$$

Since there are $\frac{1}{2}n(n-1)$ pairs in (2.7), it follows that

$$E(Q) = \tfrac{1}{4}n(n-1). \tag{2.9}$$

If Q is less than its expected value, this indicates a rising trend, whereas $Q > E(Q)$ suggests a falling trend. Q is known as the number of discordances in the series and is related to the rank correlation coefficient known as Kendall's τ (Kendall and Gibbons, 1990). We may set

$$\tau = 1 - \frac{4Q}{n(n-1)}; \tag{2.10}$$

under H_0, $E(\tau) = 0$ and

$$\mathrm{var}(\tau) = \frac{2(2n+5)}{9n(n-1)}. \tag{2.11}$$

Equivalently, we may set

$$p_{ij} = 1 - q_{ij} \quad \text{and} \quad P = \sum_{i<j}\sum p_{ij},$$

where P is the number of concordances. This leads directly to

$$\tau = \frac{4P}{n(n-1)} - 1. \tag{2.12}$$

The choice between P and Q is solely a matter of convenience.

Example 2.3   For the sheep series in Table 1.2, $n = 73$, $c = 35.5$, and $P = 530.5$; ties being scored as 0.5 by convention.

As $E(c) = \frac{1}{2}(72) = 36$, the difference-sign test fails to detect the clear trend in the sheep series. However,

$$\tau = \frac{4(530.5)}{73 \cdot 72} - 1 = -0.596$$

and var $\tau = 0.0064$. Thus, $z = \tau/(\mathrm{var}\ \tau)^{1/2} = -7.46$, leading to a decisive rejection of the null hypothesis of no trend. Strictly speaking, we should correct the variance for the presence of ties, but the correction is negligible in this case.

2.10   All the tests described in this chapter are easy to apply, so deciding which to use should depend on other considerations. It may be shown that when the alternative hypothesis is a linear trend, the most efficient tests are τ and the linear regression test. Indeed the difference-sign test has zero asymptotic efficiency against either of these alternatives (cf. Kendall *et al.* 1983, Sections 45.23–25). This is supported by the results in Example 2.3.

When it is desired to test for systematic cyclical effects and trend is absent, the turning points test can be quite effective, although the phase test is generally preferred.

Foster and Stuart (1954) considered the distribution of records in a series, a record being a value that is greater than or less than all previously noted values. As a test for trend it is less efficient than the regression coefficient or τ. The main disadvantage is, of course, that as time goes on, records tend to become sparse unless the trend is fairly marked. The principal advantage is that the test can be used when the data set consists only of 'records', as arises in sports, lists of peak floods and so on. For recent work on records tests, see Smith (1987).

2.11 It is quite common for a series to display both marked seasonality and a distinct trend. How can we test for trend in the presence of seasonality or for seasonality in the presence of a trend? One approach is to remove one component and then test the residuals for the other effect. Unfortunately, the process of removing a component usually generates correlation among the observed residuals even when the underlying error terms are independent. It is, therefore, rather dangerous to apply the foregoing tests to a set of observed residuals without some examination of the distortions induced by the removal process. We shall return to this topic in Section 5.24. For the present, we may note that residuals from a linear regression on time are not markedly affected, so this method of preliminary trend removal may be applied without seriously distorting a subsequent test for seasonality.

Testing trend in a seasonal series

2.12 For a strongly seasonal series such as the airline data in Table 1.3, the effectiveness of both the τ and linear regression tests is undermined by the strong seasonal component. However, a simple adaptation is possible. We may simply partition the observations into 12 distinct series, one for each month, and then compute 12 test statistics, say $\tau_1, \ldots, \tau_{12}$. In general, we would use r such values. The test statistic

$$\tau_S = \frac{1}{r} \sum_{j=1}^{r} \tau_j = 1 - \frac{4 \sum Q_j}{rm(m-1)}, \tag{2.13}$$

where Q_j is the number of discordances for the jth month. Under H_0, (2.13) has mean zero and variance

$$\text{var } \tau_S = \frac{2(2m+5)}{9rm(m-1)}, \tag{2.14}$$

where $n = rm$.

Surprisingly, when the alternative hypothesis is a linear trend (without seasonality), τ_S remains asymptotically as efficient as τ, although some loss of power can be expected in finite samples.

Example 2.4 For the airlines data in Table 1.3, the 12 Q_j values are, for January to December,

$$2, 1, 1, 5, 1, 3, 4, 0, 0, 0, 0, 0$$

leading to $\tau_S = 0.899$. Since $m = 8$, var $\tau_S = 1/144$ and $z = \tau_S/(\text{var } \tau_S)^{1/2}$ $= 10.79$. The null hypothesis is clearly rejected, as expected.

Testing for seasonality

2.13 The turning points test and the phase test may both be used to detect seasonal patterns, but their performance may be somewhat erratic. For example, a quarterly series with seasonal peaks in the summer and winter would produce too many turning points, whereas one with a summer peak and a winter trough would produce too few.

An alternative test, based on ranks, is a simple adaptation of the non-parametric analysis-of-variance procedure initially proposed by Friedman (1937); for a discussion, see Kendall *et al.* (1983, Sections 37.38–41). After removing a linear trend, if desired, we rank the values within each year, from 1 (smallest) to 12 (largest) for monthly data. In general, let the years represent c columns and the months r (=12) rows. Then each column represents a permutation of the integers $1, 2, ..., 12$. Summing across each row gives the monthly score M_j, $j = 1, 2, ..., 12$. Finally, under the null hypothesis of no seasonal pattern, the test statistic

$$T = 12 \sum_{j=1}^{r} \left\{ M_j - \frac{c(r+1)}{2} \right\}^2 \bigg/ cr(r+1) \qquad (2.15)$$

is approximately distributed as χ^2 with $(r-1)$ degrees of freedom.

2.14 In order to evaluate the performance of these three tests, we apply each to the airline data of Table 1.3. Any plausible test should give an unequivocal signal for this series since the seasonal pattern is so strong. In each case, we have considered the entire series, and the two sub-periods 1963–66 and 1967–70.

Example 2.5 For the turning points test we have Table 2.2. Here, as elsewhere, the z-statistic is

$$z = (\text{observed} - \text{expected under } H_0)/(\text{variance})^{1/2}. \qquad (2.16)$$

As expected, the results indicate fewer turning points than for a random pattern. However, the performance of the test is unsatisfactory for 1967–70 when the seasonal pattern is more complex.

Table 2.2

	1963–70	1963–66	1967–70
Observed	47	18	29
Expected	62.7	30.7	30.7
z-statistic	– 3.84	– 4.43	– 0.59

Example 2.6 For the phase test, we find the results shown in Table 2.3. The χ^2-value is computed using 1, 2, $\geqslant 3$ as before. Only 2 of the 3 values show significant effects. As for the turning points test, the slight change in the seasonal pattern over the later years renders the phase test ineffective, even though a strong seasonal pattern clearly persists.

Table 2.3

	1963–70		1963–66		1967–70	
d	Obs.	Exp.	Obs.	Exp.	Obs.	Exp.
1	20	38.8	5	18.8	14	18.8
2	17	16.9	4	8.1	12	8.1
3	2	4.8	1	2.3	1	2.3
4	3	1.0	2	0.5	1	0.5
5	3	0.2	3	0.1	0	0.1
6	1	0.0	1	0.0	0	0.0
All	46	61.7	16	29.7	28	29.7
χ^2		10.6		18.4		3.4

Example 2.7 The ranks within each year (no trend removed) are as shown in Table 2.4.

The rank test gives

$$T = \frac{12 \cdot 7714}{8 \cdot 12 \cdot 13} = 74.3$$

which is highly significant, based on 11 d.f.

This test has several clear advantages. It is unaffected by permuting the order of the months and it is usually more efficient than the other two tests, yet it retains their distribution-free nature. Further, it provides useful diagnostic information. For example, quick inspection of the table of ranks reveals that

(a) the peak month for travel has shifted from July to September;
(b) there are several unusual months which should be examined, notably

Table 2.4

Month	Year								Total = M_j
	63	64	65	66	67	68	69	70	
Jan.	2	2	4	3	2	1	3	2	19
Feb.	1	1	1	1	1	2	2	1	10
Mar.	5	5	5	4	5	8	5	8	45
Apr.	6	6	7	7	6	4	1	7	44
May	8	7	8	8	8	10	8	6	63
June	9	9	10	11	11	5	11	11	77
July	11	12	12	10	9	6	9	10	79
Aug.	12	11	11	9	10	11	10	9	83
Sept.	10	10	9	12	12	12	12	12	89
Oct.	7	8	6	6	7	7	7	4	52
Nov.	3	4	2	2	3	3	4	3	24
Dec.	4	3	3	5	4	9	6	5	39

March–July 1968, December 1968, April 1969, March 1970 and October 1970;

(c) generally, the later years show a less stable seasonal pattern and the reason for this should be examined.

2.15 From our investigations, we may conclude that many series are clearly non-random and formal testing may not be necessary; however, approximate tests on residuals may still be relevant, even though exact distributional results are no longer available. Other tests will be considered, in the context of stationary processes, in Chapters 6 and 7.

2.16 Throughout this chapter we have assumed that the time-series was observed at discrete points in time. The question arises, however, as to whether a process in continuous time with a continuous state space can be viewed as purely random. As early as 1827, the botanist Robert Brown observed the erratic local movements of particles suspended in a fluid, generated by random impacts from neighbouring particles. His observation of what is now known as Brownian motion led ultimately to the theory of diffusion processes; for a full account, see Cox and Miller (1968, Chapter 5).

Exercises

2.1 Apply the turning points test to the Financial Times Index series given in Table 1.6. Compute the differences $w_t = y_t - y_{t-1}$ of the series and repeat the test. Interpret your results.

2.2 Using arguments similar to those of Section 2.4, show that the probability that $(d + 3)$ consecutive terms give rise to a run-up of length d is

$$(d^2 + 3d + 1)/(d + 3)!$$

Hence find $E(N_d)$.

(*Hint*: There are $(d + 3)!$ possible orderings – which of these give rise to a run-up of length d?)

2.3 Apply the phase test to the lynx data in Appendix Table A4. Interpret the results.

2.4 Use both the difference-sign test and the τ-test to test for trend in the Financial Times Index series given in Table 1.6. Compare the results.

2.5 Table 4.4 contains data on the quarterly index of wholesale prices for vegetables in the UK, 1951–58. Test for the existence of a trend in these data, using the τ_S test.

2.6 Use the rank test to look for seasonality in the data of Table 4.4.

3

Trend

3.1 We now assume that, either by direct observation or by the application of the tests discussed in Chapter 2, it is determined that the time series exhibits certain systematic traits. In particular, in this chapter, we focus upon modelling trends.

The essential idea of trend is that it represents smooth, relatively slowly changing, features of the time series. In practice, this means that we should like to represent it by a continuous function of time. There are several classes of function which may be appropriate in particular circumstances, such as polynomials, sine waves or harmonic functions, and piecewise linear functions. For example, in the sheep series of Table 1.2, it looks as though a low-order polynomial would describe the general downward movement. The births data of Table 1.5 clearly require harmonic functions. A polynomial might also be used for the airline data of Table 1.3, after due allowance for the obvious seasonal component. Clearly, it would be unwise to extrapolate such a trend curve far into the future; the volume of air traffic must level off at some point and no polynomial can have a horizontal asymptote. We shall discuss trend models which incorporate appropriate asymptotic behaviour in Chapter 15.

Implicitly, our discussion thus far has assumed that we may wish to fit the same functional form over the entire time series. Such occasions are, however, quite rare. In addition, the procedure has at least four practical disadvantages:

(a) It may be troublesome to update when new observations become available.

(b) Adequate description of the entire series may require a model containing a considerable number of parameters. This may produce unreliable estimates of those parameters and, in turn, poor estimates of underlying trend values, particularly at the ends of the series.

(c) If the fitted model is updated, the entire set of trend values may be affected. It is undesirable that new observations should require a major reassessment of the past, particularly the distant past.

(d) As we shall see, the separation of seasonal and trend components is best
     undertaken in an iterative fashion. The use of 'global' functions to
     describe the entire length of the series may leave more 'local', but
     important, effects inadequately described. A simple example of this
     would be the use of an exponential trend to describe gross national
     product in the twentieth century, which would ignore the systematic dip
     caused by the Great Depression in the thirties.

These considerations lead us to model trends in a more local fashion. In
Sections 3.2–3.13 we consider moving averages, a method which remains
fundamental in the smoothing and seasonal adjustment of official government
series. We consider differencing in Sections 3.14–3.15; differences are often
used to eliminate trends, as we shall see in Chapters 5 and 6. In Section 3.15,
we describe one method for deciding the order of differencing, the variate
difference method; other approaches are considered in Sections 6.19–22.

Moving averages

3.2 Any smooth function can, under very general conditions, be represented
locally by a polynomial, to any desired degree of accuracy. This suggests the
following procedure. We may fit a polynomial to the first set of terms, say
$2m + 1$, and use that polynomial to determine the trend value at the $(m + 1)$th
point, the middle of the range of that set (hence the use of an odd number of
terms). We then fit the same order of polynomial to the 2nd, 3rd, ...,
$(2m + 2)$th observations and determine the trend value at the $(m + 2)$th point,
and so on, working our way along the series to the last group of $(2m + 1)$. We
do not have to fit the polynomials each time. As we now show, the procedure is
equivalent to taking linear combinations of the observations with coefficients
which can be tabulated in a standard form.

3.3 Suppose, for example, we wish to fit a polynomial of order three to sets
of seven points. Without loss of generality, we take the time points to be
$t = -3, -2, -1, 0, 1, 2, 3$. Our polynomial may be written

$$y_t = a_0 + a_1 t + a_2 t^2 + a_3 t^3. \tag{3.1}$$

We determine the coefficients a by the method of least squares; that is, we
minimise

$$\sum_{t=-3}^{3} (y_t - a_0 - a_1 t - a_2 t^2 - a_3 t^3)^2. \tag{3.2}$$

Differentiation by the as gives us the four equations:

$$\sum y_t t^j - a_0 \sum t^j - a_1 \sum t^{j+1} - a_2 \sum t^{j+2} - a_3 \sum t^{j+3} = 0, \qquad j = 0, 1, 2, 3. \tag{3.3}$$

Now the sums of the odd power of t from -3 to $+3$ vanish and the equations reduce to

$$\Sigma\, y_t = 7a_0 \qquad\qquad + 28a_2$$
$$\Sigma\, ty_t = \qquad\quad 28a_1 \qquad\qquad + 196a_3$$
$$\Sigma\, t^2 y_t = 28a_0 \qquad + 196a_2$$
$$\Sigma\, t^3 y_t = \qquad\quad 196a_1 \qquad\qquad + 1588a_3 \tag{3.4}$$

For the present, we are interested only in a_0, the values of the series at $t = 0$. We then require only the first and third of these equations to find

$$a_0 = \frac{1}{21}\left\{ \sum_{t=-3}^{3} y_t - \sum_{t=-3}^{3} t^2 y_t \right\}$$
$$= \tfrac{1}{21}\{7(y_{-3} + y_{-2} + y_{-1} + y_0 + y_1 + y_2 + y_3)$$
$$\quad - (9y_{-3} + 4y_{-2} + y_{-1} + 0 + y_1 + 4y_2 + 9y_3)\}$$
$$= \tfrac{1}{21}\{-2y_{-3} + 3y_{-2} + 6y_{-1} + 7y_0 + 6y_1 + 3y_2 - 2y_3\}. \tag{3.5}$$

That is, the trend value at time t is a weighted average of the seven points $(y_{t-3}, y_{t-2}, ..., y_{t+3})$, with weights

$$\tfrac{1}{21}[-2, 3, 6, 7, 6, 3, -2]. \tag{3.6}$$

Since the set of weights is symmetric, we abbreviate this to

$$\tfrac{1}{21}[-2, 3, 6, 7]. \tag{3.7}$$

This is known as a *moving average*, since the terms are applied successively to $(y_1, y_2, ..., y_7)$, $(y_2, y_3, ..., y_8)$ and so on.

Example 3.1 Consider the series

t	1	2	3	4	5	6	7	8	9	10
y_t	0	1	8	27	64	125	216	343	512	729

The trend value at $t = 4$ is

$$\tfrac{1}{21}\{(-2\cdot 0) + (3\cdot 1) + (6\cdot 8) + (7\cdot 27) + (6\cdot 64) + (3\cdot 125) + (-2\cdot 216)\} = 27,$$

as it should be since the original series is a cubic.

3.4 The procedure is perfectly general. If we fit $2m + 1$ points by a polynomial of order k, we have to minimize

$$\sum_{-m}^{m} (y_t - a_0 - a_1 t - \cdots - a_k t^k)^2.$$

This leads to $k + 1$ equations analogous to (3.3) and they split into two sets as in (3.4). The solution for a_0 depends on the sum $\Sigma\, t^j$ and linear functions of the ys typified by $\Sigma\, t^j y_t$. The trend value at point $t = i$ is then a weighted average of the values y_{i-m} to y_{i+m}. Using the representation in (3.7), weights for polynomials up to $k = 5$ and $n = 2m + 1 = 21$ are as follows:

Quadratic and cubic

[5] $\qquad \frac{1}{35}[-3, 12, \mathbf{17}]$

[7] $\qquad \frac{1}{21}[-2, 3, 6, \mathbf{7}]$

[9] $\qquad \frac{1}{231}[-21, 14, 39, 54, \mathbf{59}]$

[11] $\qquad \frac{1}{429}[-36, 9, 44, 69, 84, \mathbf{89}]$

[13] $\qquad \frac{1}{143}[-11, 0, 9, 16, 21, 24, \mathbf{25}]$

[15] $\qquad \frac{1}{1105}[-78, -13, 42, 87, 122, 147, 162, \mathbf{167}]$

[17] $\qquad \frac{1}{323}[-21, -6, 7, 18, 27, 34, 39, 42, \mathbf{43}]$

[19] $\qquad \frac{1}{2261}[-136, -51, 24, 89, 144, 189, 224, 249, 264, \mathbf{269}]$

[21] $\qquad \frac{1}{3059}[-171, -76, 9, 84, 149, 204, 249, 284, 309, 324, \mathbf{329}]$ $\qquad$ (3.8)

Quartic and quintic

[7] $\qquad \frac{1}{231}[5, -30, 75, \mathbf{131}]$

[9] $\qquad \frac{1}{429}[15, -55, 30, 135, \mathbf{179}]$

[11] $\qquad \frac{1}{429}[18, -45, -10, 60, 120, \mathbf{143}]$

[13] $\qquad \frac{1}{2431}[110, -198, -135, 110, 390, 600, \mathbf{677}]$

[15] $\frac{1}{46189}[2\,145, -2\,860, -2\,937, -165, 3\,755, 7\,500, 10\,125, \mathbf{11\,063}]$

[17] $\qquad \frac{1}{4199}[195, -195, -260, -117, 135, 415, 660, 825, \mathbf{883}]$

[19] $\qquad \frac{1}{7429}[340, -255, -420, -290, 18, 405, 790, 1\,110, 1\,320, \mathbf{1\,393}]$

[21] $\frac{1}{260015}[11\,628, -6\,460, -13\,005, -11\,220, -3\,940, 6\,378, 17\,655, 28\,190,$

$$36\,660, 42\,120, \mathbf{44\,003}] \quad (3.9)$$

Such formulae may also be expressed in terms of differences: see Kendall *et al.* (1983, Section 46.7).

3.5 Certain properties of such moving averages are easily derived:

(a) The weights sum to unity. This must be so because, if we apply them to a series consisting simply of the same constant repeated, the average must be the same constant.

(b) The weights are symmetric about the middle value, by construction.

(c) It follows from (b) that we get the same trend values whether we fit forwards or backwards in time.

(d) Since the estimating equations such as (3.4) split into two groups, we get the same value for a_0 whether there is a term in $a_3 t^3$ or not. In other words, the formulae are the same for a polynomial of even order $2k$ as for the polynomial of order $2k + 1$.

(e) As we have derived the formulae there are no trend values for the first and the last m values of the series. We shall remedy this deficiency in Section 3.9.

(f) Although formulae could be derived for fitting trends to an even number of points, the result would be to give trend values half-way between the intervals of observation, which would clearly be inconvenient. See Section 3.11 for further discussion.

3.6 The moving averages developed thus far require that the calculations be repeated for each set of $(2m + 1)$ observations. A convenient simplification is to use simple averages repeated several times. For example, if we take a simple moving average of threes and then another simple moving average of fives of the result, we have weights given by $\frac{1}{15}$:

$$
\begin{array}{c}
1, 1, 1 \\
\quad 1, 1, 1 \\
\qquad 1, 1, 1 \\
\qquad\quad 1, 1, 1 \\
\qquad\qquad 1, 1, 1 \\
\hline
[1, 2, 3, 3, 3, 2, 1]
\end{array}
$$

This is expressed more compactly as $\frac{1}{15}$ [3] [5]; that is, a simple average of three overlaid by a simple average of five terms. The order of the operations is immaterial, so that [3] [5] $\equiv$ [5] [3], and so on.

Example 3.2 Spencer's formulae   In 1904, Spencer, an English actuary, introduced two smoothing formulae which used simple averages but closely approximated the expressions developed earlier. Spencer's 15-point formula may be expressed as

$$\tfrac{1}{320} [4] [4] [5] [-3, 3, 4]$$
$$\equiv \tfrac{1}{320} [-3, -6, -5, 3, 21, 46, 67, \mathbf{74}]. \tag{3.10}$$

This compares with the 15-point form based on fitting a cubic

$$\tfrac{1}{1105} [-78, -13, 42, 87, 122, 147, 162, \mathbf{167}], \tag{3.11}$$

which is approximately

$$\tfrac{1}{320} [-23, -4, 12, 25, 35, 43, 47, \mathbf{48}]. \tag{3.12}$$

The patterns of weights are clearly quite similar. Spencer's 21-point formula may be written as

$$\tfrac{1}{350} [5] [5] [7] [-1, 0, 1, 2]$$
$$\equiv \tfrac{1}{350} [-1, -3, -5, -5, -2, 6, 18, 33, 47, 57, \mathbf{60}]. \tag{3.13}$$

Given the availability of modern computing power, the need for such simplifying expressions is less apparent, but these flexible formulae are still popular in some areas of application, such as actuarial work. One practical advantage of such methods is that the component parts of (3.10) or (3.12) are easier to understand than the final forms.

3.7 One question which may be raised is how do we assess the smoothing performance of these various formulae. One simple answer is to consider a random series of observations $\{y_t\}$; that is, such that successive values are independent and identically distributed with mean μ and variance σ^2. Then the smoothing effect of

$$\bar{y}_t = \sum_{j=-m}^{m} \alpha_j y_{t-j} \tag{3.14}$$

may be assessed by evaluating the variance of $\tilde{y}_t$, which is

$$\text{var}(\tilde{y}_t) = \sigma^2 \Sigma \alpha_j^2, \qquad (3.15)$$

given the assumptions; see Section 5.21. In making comparisons between formulae, we may set $\sigma^2 = 1$ without loss of generality. Thus, Spencer's 15-point average has $\text{var}(\tilde{y}) = 0.193$, whereas the 15-point formula in (3.11) has $\text{var}(\tilde{y}) = 0.151$. A purely random series smoothed by each of (3.10) and (3.11) would produce two correlated 'smoothed series', the correlation being $(0.151/0.193)^{1/2} = 0.885$; the reason for this correlation structure is given by Kendall and Stuart (1979, Sections 17.28–29). The variances for the 21-point formulae are 0.143 and 0.108, respectively, with a correlation of 0.870. The variance may be interpreted as the *error-reduction power* of the smoothing formulae. The error-reduction effect is clearly illustrated in Table 3.1, which shows the effects of applying Spencer's 21-point formula to a series of 51 observations drawn at random from a discrete uniform distribution, defined on the integers $0, 1, 2, \ldots, 99$.

Appendix C includes, as the last row in each table, the standard errors for the various moving averages. The element in the last column corresponds to the symmetric moving average and the remaining elements cover the ends of the series. The value in column '0' gives the standard error corresponding to extrapolating the polynomial one period beyond the end of the series; it can be seen that this value increases substantially when n is small.

3.8 As may be seen from Fig. 1.4, the sheep series shows a generally declining trend, with some evidence of oscillations about the trend. We now attempt to remove the trends so that cyclical effects, if any, may be examined.

Table 3.1 Spencer smoothing of a rectangular random series

Number of term	Series	Spencer 21-point 'trend'	Number of term	Series	Spencer 21-point 'trend'	Number of term	Series	Spencer 21-point 'trend'
1	23		18	3	43	35	10	39
2	15		19	67	40	36	96	38
3	75		20	44	39	37	22	37
4	48		21	5	39	38	13	36
5	59		22	54	39	39	43	35
6	1		23	55	40	40	14	34
7	83		24	50	41	41	87	34
8	72		25	43	42	42	16	
9	59		26	10	43	43	3	
10	93		27	74	44	44	50	
11	76	67	28	35	44	45	32	
12	24	66	29	8	45	46	40	
13	97	63	30	90	44	47	43	
14	8	60	31	61	44	48	62	
15	86	55	32	18	43	49	23	
16	95	51	33	37	42	50	50	
17	23	47	34	44	41	51	5	

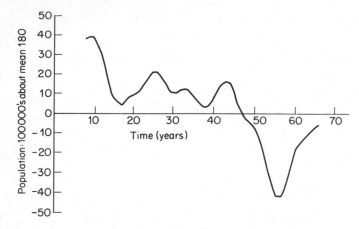

Fig. 3.1 Spencer 15-point fitted to the sheep series of Fig 1.4

Figure 3.1 shows the Spencer 15-point average. It appears that longer-term trends are not separated out from short-term effects as the fitted curve follows the observations too faithfully. By way of contrast, Fig. 3.2 shows the result of applying a simple moving average of order 11 to the series; it appears to capture the main features of the series without excessive detail. Figure 3.3 shows a simple average of order 5, which again seems too close to the original data. A simple average of order 9 seems to achieve a good balance and the residuals (= actual-trend) are given Table 3.2.

This 65-term series contains only 21 turning points as opposed to an expected number of $E(p) = \frac{2}{3}(63) = 42$ and $\text{var}(p) = 11.23$. This gives a value of -6.27 for the test statistic, clearly indicating the presence of oscillatory behaviour, which, as previously noted, could be partially induced by the trend-removal procedure.

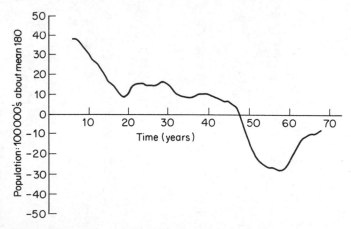

Fig. 3.2 Moving average of 11 fitted to the sheep series

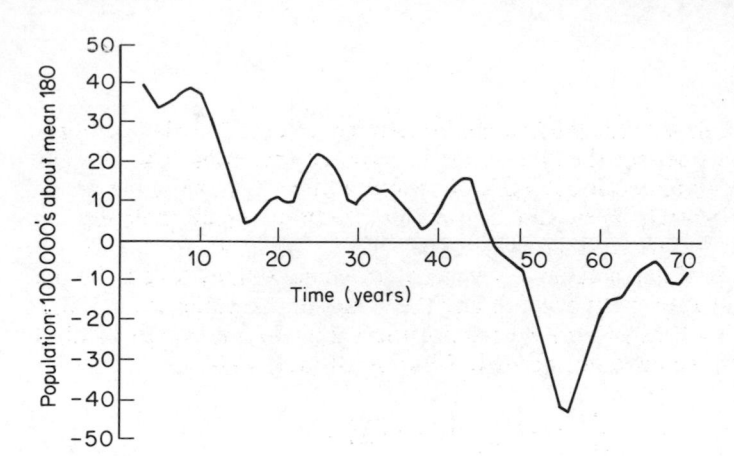

Fig. 3.3   Moving average of 5 fitted to the sheep series

Table 3.2 Residual values of the sheep series of Table 1.2 after removal of trend by a simple 9-point moving average

Year	Residual (10 000)	Year	Residual (10 000)	Year	Residual (10 000)
1871	− 176	1893	+ 34	1915	+ 19
72	− 112	94	− 103	16	+ 128
73	+ 50	95	− 104	17	+ 97
74	+ 141	96	− 15	18	+ 69
75	+ 60	97	− 23	19	− 29
76	− 20	98	+ 17	20	− 174
77	+ 12	99	+ 71	21	− 107
78	+ 82	1900	+ 35	22	− 142
79	+ 130	01	+ 16	23	− 109
80	− 14	02	− 27	24	− 23
81	− 166	03	− 32	25	+ 60
82	− 179	04	− 49	26	+ 121
83	− 84	05	− 61	27	+ 94
84	+ 38	06	− 52	28	− 25
85	+ 97	07	− 24	29	− 90
86	+ 8	08	+ 68	30	− 75
87	− 5	09	+ 141	31	+ 72
88	− 105	10	+ 119	32	+ 152
89	− 99	11	+ 66	33	+ 112
90	+ 35	12	− 52	34	− 64
91	+ 159	13	− 117	35	− 87
92	+ 167	14	− 61		

End-effects

3.9   As we have derived them and as can be seen from Table 3.1, the formulae provide no trend values for the first or last m terms of the series. The absence of values at the beginning is usually of minor importance; to have values at the end is usually essential. They can be obtained without great difficulty by extending the method we have already used.

In Section 3.3, we fitted a cubic to seven points which we may take to be the last seven points in a series. In order to find the values of this cubic at $t = 1, 2, 3$ (measured from $t = 0$ as origin) we require the values of a_1, a_2, a_3 in (3.1) which, up to now, we have not needed. Solution of (3.4) yields

$$a_1 = \frac{1}{1512}\left\{397 \sum_{-3}^{3} t y_t - 49 \sum_{-3}^{3} t^3 y_t\right\}$$

$$a_2 = \frac{1}{84}\left\{-4 \sum_{-3}^{3} y_t + \sum_{-3}^{3} t^2 y_t\right\}$$

$$a_3 = \frac{1}{216}\left\{-7 \sum_{-3}^{3} t y_t + \sum_{-3}^{3} t^3 y_t\right\}. \tag{3.16}$$

Expressing these as moving averages of the last seven terms, we have, in an obvious notation,

$$y_t = \tfrac{1}{21}[-2, 3, 6, 7, 6, 3, -2] + \tfrac{1}{252}[22, -67, -58, 0, 58, 67, -22]t$$
$$+ \tfrac{1}{84}[5, 0, -3, -4, -3, 0, 5]t^2 + \tfrac{1}{36}[-1, 1, 1, 0, -1, -1, 1]t^3. \tag{3.17}$$

For example, with $t = 1, 2, 3$, these reduce to the following averages, centred on y_{n-3}:

$$\bar{y}_{n-2} = \tfrac{1}{42}[1, -4, 2, 12, 19, 16, -4]y_{n-3} \tag{3.18}$$
$$\bar{y}_{n-1} = \tfrac{1}{42}[4, -7, -4, 6, 16, 19, 8]y_{n-3} \tag{3.19}$$
$$\bar{y}_n = \tfrac{1}{42}[-2, 4, 1, -4, -4, 8, 39]y_{n-3}. \tag{3.20}$$

If the last seven terms were 0, 1, 8, 27, 64, 125, 216 we should have for the third from the end

$$\bar{y}_{n-2} = \tfrac{1}{42}[(1 \cdot 0) + (-4 \cdot 1) + (2 \cdot 8) + \cdots + (4 \cdot 216)] = 64.$$

Likewise, $\bar{y}_{n-1} = 125$ and $\bar{y}_n = 216$ as expected. The coefficients sum to one, as they must, but they are no longer symmetrical. Moreover, since there are different formulae for different terms, a complete set of coefficients corresponding to (3.8) and (3.9) occupy rather a lot of space. They were tabulated for $p \leqslant 5$, $n \leqslant 25$ by Cowden (1962) with whose permission we reproduce them in Appendix C for $p \leqslant 3$ and $n \leqslant 15$. It is to be noted that we can no longer use the same formulae for polynomials of order $2m$ as for those of order $2m + 1$.

3.10   As we get nearer to the end of the series, the coefficients become more and more unequal and hence their sums of squares tend to increase. For example, the sum of squares of the coefficients of (3.18)–(3.20) are 0.452, 0.452, 0.928 compared to the central value of 0.333. This is as we might expect: the nearer the tails, the less reliable is the trend point, as measured by the error-reducing power at that point. The fitted curve, it has been said, tends to wag its tail.

Centred averages

3.11 We have noticed that it is convenient to use an odd number of points when constructing an average. But it so happens that many of the time spans over which we wish to average comprise an even number of points – the twenty-four hours of the day, the four weeks of the month, the four quarters and twelve months of the year. It is desirable in such cases to bring the trend estimates into line with the time points of the observations. This is usually done by taking a simple arithmetic mean of two adjacent trend values.

For example, suppose we have observations on the last day of each month, say January through December. A simple moving average of 12, with $\frac{1}{12}$ [12], would give us a trend value for the middle of July. We therefore take the mean of the values at the middle of July and middle of August to provide a trend value at the end of July for comparison with the observed value at that point. This is easily seen to be equivalent to taking an average over thirteen months with weights

$$\tfrac{1}{24}\,[2]\,[12] \equiv \tfrac{1}{24}\,[1, 2, 2, ..., 2, 2, 1]. \tag{3.21}$$

There are two Januaries (and in general two identical months) in the average, but each has only half the weight of the other months.

The effect of moving averages on other components

3.12 As yet no advice has been offered as to the best choice of the numbers $2m + 1$ and k, the extent of the average and the degree of polynomial embodied in it. Before we can take up these points we must consider the effect of moving averages on the other constituents in an additive model.

Consider first of all what happens if we apply a simple moving average of $k = 2m + 1$ terms to the purely deterministic time series given by the sine wave

$$y_t = \sin \lambda t. \tag{3.22}$$

Since

$$\sum_{j=-m}^{m} \sin \lambda(t + j) = \frac{\sin (m + \frac{1}{2}\lambda)\sin \lambda t}{\sin \frac{1}{2}\lambda}, \tag{3.23}$$

it follows that

$$\frac{1}{k}\,[k]\,y_t = \left\{ \frac{\sin(m + \frac{1}{2})\lambda}{(2m + 1)\sin \frac{1}{2}\lambda} \right\} y_t. \tag{3.24}$$

It follows from (3.24) that the trend value at each time t retains the term $\sin \lambda t$ but that the magnitude, or amplitude, of the sine wave is modified by the factor in curly brackets. When λ is small, this term is close to one, but when λ is large, the sine wave may be almost obliterated. On reflection, this is

understandable. If λ is small, the sine wave is long compared to the interval of observation and is treated as a trend. If λ is large, the wave repeats itself several times in the range of the average, which therefore comes close to zero.
3.13 For series which consist of, or can be represented as, the sum of a number of sine waves, using $y_t - \bar{y}_t$ to represent the detrended series will tend to emphasize the shorter cycles at the expense of the longer ones. Thus there is some danger that cyclical movements may be distorted by trend removal. For oscillations which are not cyclical in the strict sense the same kind of danger exists. A long swing, even if irregular, may be mistaken for trend by the moving average and included in the trend, so that detrended series has lost some of the purely oscillatory component.

This discussion suggests that any moving average may distort the cyclical and shorter-term component in the series. Such effects are inherent in any trend-removal method, and the best we can do is to understand the nature of these effects and not be misled by our analyses. We return to these topics in Sections 10.31–35, and now turn to other approaches to trend removal.

Differencing

3.14 A quite different approach to trend removal and one which has become very popular, is the use of differences. We define the (first) *backward difference* of the series $\{y_t\}$ as

$$\nabla y_t = y_t - y_{t-1}, \tag{3.25}$$

so that

$$\begin{aligned}
\nabla^2 y_t &= \nabla(\nabla y_t) \\
&= (y_t - y_{t-1}) - (y_{t-1} - y_{t-2}) \\
&= y_t - 2y_{t-1} + y_{t-2}, \tag{3.26}
\end{aligned}$$

and so on. Later, we shall find it useful to introduce seasonal differences of the form

$$\nabla_s y_t = y_t - y_{t-s}. \tag{3.27}$$

For example, for monthly data, we have $S = 12$ and

$$\nabla_{12} y_t = y_t - y_{t-12} \tag{3.28}$$

may be used to remove the seasonal component by considering (January, year k)–(January, year $k - 1$) and so on. It may be shown that the sequence of differencing operations is immaterial; that is,

$$\nabla \nabla_S \equiv \nabla_S \nabla. \tag{3.29}$$

A further item of notation that is worth introducing here is the *backshift operator*, B, where

$$By_t = y_{t-1}, \qquad B^s y_t = y_{t-s}$$

and

$$\nabla \equiv 1 - B;$$

properties of these operators are described in Appendix 3A at the end of this chapter. These operators are not only a useful notation, they also help in the derivation of some useful structural properties.

Suppose that

$$y_t = \beta_0 + \beta_1 t + \beta_2 t^2;$$

it follows that

$$\nabla y_t = \beta_1 - \beta_2 + 2\beta_2 t, \qquad \nabla^2 y_t = 2\beta_2,$$

and

$$\nabla^k y_t = 0, \qquad k > 2.$$

In general, if the trend follows a polynomial of degree d, we have

$$\nabla^d y_t = \text{constant and } \nabla^k y_t = 0, \qquad k > d.$$

This provides a straightforward way of removing polynomial trends. Exponential trends such as

$$y_t = \beta_0 e^{\beta_1 t + \beta_2 t^2}$$

can be removed by taking logarithms and then differencing:

$$z_t = \log_e y_t,$$
$$\nabla^2 z_t = 2\beta_2, \quad \nabla^k z_t = 0, \qquad k > 2$$

3.15 In order to use differencing to remove trends, we must decide how many differences to take. Suppose it is reasonable to consider the model

$$y_t = \beta_0 + \beta_1 t + \cdots + \beta_k t^k + \varepsilon_t, \tag{3.30}$$

where the errors are independent and identically distributed with zero mean and variance σ^2. Since the errors are independent, we obtain, for successive values of k

$$k = 0: \sigma_0^2 = \text{var}(y_t) = \sigma^2$$
$$k = 1: \sigma_1^2 = \text{var}(\nabla y_t) = \text{var}(\nabla \varepsilon_t)$$
$$= \text{var}(\varepsilon_t) + \text{var}(\varepsilon_{t-1})$$
$$= 2\sigma^2,$$
$$k = 2: \sigma_k^2 = \text{var}(\nabla^2 y_t) = \text{var}(\nabla^2 \varepsilon_t)$$
$$= \text{var}(\varepsilon_t) + 4\,\text{var}(\varepsilon_{t-1}) + \text{var}(\varepsilon_{t-2})$$
$$= 6\sigma^2.$$

Generally, we find

$$\sigma_k^2 = \text{var}(\nabla^k y_t) = \binom{2k}{k}\sigma^2. \tag{3.31}$$

Working with model (3.30), we might select that value of k for which the ratio

$$V_k = \text{sample variance of } (\nabla^k y_t) \Big/ \binom{2k}{k} = \hat{\sigma}_k^2 \Big/ \binom{2k}{k}$$

is minimized. This approach to selecting k is known as *variate differencing*; see Kendall *et al* (1983, Sections 46.24–32) for further details. Unfortunately, model (3.30) is generally unrealistic as the errors are often correlated. An alternative approach is to consider the sequence of alternative models

$$\nabla^k y_t = \varepsilon_t \tag{3.32}$$

and then choose the value of k which minimizes $\hat{\sigma}_k^2 = \text{var}(\nabla^k y_t)$. Again, the procedure is affected by correlation among the error terms but gives a lower and usually more realistic value for k.

Example 3.3   From the sheep data of Table 1.2 we obtain Table 3.3.

The two criteria give strikingly different results. As we shall see in Chapter 7, there is considerable evidence to favour the analysis based upon model (3.32); the discussion in Sections 6.19–22 provides a more formal criterion for selecting k.

3.16   The discussion in this chapter serves to demonstrate that trend fitting and trend removal need to be approached with some care. The choice of method and the mode of application require an appreciation of the subject matter of the series being analysed and an element of personal judgement. To a scientist it is always felt as a departure from correctness to incorporate subjective elements into the analysis. The student of time series cannot be a purist in that sense. What can be done is to make available the primary data and to explain unambiguously how the analysis was performed. Anyone who disagrees with what has been done can then carry out his or her own investigation.

Table 3.3

k	V_k	$\hat{\sigma}_k^2$
0	49 640	49 640
1	3 500	7 001
2	1 463	8 780
3	866	17 319
4	637	44 563
5	526	132 569

Exercises

3.1   Derive the weights given in (3.8) for fitting a quadratic to five points.

3.2   Compare the effects of smoothing the barley data, given in Table 1.1, with (a) a cubic fitted to seven points and (b) a simple $\frac{1}{15}$ [3] [5] average.

3.3   Using (3.15), calculate the variance for a random series using each of the moving averages used in Exercise 3.2. Compare these theoretical values with the variances for the smoothed barley series.

3.4   Verify the weights for Spencer's 15-point formula, given in (3.10).

3.5   Calculate the variance of y_t, ∇y_t and $\nabla^2 y_t$ for the Financial Times Index data given in Table 1.6. What order of differencing is appropriate? Compare your results with those from Exercise 2.1.

(*Note*:   The random walk model $\nabla y_t = \varepsilon_t$, described in Section 5.12, is the basis of models for an efficient stock market. Why? The model was first proposed by Bachelier in 1900 but his work was not followed up until the 1950s and 1960s. The model is now a cornerstone of financial theory.)

3.6   Consider the series

$$1, 2, 4, 8, 16, 32, 64, 128, \ldots .$$

What happens if you difference to try to remove the trend? Take logarithms ($\log_2(2^k) = k$) and try differencing again.

3.7   Show that

$$\nabla \nabla_s y_t = \nabla_s \nabla y_t.$$

Appendix 3A: Properties of the ∇ and B Operators

We assume that c is a constant and $\{y_t\}$ is a time series. The following properties hold for B:

$$Bc = c;$$
$$Bcy_t = cy_{t-1} = cBy_t$$
$$(a_1 B^i + a_2 B^j)y_t = a_1 B^i y_t + a_2 B^j y_t = a_1 y_{t-i} + a_2 y_{t-j}$$
$$B^i B^j y_t = B^{i+j} y_t = y_{t-i-j}$$
$$\frac{1}{(1 - aB)} y_t = \{1 + aB + a^2 B^2 + \cdots\}y_t$$
$$= y_t + ay_{t-1} + a^2 y_{t-2} + \cdots, \text{provided } |a| < 1.$$

An analogous set of properties hold for ∇, save that $\nabla c = 0$.

4

Seasonality

4.1 Seasonal effects, although they may vary somewhat in their average time of occurrence during the year, have a degree of regularity which other elements of time series usually do not. When we discuss spectrum analysis, we shall see that it is sometimes possible to isolate the seasonal component. However, in the time domain it is impossible to determine the seasonal effects without some prior adjustments for the trend. Consider, for example, a monthly series consisting of a slowly rising trend, 100 for January 1970, 101 for February 1970, ..., 112 for December 1970, 113 for January 1971 and so on. In any year January is the lowest month and December is the highest. Yet these are not seasonal effects which, in this case, do not exist. The problem is to distinguish such cases from, for example, the monthly sales of Christmas cards, which presumably also have their highest value in December, such variation being seasonal in any ordinary sense of the word.

4.2 There are several different reasons for wanting to examine seasonal effects, just as there were various reasons for looking at residual effects after the removal of trend:

(a) To compare a variable at different points of the year as a purely intra-year phenomenon; for example, in deciding how many hotels to close out of season, or at what points to allow stocks to run down.

(b) To remove seasonal effects from the series in order to study its other constituents uncontaminated by the seasonal component.

(c) To 'correct' a current figure for seasonal effects, e.g. to state what the unemployment figures in a winter month would have been if customary seasonal influences had not increased them.

These objectives are not the same, and it follows that one single method of seasonal determination may not be suitable to meet them all. This is, perhaps, the reason why different agencies (especially in government) favour different techniques for dealing with the seasonal problem.

Types of model

4.3   We shall consider three types of model, depending on whether the seasonal effect is additive or multiplicative. If m_t is the smooth component of the series (trend and cyclical effects), s_t is the seasonal component and ε_t the error term, we may have

$$y_t = m_t + s_t + \varepsilon_t, \tag{4.1}$$

$$y_t = m_t s_t \varepsilon_t, \tag{4.2}$$

or, the multiplicative–seasonal model

$$y_t = m_t s_t + \varepsilon_t. \tag{4.3}$$

The purely multiplicative model (4.2) may be converted to linear form by taking logarithms

$$\log y_t = \log m_t + \log s_t + \log \varepsilon_t. \tag{4.4}$$

In making the transformation (4.4) we assume that ε_t in (4.2) can only take on positive values; otherwise $\log \varepsilon_t$ is undefined. Thus we may choose to write $\eta_t = \log \varepsilon_t$, where η_t is a random variable with zero mean. Then (4.4) becomes

$$\log y_t = \log m_t + \log s_t + \eta_t \tag{4.5}$$

corresponding to

$$y_t = m_t s_t e^{\eta_t}. \tag{4.6}$$

4.4   We shall now concentrate upon the additive model, returning to the mixed model in Section 4.6. Since we wish to separate out the seasonal and trend components, it is reasonable to impose the condition that the sum of the seasonal effects is zero. Thus, for monthly data, the subscript t may be written as $t = 12(i - 1) + j$, corresponding to the jth month of the ith year. Assuming the seasonal effects are the same in different years, we would impose the condition

$$\sum_{j=1}^{12} s_t = \sum_{j=1}^{12} s_j = 0 \tag{4.7}$$

since $s_{12i+j} = s_j$ for all i and j. To determine the trend we may now take a 12-month centred moving average (as suggested in Section 3.11) with weights

$$\tfrac{1}{24} [2][12] = \tfrac{1}{24} [1, 2, 2, ..., 2, 1]. \tag{4.8}$$

For quarterly data, (4.7) is summed over the four components of s_j and (4.8) becomes $\tfrac{1}{8} [2][4]$. This moving average removes seasonality as Example 4.1 illustrates. The values (y_t − trend) yield $s_j + \varepsilon_t$, or simply s_j for the error-free series in the Example.

Example 4.1   Consider the quarterly series with values

$$y_t = 10 + t + s_j, \qquad t = 4(i - 1) + j,$$
$$s_1 = -3, \quad s_2 = 1, \quad s_3 = 4, \quad s_4 = -2.$$

The effect of applying the moving-average is as shown in Table 4.1

Table 4.1

(1) Year	(2) Quarter	(3) Series	(4) $\frac{1}{8}$ [2] [4]	(5) Col. (3) – Col. (4)
1	1	8		
	2	13		
	3	17	13	4
	4	12	14	– 2
2	1	12	15	– 3
	2	17	16	1
	3	21	17	4
	4	16	18	– 2
3	1	16	19	– 3
	2	21	20	1
	3	25	21	4
	4	20	22	– 2
4	1	20	23	– 3
	2	25	24	1
	3	29		
	4	24		

The use of a simple moving average, combined with the restriction (4.7), serves to eliminate a quadratic trend rather than a linear one. In general, the order of polynomial removal is increased by one when restriction (4.7) is imposed.

Example 4.2 Consider a quarterly series with a quadratic trend as shown in Table 4.2.

If a fixed seasonal effect was added to the series in column (3), its mean under averaging would be zero and it would be added to column (5). Since we have constrained the seasonal components to sum to zero, the values in column (5) should be adjusted to sum to zero. This gives the same result for seasonality as if we had added – 1.5 to the centred average in column (4), which would then reproduce the original quadratic exactly.

The procedure for computing the deseasonalised series may be summarised as follows: we assume that $(p + 1)$ complete years of monthly data are available, giving $12(p + 1)$ observations in all.

(1) Estimate the trend, $\hat{m}_t$, using an appropriate moving average.
(2) Compute

$$x_t = y_t - \hat{m}_t \tag{4.9}$$

and estimate the seasonal component by

$$s_j = \bar{x}_j - \bar{x} \tag{4.10}$$

where

$$\bar{x}_j = \sum_{i=1}^{p} x_t/p \quad \text{and} \quad \bar{x} = \sum_j \bar{x}_j/12 \tag{4.11}$$

Table 4.2

(1) Year	(2) Quarter	(3) Series	(4) $\frac{1}{8}$ [2] [4]	(5) Col. (3) − Col. (4)
1	1	0		
	2	1		
	3	4	5.5	− 1.5
	4	9	10.5	− 1.5
2	1	16	17.5	− 1.5
	2	25	26.5	− 1.5
	3	36	37.5	− 1.5
	4	49	50.5	− 1.5
3	1	64	65.5	− 1.5
	2	81	82.5	− 1.5
	3	100	101.5	− 1.5
	4	121	122.5	− 1.5
4	1	144	145.5	− 1.5
	2	169	170.5	− 1.5
	3	196		
	4	225		

The t subscript is defined as

$$t = 12(i - 1) + j, \qquad j = 7, \ldots, 12$$
$$t = 12i + j, \qquad\qquad j = 1, \ldots, 6,$$

the different subscripts being to allow for the observations omitted at the ends of the series.

(3) The deseasonalized series is then $y_t - s_j$. The following example illustrates the method.

Example 4.3 Table 4.3 gives the quarterly index numbers of the wholesale price of vegetable food in the United Kingdom for the years 1951–58. For arithmetic convenience the scale is multiplied by 10 and the series then transferred to origin 300 in Table 4.4.

Table 4.3 Quarterly index numbers of the wholesale price of vegetable food in the United Kingdom, 1951–58 (data from the Journal of the Royal Statistical Society for appropriate years; 1867–77 = 100)

	1951	1952	1953	1954	1955	1956	1957	1958
First quarter	295.0	324.7	372.9	354.0	333.7	323.2	304.3	312.5
2nd quarter	317.5	323.7	380.9	345.7	323.9	342.9	285.9	336.1
3rd quarter	314.9	322.5	353.0	319.5	312.8	300.3	292.3	295.5
4th quarter	321.4	332.9	348.9	317.6	310.2	309.8	298.7	318.4

Table 4.4 Data of Table 4.3 with origin 300, values multiplied by 10

	1951	1952	1953	1954	1955	1956	1957	1958
First quarter	− 50	247	729	540	337	232	43	125
2nd quarter	175	237	809	457	239	429	− 141	361
3rd quarter	149	225	530	195	128	3	− 77	− 45
4th quarter	214	329	489	176	102	98	− 13	184

Table 4.5 gives the residuals after elimination of trend by a centred average of fours. The mean values for each quarter (over seven years) are shown in the last column. These means sum to 24.01 with an overall mean of 6.00. Thus the seasonal effects are measured by subtracting 6.00 from the last column, e.g. 68.46 − 6.00. After division by 10 to restore the original scale we have for the seasonal factors in the four quarters

$$6.25, \quad 8.62, \quad -8.84, \quad -6.03,$$

which sum to zero as required. The seasonally adjusted series is given by subtracting these values from y_t in Table 4.3.

Table 4.6 gives the similar residuals obtained by fitting a seven-point cubic $\frac{1}{21}$ $[-2, 3, 6, 7]$. The seasonal adjustments are found to be

$$6.81, \quad 6.87, \quad -8.07, \quad -5.61.$$

The difference between the two sets of results are not very great. The seasonal effect is perceptible but not very marked.

4.5 Our development has assumed that the trend values should be determined at every point of the series, except the ends, in order to compute the seasonal effects. In fact, Durbin (1963) showed that seasonal effects can be computed directly from the original series; details are given in Kendall *et al.* (1983, Section 46.40).

4.6 We now consider the multiplicative-seasonal model given in (4.3). Although the model is non-linear, the component, $\hat{m}_t$, is usually estimated by a moving average, as before. Indeed, the only changes in steps (1)–(3) given in Section 4.4 are that the detrended series is now estimated by

$$x_t = y_t / \hat{m}_t \tag{4.12}$$

and the seasonal components are defined as

$$s_j = \bar{x}_j / \bar{x}, \tag{4.13}$$

where $\bar{x}_j$ and $\bar{x}$ are as in (4.11). Finally, the deseasonalized series is given by y_t / s_j.

There is an element of arbitrariness here in that it seems odd to set the *mean* of the seasonal effects equal to one, rather than their product. However, no method is likely to resolve the problem completely, and this rule seems to work well.

4.7 Up to this point we have assumed that the seasonal effects do not change over time. If, for any reason, the seasonal pattern is thought to be changing, the estimates of the effects must also change over time. Several methods have been developed to handle this problem, the best known being the Census

Table 4.5 Residuals in the data of Table 4.4 after removal of trend by a centred moving average of fours

	1951	1952	1953	1954	1955	1956	1957	1958	Totals	Means
First quarter		25.750	167.875	77.875	108.625	24.875	52.250	22.000	479.250	68.46
2nd quarter		− 8.125	189.750	75.875	28.250	238.000	107.875	229.375	645.250	92.18
3rd quarter	− 10.125	− 94.750	− 85.625	− 121.625	− 60.375	− 163.875	− 40.250		− 576.625	− 82.38
4th quarter	10.000	− 122.500	− 59.000	− 88.000	− 97.000	26.000	− 49.250		− 379.750	− 54.25

Table 4.6 Residuals in the data of Table 4.4 after removal of trend by a cubic fitted to 7 points

	1951	1952	1953	1954	1955	1956	1957	1958	Totals	Means
First quarter		+ 30.38	122.14	79.95	113.19	16.33	91.14	5.43	458.56	65.51
2nd quarter		+ 29.19	135.62	82.24	27.43	206.57	− 84.24		396.81	66.13
3rd quarter		− 53.71	− 123.95	− 106.81	− 35.76	− 191.57	12.38		− 499.42	− 83.24
4th quarter	− 12.67	− 128.67	− 72.57	− 70.48	− 97.90	25.81	− 54.00		− 410.48	− 58.64

Method II, Version X-11 devised by Julius Shiskin for the US Bureau of the Census and widely used throughout the world (Shiskin, 1967). This method is often known simply as X-11 and is accessible through the SAS/ETS package (Statistical Analysis System/Econometrics and Time Series). The properties of the current version are discussed by Wallis (1974) and Burman (1979) among others.

The X-11 method

4.8 Several implementations of X-11 are in circulation, but the main features are as summarized below. The description is in terms of the multiplicative-seasonal model (4.3) for monthly data, but the method also can handle quarterly data. Also, the additive-seasonal model (4.1) may be used in place of (4.3).

(1) An option is available to adjust the series for the number of trading or working days. If it is adopted, all subsequent operations are on the adjusted series; cf. Bell and Hillmer (1983a,b).

(2) A moving average is taken. There are a number of options with respect to the extent and weighting of the average.

(3) The series is divided by the moving average to give a first estimate of the seasonal-plus-irregular component. End values are estimated, usually by the nearest value for the same month. Extreme values are replaced by the mean of the two values for that month lying on either side of it.

(4) To decide the relative importance of seasonal and irregular components, an analysis of variance is carried out between years and between months. An *F*-test is used to test whether the variance between months is significantly larger than the residual variance, which would indicate seasonal effects.

(5) For any month the ratio of within-month to residual variance for that month is used to decide from among a number of options what moving average shall be used to smooth the residual term. A different average may be used for different months.

(6) In some cases the seasonal effect is divided into the primary series to get a preliminary deseasonalized series, and another moving average taken to get a second estimate of the trend. The seasonal factors are adjusted so as to sum to 12.

(7) These results provide estimates of the smooth component and of a moving seasonal component. The residual is obtained by subtraction (or sometimes by dividing the primary series by the smooth-plus-seasonal component if the error is regarded as multiplicative).

(8) Various subsidiary statistics such as the error variance are computed.

For further details of the procedure, see Wallis (1974), Kendall *et al.* (1983, Section 46.41) or the SAS/ETS manual. As illustrated by the account given above, the method has many *ad hoc* features, but it appears to perform remarkably well in practice.

Example 4.4   Figure 4.1 shows the smooth, seasonal, and irregular factors for the airline data of Table 1.3. The analysis on which these diagrams were based was carried out on a Univac 1108 by a program adapted from the X-11 variant of the Census Mark II. No adjustment was made for trading days or length of month. An initial 15-term moving average was fitted. The resulting values were divided into the original series to get an estimate of the irregular-plus-seasonal component. From the trend values an average from month to month, regardless of sign, was compared with the average from month to month of the estimated irregular-plus-seasonal component. This comparison dictated a second moving average of the original series, also a 15-term average, and again the trend estimates were divided into the original series to give the seasonal-plus-irregular component. The latter were smoothed by a seven-term moving average in order to yield preliminary estimates of the seasonal and irregular components. For each month separately these preliminary estimates were used to select a further smoothing, in this case a five-term average. This provided an estimate of the seasonal components. They were divided into the seasonal-plus-irregular component to give the irregular component, which is treated as multiplicative in this case.

An extension to the X-11 method, known as X11-ARIMA, uses the time series models described in Chapter 5. We shall discuss this in Chapter 15.

4.9   Several other seasonal adjustment procedures are available and are briefly noted below.

The SABL package developed by Cleveland *et al*. (1981) uses robust methods and appears to have better statistical foundations than X-11. Adjustments for calender effects have also been developed (Cleveland and Devlin, 1982).

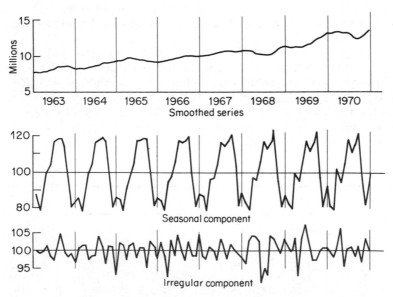

Fig. 4.1   Smooth, seasonal and irregular factors for the airline data of Table 1.3

Durbin and Murphy (1975) consider the mixed-scasonal model

$$y_t = m_t(1 + \beta_t) + \alpha_t + \varepsilon_t. \qquad (4.14)$$

As in Section 4.4, m_t may be estimated by a moving average; α_t and β_t are then estimated by regressing y_t on $\hat{m}_t$. The seasonal effects satisfy the conditions

$$\alpha_t = \alpha_{t+12}, \qquad \beta_t = \beta_{t+12}$$
$$\Sigma \, \alpha_j = 0, \qquad \Sigma \, \beta_j = 0.$$

In order to reduce the total number of parameters, harmonic components may be used to describe the seasonal effects. Significant terms may then be selected by stepwise regression; for details see Durbin and Murphy (1975).

Various regression methods have also been considered; the interested reader is referred to Kendall *et al.* (1983, Sections 46.42–45).

4.10 In practical time series analysis there are few golden rules, and few general rules which can be applied without detailed thought about the nature of the series and the purpose for studying it. Experience would indicate that it is often better to remove known effects before proceeding to further study. Indeed, trend removal in one form or another is usually carried out at a preliminary stage in the analysis. Opinions differ about the desirability of removing seasonal effects as opposed to modelling them directly. Interestingly, as we shall see in Section 5.34, these two approaches are not so far removed as they seem.

What is evident is that, after removing trends, there will often be systematic movements in the irregular component which suggest that further structural modelling is possible. It is this issue which we shall address in the next three chapters.

Exercises

4.1 Consider the artificial quarterly time series generated by the model

$$y_t = (10 + t)s_j, \qquad t = 1, ..., 20,$$

where $s_1 = s_4 = 0.7$ and $s_2 = s_3 = 1.3$ (see Table 4.7).

Find the seasonally adjusted series using the multiplicative–seasonal model (4.3) with a $\frac{1}{8}$ [4] [2] moving average for trend. (Note that the underlying trend, $10 + t$, is almost recovered although small systematic biases remain.)

Table 4.7

		Year				
		1	2	3	4	5
Quarter	1	7.7	10.5	13.3	16.1	18.9
	2	15.6	20.8	26.0	31.2	36.4
	3	16.9	22.1	27.3	32.5	37.7
	4	9.8	12.6	15.4	18.2	21.0

4.2 Use the multiplicative–seasonal model to provide a seasonally adjusted series for the data in Table 4.3. Use the $\frac{1}{8}$ [4] [2] moving average for trend.

4.3 Using the $\frac{1}{8}$[4] [2] moving average to represent $\hat{m}_t$ for the data in Table 4.3, estimate the parameters of the mixed model (4.14) for the data in Table 4.3. (Use a linear regression package and formulate the model as in Section 4.9.)

4.4 Develop a seasonally adjusted series for the airline data of Table 1.3 using (a) additive model (4.1) and (b) multiplicative model (4.3).

4.5 Develop a seasonally adjusted series for the airline data using X-11 or other available computer packages. Check the various stages of development as described in Section 4.8.

5

Stationary series

5.1   In most of statistics, random sampling procedures enable us to obtain replicated observations under identical conditions. Further, these observations are independent. By contrast, in time-series analysis, we are faced with only a single realization at each point in time and observations which are dependent over time. In order to develop inferential procedures, we must recreate some notion of replicability. This is achieved by the assumption of *stationarity*. Intuitively, this means we assume that the mean and variance of the series are constant over time and that the structure of the series depends only upon the relative position in time of the two observations.

In this chapter we define two versions of stationarity and show how they are used in time-series analysis. We then develop some basic properties of stationary processes and introduce three major classes of stationary model: autoregressive (AR), moving average (MA) and mixed, or autoregressive-moving average (ARMA). We then extend these ideas to cover seasonal schemes and mixed regular and seasonal models.

Stationarity

5.2   For many purposes, all that is needed is knowledge of the mean, variance and autocorrelation structure of the series. If y_t, $t = 1, 2, \ldots, n$ denotes the time series, we have the general definitions

$$\text{mean:} E(y_t) = \mu(t) \text{ or } \mu_t \tag{5.1}$$

$$\text{variance:} \text{var}(y_t) = \gamma(t) = E(y_t - \mu_t)^2 \tag{5.2}$$

$$\text{autocovariance:} \text{cov}(y_t, y_{t-k}) = \gamma(t, t-k)$$
$$= E\{(y_t - \mu_t)(y_{t-k} - \mu_{t-k})\}. \tag{5.3}$$

A time series is *weakly*, *second-order* or *covariance* stationary if

(i) $\mu(t) = \mu$ and $\gamma(t) = \gamma_0$,   for all t

(ii) $\gamma(t, t-k) = \gamma_k$,   for all t and k. \tag{5.4}

Condition (ii) implies that two observations, k time periods apart, have the same covariance no matter where they occur in the series. Further, conditions (i) and (ii) imply that we can define the *autocorrelation* between y_t and y_{t-k} as

$$\text{corr}(y_t, y_{t-k}) = \frac{\text{cov}(y_t, y_{t-k})}{\{\text{var}(y_t)\text{var}(y_{t-k})\}^{1/2}}$$

$$= \gamma_k/\gamma_0, \tag{5.5}$$

which we shall denote by ρ_k. Note that $\rho_k = \rho_{-k}$.

A time series is said to be *strongly*, or *strictly*, stationary if the joint density functions depend only upon the relative locations of the observations. That is, if $f\{y(t_1), y(t_2), \ldots, y(t_h)\}$ denotes the joint density of the observations at times $t_1, t_2, \ldots, t_h$, we require that

$$f\{y(t_1 + k), y(t_2 + k), \ldots, y(t_h + k)\} = f\{y(t_1), y(t_2), \ldots, y(t_h)\}, \tag{5.6}$$

for all k and for all choices of the $\{t_i\}$. In particular, this means that

$$f\{y(t)\} = f\{y(t + k)\} \tag{5.7}$$

$$f\{y(t_1), y(t_2)\} = f\{y(t_1 + k), y(t_2 + k)\} \tag{5.8}$$

and so on. In this book, we shall always make the assumption that the mean and variance exist, then conditions (5.6) imply conditions (5.4).

5.3   The observed series is just one out of the infinite number of possible realizations; in effect, it is a sample of size one. The stationarity conditions make it possible to use the corresponding sample statistics to estimate μ, γ_0 and the $\{\gamma_k\}$. The precise conditions under which this is possible are given by *ergodic* theorems, first proved by Birkhoff (1936) and Khintchine (1932). Given (5.4) or (5.6), the basic requirement is that

$$\lim_{n \to \infty} \frac{1}{n} \sum_{j=1}^{n} |\rho_j| = 0. \tag{5.9}$$

In general, we shall find that either $\rho_k = 0$ beyond some point k_0, or that the ρ_k converge towards zero sufficiently quickly for their mean to approach zero.

Variance stabilization

5.4   The condition of stationarity is clearly fundamental to the statistical analysis of time series, but it is not an assumption that can be made automatically. For any particular series we must ensure that conditions (5.4) are satisfied, at least to a reasonable degree of approximation.

In Chapter 3 we saw how to de-trend a series and that some form of deviation from a moving average could be used to achieve this purpose. As we shall see in Section 5.23, such procedures may induce extraneous autocorrelations in the series so that modern practice favours the use of differencing, as outlined in Sections 3.14–15. Thus, for the purposes of the rest of this chapter, we shall assume that the series has been, or can be, rendered stationary in the mean by use of differencing.

The next question which arises is that of variance stability. Box and Cox

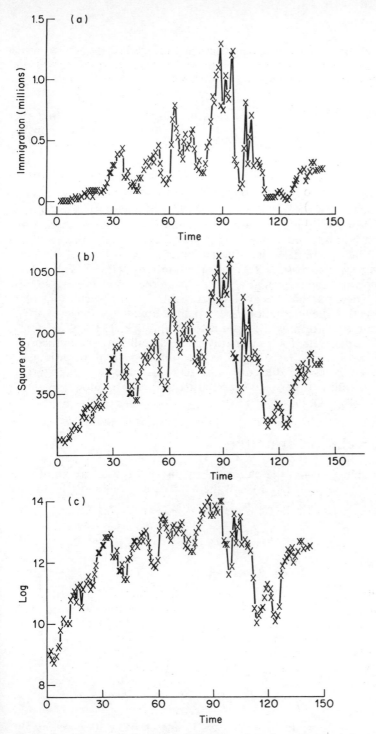

Fig. 5.1 (a) plot of the US immigration data from Table 1.4; (b) plot of square roots; (c) plot of logarithms

(1964) introduced the class of variance stabilizing transformations

$$y^{(\lambda)} = \begin{cases} (y^\lambda - 1)/\lambda, & \lambda \neq 0 \\ \log_e y, & \lambda = 0, \end{cases} \tag{5.10}$$

where, typically, $-1 \leqslant \lambda \leqslant 1$ and the random variable y is such that the probability of a negative value is negligible. Provided $\mu^2 \gg \gamma_0$, an approximation for the variance of the transformed variable is

$$\text{var } y^{(\lambda)} \doteq \mu^{2\lambda - 2}\text{var}(y). \tag{5.11}$$

Thus if the variance of y appears to increase linearly with the mean, we should use $\lambda = 0.5$. If the variance increases quadratically with the mean, $\lambda = 0$ or the logarithmic transformation is appropriate. Box and Cox (1964) estimate λ by maximum likelihood, but a graphical check for the values $\lambda = -1, -0.5, 0, 0.5, 1$ will often be sufficiently accurate. In Fig. 5.1, we give the plots of y_t, $\sqrt{y_t}$ and log y_t for the immigration data of Table 1.4. Fluctuations appear greatest just after the turn of the century when immigration peaked. These fluctuations are less noticeable for $\sqrt{y_t}$, whereas the plot for log y_t suggests that the fluctuations are virtually unrelated to the level of y_t. Accordingly, a log transform would appear to be appropriate for these data.
5.5   The third requirement for stationarity is the stability of the autocorrelations. There are no formal tests of this assumption in general use although, when the observed series is sufficiently long to justify the step, it is worth dividing the series into two parts and comparing the sample autocorrelations computed separately from each part.

The autocorrelation function

5.6   The set of values ρ_k and the plot of ρ_k against $k = 1, 2, \ldots$ are known as the *autocorrelation function* (ACF). All the autocorrelations must lie in the range $[-1, +1]$ but, in addition, they must satisfy the conditions

$$V_k = \text{var}\left(\sum_{j=0}^{k} a_j y_{t-j}\right) \geqslant 0, \tag{5.12}$$

for all k and all choices of the constants $\{a_j\}$; $a_0 = 1$ without loss of generality. When $k = 1$, (5.12) implies $|\rho_1| \leqslant 1$. For $k = 2$, it may be shown that V_2 is minimized when a_1 and a_2 are given by

$$a_1 = \frac{-\rho_1(1 - \rho_2)}{1 - \rho_1^2} \quad \text{and} \quad a_2 = \frac{-(\rho_2 - \rho_1^2)}{1 - \rho_1^2}. \tag{5.13}$$

Substituting back into (5.12) we obtain the constraint

$$\rho_2 - 2\rho_1^2 + 1 \geqslant 0. \tag{5.14}$$

Thus if $\rho_1 = 0.8$, we must have $\rho_2 \geqslant 0.28$. If successive values are highly correlated, there is still a strong correlation between values two time periods apart.

The partial autocorrelation function

5.7 The carry-over effect just observed suggests that it would be useful to look at the *partial autocorrelation* between y_t and y_{t-k}, after allowing for the effect of the intervening values $y_{t-1}, \ldots, y_{t-k+1}$. It follows from the general expression for partial correlations (Kendall and Stuart, 1979, Section 27.3–5) that the partial autocorrelation between y_t and t_{t-2}, allowing for y_{t-1}, is

$$\text{corr}(y_t, y_{t-2} \mid y_{t-1}) = \frac{\rho_2 - \rho_1^2}{1 - \rho_1^2} \tag{5.15}$$

which is exactly the coefficient a_2 in (5.13). As we shall see in Section 5.14, this is no accident. Higher-order partial autocorrelations may be defined similarly, but we prefer to develop these expressions in a more intuitively appealing manner in Section 5.11.

Example 5.1
(a) If $\rho_1 = 0.8$ and $\rho_2 = 0.28$, $\text{cov}(y_t, y_{t-2} \mid y_{t-1}) = -1$, confirming the constraint given by (5.14).
(b) If $\rho_2 = \rho_1^2$, $\text{cov}(y_t, y_{t-2} \mid y_{t-1}) = 0$, a property which arises naturally in Section 5.9.

Autoregressive processes

5.8 An important classs of linear time-series models is the set of autoregressive schemes. The pth-order scheme, denoted by $AR(p)$, may be written as

$$y_t = \delta + \phi_1 y_{t-1} + \cdots + \phi_p y_{t-p} + \varepsilon_t; \tag{5.16}$$

the coefficients $\phi_1, \ldots, \phi_p$ being the (auto)regressive coefficients for y_t on $y_{t-1}, \ldots, y_{t-p}$ with δ denoting the constant term. The error term, ε_t, is assumed to have the properties

$$E(\varepsilon_t) = 0, \qquad \text{var}(\varepsilon_t) = \sigma^2, \tag{5.17a}$$

$$\text{cov}(\varepsilon_t, \varepsilon_{t-k}) = 0, \qquad k \neq 0, \tag{5.17b}$$

and

$$\text{cov}(\varepsilon_t, y_{t-k}) = 0, \qquad k > 0. \tag{5.17c}$$

Condition (5.17c) states that the new error is independent of past values of the process; this assumption is critical to later developments.

We shall now examine two important special cases of (5.16), the Markov scheme ($p = 1$) and the Yule scheme ($p = 2$).

The Markov scheme

5.9 The Markov, or AR(1), scheme may be written as

$$y_t = \delta + \phi_1 y_{t-1} + \varepsilon_t. \tag{5.18}$$

Taking expected values and recalling that $E(y_t) = \mu$ given stationarity, we obtain

$$E(y_t) = \mu = \delta + \phi_1\mu$$

or

$$\mu = \delta/(1 - \phi_1) \tag{5.19}$$

provided $\phi_1 \neq 1$. In what follows we shall use

$$u_t = y_t - \mu$$

so that (5.18) and (5.19) give

$$u_t = \phi_1 u_{t-1} + \varepsilon_t. \tag{5.20}$$

In effect, expression (5.20) corresponds to setting the mean to zero; this simplifies further algebraic developments without any loss of generality. To obtain the variance, we square both sides and take expectations to obtain

$$\begin{aligned}
\gamma_0 = E(u_t^2) &= E\{(\phi_1 u_{t-1} + \varepsilon_t)^2\} \\
&= \phi_1^2 E(u_{t-1}^2) + 2\phi_1 E(u_{t-1}\varepsilon_t) + E(\varepsilon_t^2) \\
&= \phi_1^2 \gamma_0 + \sigma^2,
\end{aligned} \tag{5.21}$$

using condition (5.17c). Thus, from (5.21) we have

$$\gamma_0 = \sigma^2/(1 - \phi_1^2). \tag{5.22}$$

Since the right-hand side of (5.22) must be strictly positive, this implies that

$$1 - \phi_1^2 > 0 \quad \text{or} \quad |\phi_1| < 1. \tag{5.23}$$

As we shall see below, (5.23) is precisely the condition for the Markov scheme to be stationary.

To find ρ_1 we multiply both sides of (5.20) by u_{t-1} and take expectations to arrive at

$$\begin{aligned}
E(u_t u_{t-1}) = \gamma_1 &= E\{(\phi_1 u_{t-1} + \varepsilon_t)u_{t-1}\} \\
&= \phi_1 E(u_{t-1}^2) \\
&= \phi_1\gamma_0,
\end{aligned} \tag{5.24}$$

using (5.17c) again. Thus (5.24) yields

$$\rho_1 = \gamma_1/\gamma_0 = \phi_1. \tag{5.25}$$

Similarly, on multiplying (5.20) by u_{t-k} and taking expectations, we have

$$\begin{aligned}
\text{cov}(u_t, u_{t-k}) = \gamma_k &= \phi_1 \, \text{cov}(u_{t-1}, u_{t-k}) \\
&= \phi_1\gamma_{k-1}, \quad k = 1, 2, \ldots
\end{aligned}$$

Hence, dividing through by γ_0,

$$\begin{aligned}
\rho_k &= \phi_1\rho_{k-1} = \phi_1^2\rho_{k-2} = \cdots \\
&= \phi_1^k.
\end{aligned} \tag{5.26}$$

The typical form of the autocorrelation function for $\phi_1 > 0$ is given in Fig. 5.2(a).

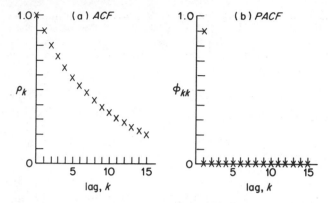

Fig. 5.2   ACF and PACF for AR(1) scheme

5.10   For the AR(1) scheme, the basic condition for stationarity given in (5.9) becomes a condition on

$$\frac{1}{n} \sum_{j=1}^{n} |\phi_1|^j. \tag{5.27}$$

From Appendix 5A (5.27) may be written as

$$\frac{1 - |\phi_1|^{n+1}}{n(1 - |\phi_1|)}$$

which tends to zero as $n \to \infty$, provided $|\phi_1| < 1$.

5.11   Referring back to (5.15) and Example 5.1, we see that the first-order partial autocorrelation is zero since $\rho_2 = \rho_1^2$. More generally, since $\mathrm{var}(y_t) = \sigma_y^2$ for all t, the (partial) regression coefficient for y_t on y_{t-p} in (5.16) corresponds exactly to the partial autocorrelation for y_t on y_{t-p}, given $y_{t-1}, ..., y_{t-p+1}$. That is, if we write

$$\phi_{jj} = \mathrm{corr}(y_t, y_{t-j} | y_{t-1}, ..., y_{t-j+1}), \tag{5.28}$$

we may compute ϕ_{jj} by performing the autoregression of y_t on $(y_{t-1}, ..., y_{t-j})$ consecutively for $j = 1, 2, ...$. A more efficient numerical procedure is given in Section 5.19. Since the Markov scheme is given by (5.20), it follows immediately that $\phi_{jj} = 0, j \geq 2$ and so.

$$\phi_{11} = \phi_1, \qquad \phi_{jj} = 0, \quad j \geq 2; \tag{5.29}$$

as illustrated in Fig. 5.2(b).

5.12   If $\phi_1 = 1$ in (5.20), we have

$$u_t = u_{t-1} + \varepsilon_t$$

or

$$u_t - u_{t-1} = \nabla u_t = \varepsilon_t, \tag{5.30}$$

known as the *random walk* model since the mean no longer exists but the conditional expectation for u_t, given u_{t-1}, is

$$E(u_t | u_{t-1}) = u_{t-1}. \tag{5.31}$$

Roughly speaking, (5.31) tells us that the 'best guess' that we can make for u_t at time t_{-1} is that its conditional expectation is u_{t-1}. As we mentioned in Exercise 3.5 this makes it a natural starting point for models of stock prices. If we use (5.18) with $\phi_1 = 1$, we can obtain

$$\nabla y_t = \delta + \varepsilon_t, \tag{5.32}$$

known as the *random walk with drift* (δ). It is interesting to note that the differencing in (5.32) has arisen from purely stochastic considerations, rather than to eliminate trends, as in Sections 3.14–15. Perhaps for this reason, random walk processes are sometimes said to exhibit a 'stochastic trend', something of a contradiction in terms.

Example 5.2   *The effects of aggregation; Working (1960)*   It follows from (5.30) that the successive terms $u_1, u_2, ..., u_m$ may be rewritten as

$$u_1, \quad u_1 + \varepsilon_2, \quad u_1 + \varepsilon_2 + \varepsilon_3, \quad u_1 + \varepsilon_2 + \varepsilon_3 + \cdots + \varepsilon_m.$$

A second set of the next m terms may be written as

$$u_1 + \varepsilon_2 + \varepsilon_3 + \cdots + \varepsilon_{m+1}, ..., u_1 + \varepsilon_2 + \varepsilon_3 + \cdots + \varepsilon_{2m}.$$

If we compute the average values for each set as

$$\bar{u}_{(1)} = \sum_{i=1}^{m} u_i/m \quad \text{and} \quad \bar{u}_{(2)} = \sum_{i=m+1}^{2m} u_i/m,$$

it follows that

$$d_{(1)} = \bar{u}_{(2)} - \bar{u}_{(1)}$$

$$= \frac{1}{m} \{\varepsilon_2 + 2\varepsilon_3 + \cdots + (m-1)\varepsilon_m + m\varepsilon_{m+1} + (m-1)\varepsilon_{m+2} + \cdots + \varepsilon_{2m}\}, \tag{5.33}$$

which has mean zero and variance

$$\text{var}(d_{(1)}) = \frac{\sigma^2}{m^2} \left\{ m^2 + 2 \sum_{j=1}^{m-1} j^2 \right\} = \frac{\sigma^2}{3m}(2m^2 + 1). \tag{5.34}$$

Likewise if we set $d_{(2)} = \bar{u}_{(3)} - \bar{u}_{(2)}$, it follows that

$$\text{cov}(d_{(1)}, d_{(2)}) = \sigma^2(m^2 - 1)/6m. \tag{5.35}$$

Finally, the first-order correlation for $d_{(t)}$ is

$$(m^2 - 1)/(4m^2 + 2)$$

which approaches 0.25 as m increases. Thus, although the first differences of the original series are uncorrelated, those of the aggregated series are not. For this reason, stock prices are now quoted as closing prices rather than daily averages.

5.13   There has been considerable debate in the econometric literature (e.g., Chan *et al.* 1977; Nelson and Kang, 1981) about whether trends should be modelled by including a polynomial in time in the regression model or by differencing. The evidence appears to be running strongly in favour of differencing. The same conclusion had been reached earlier for more pragmatic reasons by time-series analysts. Henceforth, we shall concentrate upon the use of differences to remove trends.

The Yule scheme

5.14 The Yule, or AR(2), process may be written as

$$y_t = \delta + \phi_1 y_{t-1} + \phi_2 y_{t-2} + \varepsilon_t, \tag{5.36}$$

with the same assumptions on ε as before. Assuming that the process is stationary, we take expectations and find

$$\mu = \delta + \phi_1 \mu + \phi_2 \mu$$

or

$$\delta = (1 - \phi_1 - \phi_2)\mu. \tag{5.37}$$

Again, we may set $u_t = y_t - \mu$ so that (5.36) becomes

$$u_t = \phi_1 u_{t-1} + \phi_2 u_{t-2} + \varepsilon_t. \tag{5.38}$$

The analysis now follows the same form as for the Markov scheme. Squaring both sides and adding gives

$$\gamma_0 = \gamma_0(\phi_1^2 + \phi_2^2 + 2\phi_1\phi_2\rho_1) + \sigma^2. \tag{5.39}$$

Likewise, multiplying (5.38) through by u_{t-1} and u_{t-2} in turn, and taking expectations, gives the pair of equations

$$\rho_1 = \phi_1 + \phi_2\rho_1$$
$$\rho_2 = \phi_1\rho_1 + \phi_2 \tag{5.40}$$

after dividing through by γ_0. This pair of equations enables us to express the ρs in terms of the ϕs as

$$\rho_1 = \frac{\phi_1}{1 - \phi_2} \quad \text{and} \quad \rho_2 = \phi_2 + \frac{\phi_1^2}{1 - \phi_2}. \tag{5.41}$$

Conversely,

$$\phi_1 = \frac{\rho_1(1 - \rho_2)}{1 - \rho_1^2} \quad \text{and} \quad \phi_2 = \frac{\rho_2 - \rho_1^2}{1 - \rho_1^2}. \tag{5.42}$$

Expressions (5.42) correspond exactly to those given in (5.13) since the coefficients of the Yule scheme are determined by choosing ϕ_1 and ϕ_2 to minimize $\mathrm{var}(y_t - \phi_1 y_{t-1} - \phi_2 y_{t-2})$. Finally, substituting (5.41) into (5.39) we obtain

$$\gamma_0\left(1 - \phi_1^2 - \phi_2^2 - \frac{2\phi_1^2\phi_2}{1 - \phi_2}\right) = \sigma^2$$

or

$$\gamma_0 = \frac{\sigma^2(1 - \phi_2)}{(1 - \phi_1 - \phi_2)(1 - \phi_2 + \phi_1)(1 + \phi_2)}. \tag{5.43}$$

To ensure that $\gamma_0 > 0$, (5.43) suggests that

$$1 - \phi_1 - \phi_2 > 0, \quad 1 - \phi_2 + \phi_1 > 0 \quad \text{and} \quad |\phi_2| < 1. \tag{5.44}$$

These are precisely the conditions required for stationarity. In general, the

Yule scheme (5.38) gives rise to the auxiliary equation

$$\phi(x) = 1 - \phi_1 x - \phi_2 x^2 = 0, \tag{5.45}$$

and the conditions for stationarity are that the roots of (5.45) should be greater than one in absolute value. These conditions are equivalent to those in (5.44) and are often stated as the requirement that the roots of $\phi(x)$ lie outside the unit circle.

Yule–Walker equations

5.15   The pair of equations (5.40) are known as the *Yule–Walker equations*. It is evident from (5.40) that the first two autocorrelations determine the autocorrelation structure of the Yule process completely. In general, the first p autocorrelations determine uniquely the coefficients of an $AR(p)$ scheme, see Section 5.19.

If we multiply (5.38) by u_{t-k} and take expectations, we obtain

$$\rho_k = \phi_1 \rho_{k-1} + \phi_2 \rho_{k-2}, \tag{5.46}$$

recalling that $\rho_0 \equiv 1$ and $\rho_1 = \rho_{-1}$. Equation (5.46) is a second-order difference equation, whose general solution is described in Appendix 5B. When $\phi_1^2 + 4\phi_2 \geqslant 0$, the roots are real and the autocorrelation function is of damped exponential form, with a steady or oscillatory pattern depending on the signs of the roots. When $\phi_1^2 + 4\phi_2 < 0$, the roots are complex and the *ACF* is a damped sine wave.

Example 5.3   What is the *ACF* of the Yule scheme

$$u_t = 0.8 u_{t-1} - 0.64 u_{t-2} \ ?$$

The model has auxiliary equation

$$1 - 0.8x + 0.64x^2 = 0$$

with roots

$$x = 1.25 \left(\frac{1}{2} \pm i \frac{\sqrt{3}}{2} \right) = 1.25 \left(\cos \frac{\pi}{3} \pm i \sin \frac{\pi}{3} \right).$$

From Appendix 5B we obtain the solution

$$\rho_j = (0.8)^j \sin \left(\frac{\pi j}{3} + \omega \right) / \sin \omega,$$

where $\omega \doteq 0.46\pi$. Hence,

$$\rho_1 = 0.49, \qquad \rho_2 = -0.25, \qquad \rho_3 = -0.51$$

as may be checked directly from (5.41) and (5.46).

A typical *ACF* for the Yule scheme with complex roots is given in Fig. 6.4. As we shall see in Chapter 6, the general shape of the *ACF* is more important than the detailed algebraic solution.

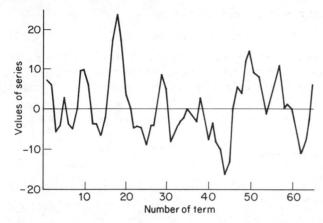

Fig. 5.3 Plot of 65 terms of a second order autoregressive Yule scheme

5.16 From (5.42), the *PACF* for the Yule scheme is

$$\phi_{11} = \rho_1 = \frac{\phi_1}{1 - \phi_2}, \qquad \phi_{22} = \phi_2, \qquad \phi_{jj} = 0, \quad j \geqslant 3,$$

so that the plot of the *PACF* consists only of two spikes.

5.17 In Fig. 5.3 we plot 65 terms from the Yule scheme with $\phi_1 = 1.1$ and $\phi_2 = -0.5$. The error term is uniformly distributed in range $[-9.5, 9.5]$. For comparison, Fig. 5.4 gives the values of 60 terms of the harmonic series

$$u_t = 10 \sin \frac{\pi t}{5} + \varepsilon_t, \tag{5.47}$$

where ε_t is uniformly distributed on $[-5, 5]$. The sine wave has regular peaks every 10 terms, but the Yule scheme is clearly less regular in its fluctuations.

5.18 In general, it may be shown that the mean distance between peaks is $2\pi/\theta$, where $0 < \theta < 2\pi$ and

$$\cos \theta = \text{corr}\{\nabla u_t, \nabla u_{t-1}\}; \tag{5.48}$$

see Kendall *et al.* (1983, Section 46.19–20) for details.

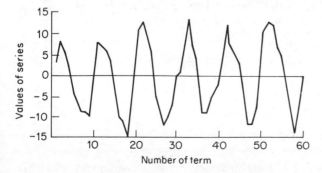

Fig. 5.4 Plot of 60 terms of a harmonic series

For the Markov scheme

$$\cos \theta = -\tfrac{1}{2}(1 - \phi_1)$$

and for the Yule scheme

$$\cos \theta = -\tfrac{1}{2}(1 - \phi_1 - \phi_2).$$

Thus for a purely random process, $\theta = 2\pi/3$, and the mean difference between peaks is 3. Hence, the mean distance between turning points is 1.5, as noted in Section 2.4.

For the Yule series plotted in Fig. 5.3, $\theta \doteq 0.4\pi$, and the mean distance between peaks is about five time periods. From Fig. 5.3, we see that there are 12 peaks including a pair of adjacent peaks at $t = 9$ and 10. The first peak occurs at $t = 4$ and the last at $t = 59$, giving an observed mean distance of $(59 - 3 - 1)/12 = 4.6$, where 1 is subtracted for the tie. Even though the distribution is not normal, the observed and theoretical distances between peaks are in close agreement.

General autoregressive schemes

5.19   Analysis for the general $AR(p)$ scheme follows along the same lines as for the Markov and Yule schemes. Thus, we obtain the set of Yule–Walker equations

$$\rho_k = \phi_1\rho_{k-1} + \phi_2\rho_{k-2} + \cdots + \phi_p\rho_{k-p}, \tag{5.49}$$

for $k = 1, 2, ..., p$. The p equations (5.49) may be used to express the ϕ_k in terms of the ρ_k or vice versa. For $k > p$, we obtain the same recurrence relation (5.49) so that higher-order autocorrelations may be calculated. The partial autocorrelation are obtained by solving the sets of equations successively for $p = 1, 2, 3, ...$; that is,

$$
\begin{array}{ll}
p = 1 & \rho_1 = \phi_{11} \\
p = 2 & \rho_1 = \phi_{21} + \phi_{22}\rho_1 \\
& \rho_2 = \phi_{21}\rho_1 + \phi_{22} \\
p = 3 & \rho_1 = \phi_{31} + \phi_{32}\rho_1 + \phi_{33}\rho_2 \\
& \rho_2 = \phi_{31}\rho_1 + \phi_{32} + \phi_{33}\rho_1 \\
& \rho_3 = \phi_{31}\rho_2 + \phi_{32}\rho_1 + \phi_{33}
\end{array}
$$

and so on. The *PACF* is given by ϕ_{jj}, $j = 1, 2, ...$. Rather than solve the complete set of equations at each stage Durbin (1960) showed that these could be solved iteratively as

$$\phi_{k+1,k+1} = \frac{(\rho_{k+1} - \sum_{j=1}^{k} \phi_{kj}\rho_{k+1-j})}{(1 - \sum_{j=1}^{k} \phi_{kj}\rho_j)} \tag{5.50}$$

$$\phi_{k+1,j} = \phi_{kj} - \phi_{k+1,k+1}\phi_{k,k+1-j}, \qquad j = 1, 2, ..., k. \tag{5.51}$$

For further details, see Kendall *et al.* (1983, Section 47.15). This procedure is generally known as the Durbin–Levinson algorithm; for a history and description of its other uses in statistics, see Morretin (1984).

5.20   The stationarity conditions for the $AR(p)$ scheme follow from (5.49) as the requirement that the roots of the auxiliary equation

$$\phi(x) = 1 - \phi_1 x - \phi_2 x^2 - \cdots - \phi_p x^p = 0 \qquad (5.52)$$

should be greater than one in absolute value. If any root is $\leqslant 1$ in absolute value, the process is non-stationary. As before, differencing and transformations may be used to induce stationarity.

Moving average schemes

5.21   In Chapter 2 we used moving averages to represent trends. We shall now use them in a rather different way to model the persistence of random effects over time. Suppose, for example, y_t denotes weekly sales figures and some sales are not recorded until the following week. A possible model would be

$$y_t = \mu + \varepsilon_t - \theta_1 \varepsilon_{t-1}, \qquad (5.53)$$

where the term in ε_{t-1} reflects the carryover from one week to the next. Model (5.53) represents a first-order moving average scheme, or $MA(1)$. The general $MA(q)$ form is

$$y_t = \mu + \varepsilon_t - \theta_1 \varepsilon_{t-1} - \cdots - \theta_q \varepsilon_{t-q}. \qquad (5.54)$$

5.22   Since $E(\varepsilon_t) = 0$, we see immediately that

$$E(y_t) = \mu. \qquad (5.55)$$

Further, since the ε_t are uncorrelated, it follows that

$$\begin{aligned}
\mathrm{var}(y_t) &= E\{(y_t - \mu)^2\} \\
&= E\{(\varepsilon_t - \theta_1 \varepsilon_{t-1} - \cdots - \theta_q \varepsilon_{t-q})^2\} \\
&= E(\varepsilon_t^2 + \theta_1^2 \varepsilon_{t-1}^2 + \cdots + \theta_q^2 \varepsilon_{t-q}^2) \\
&= \sigma^2(1 + \theta_1^2 + \cdots + \theta_q^2).
\end{aligned} \qquad (5.56)$$

By implication, *any $MA(q)$* scheme is stationary.

Example 5.4   Find the *ACF* and the *PACF* for the $MA(1)$ scheme. Let $u_t = y_t - \mu$ and write the $MA(1)$ scheme (5.53) as

$$u_t = \varepsilon_t - \theta_1 \varepsilon_{t-1}. \qquad (5.57)$$

If we multiply through by u_{t-k} and take expectations, we obtain

$$E(u_t u_{t-k}) = \gamma_k = E\{u_{t-k}(\varepsilon_t - \theta_1 \varepsilon_{t-1})\}.$$

For $k = 1$,

$$E(u_t u_{t-1}) = E\{(\varepsilon_{t-1} - \theta_1 \varepsilon_{t-2})(\varepsilon_t - \theta_1 \varepsilon_{t-1})\} \qquad (5.58)$$

yielding

$$\gamma_1 = -\theta_1 \sigma^2,$$

and

$$\gamma_k = 0, \qquad k \geqslant 2.$$

From (5.56), $\gamma_0 = \sigma^2(1 + \theta_1^2)$ so that

$$\rho_1 = -\theta_1/(1 + \theta_1^2); \tag{5.59}$$

clearly, $\rho_k = 0$, $k \geqslant 2$. Thus the *ACF* has a single non-zero term at lag one. The *PACF* may be found from (5.50) and (5.51) as

$$\phi_{11} = \frac{-\theta_1}{1 + \theta_1^2}; \quad \phi_{22} = \frac{-\theta_1^2}{1 + \theta_1^2 + \theta_1^4}; \quad \phi_{33} = \frac{-\theta_1^3}{1 + \theta_1^2 + \theta_1^4 + \theta_1^6}, \tag{5.60}$$

and so on. The partial autocorrelations decay at an exponential rate towards zero, all being negative if $\theta_1 > 0$ or with alternating signs if $\theta_1 < 0$.

This example provides a general idea of the structure of *MA* schemes. The *MA*(q) process possesses an *ACF* with q non-zero values but an exponentially decaying, possibly harmonic, *PACF*. In this regard its behaviour is the converse of that for *AR* schemes and we shall exploit this fact in our model selection procedures in the next chapter.

5.23   One final point worth noting about the *MA*(1) scheme is that the maximum value of $|\rho_1| = 0.5$. This is another manifestation of the effect noted in Section 5.6; if observations two time periods apart are uncorrelated, adjacent values cannot be 'too strongly' correlated. Various upper limits are available for higher-order *MA* schemes; see Chanda (1962) and O. D. Anderson (1975).

5.24   Following the approach of Example 5.4, we find that the kth autocorrelation for an *MA*(q) scheme is

$$\rho_k = \frac{\theta_0\theta_k + \theta_1\theta_{k+1} + \cdots + \theta_{q-k}\theta_q}{1 + \theta_1^2 + \cdots + \theta_q^2}, \quad k \leqslant q$$

$$= 0, \quad\quad\quad\quad\quad\quad k > q, \tag{5.61}$$

where $\theta_0 = -1$ by convention.

Example 5.5   (*The Slutzky–Yule effect*)   Consider the purely random series

$$y_t = \varepsilon_t$$

and suppose that we form a moving average, in the sense of Chapter 2, as

$$u_t = \sum_{j=0}^{q} \alpha_j \varepsilon_{t-j}; \tag{5.62}$$

nothing is lost by scaling the α_j so that $\alpha_0 = 1$. Clearly (5.62) is an *MA*(q) scheme and its *ACF* is given by (5.61) with $\theta_j = -\alpha_j$. That is, when we apply a $(q + 1)$-order moving average to a purely random series we induce autocorrelation up to and including lag q. Indeed, systematic effects in the smoothed series may be nothing more than artifacts produced by the moving average. This is known as the *Slutzky–Yule effect* after its discoverers.

Example 5.6   Table 5.1 shows the autocorrelations generated in a random series by the Spencer 21-point formula whose weights were given in (3.13). The *ACF* is plotted in Fig. 5.5. From $k = 13$ the autocorrelations are small and from $k = 21$ onwards they are zero, as expected from (5.61). For small k, the

Table 5.1 Autocorrelations generated by a Spencer 21-point average

k	$\Sigma\, \alpha_j\alpha_{j+k}$	ρ_k	k	$\Sigma\, \alpha_j\alpha_{j+k}$	ρ_k
0	17 542	1.000	11	− 930	− 0.053
1	16 786	0.957	12	− 528	− 0.030
2	14 667	0.836	13	− 214	− 0.012
3	11 584	0.660	14	− 27	− 0.002
4	8 085	0.461	15	50	0.003
5	4 726	0.269	16	59	0.003
6	1 951	0.111	17	40	0.002
7	6	0.000	18	19	0.001
8	− 1 074	− 0.061	19	6	0.000
9	− 1 430	− 0.082	20	1	0.000
10	− 1 298	− 0.074	21	0	0.000

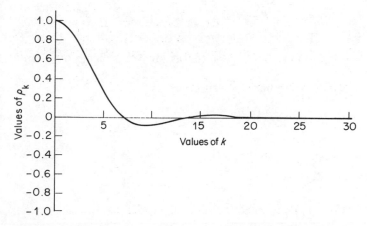

Fig. 5.5   ACF of the series generated by the Spencer 21-point formula

autocorrelations are substantial and we might expect that the formula would induce a considerable amount of smoothing as, indeed, we saw in Section 3.7.

5.25   The realization that apparently systematic fluctuations can be generated merely as the average of random events came as something of a shock when Slutzky (1927) and Yule (1927) first called attention to the fact, especially as Slutzky was able to mimic an actual trade 'cycle' of the nineteenth century very closely by a moving-average process.

It is for this reason that many econometricians now prefer to work with data that have not been seasonally adjusted.

Duality between AR and MA schemes

5.26   We now have two classes of process suitable for describing time series − the *AR* and *MA* schemes. Indeed, it may be shown that any stationary process may be approximated by either an *AR* or an *MA* process of sufficiently high order (Fuller 1976, pp. 148–51). We shall now show how AR schemes may be written in MA form and vice versa.

Consider the $AR(1)$ scheme with $u_t = y_t - \mu$

$$u_t = \phi u_{t-1} + \varepsilon_t, \tag{5.63}$$

where $|\phi| < 1$. Using the *B*-operator (Appendix 3A), $u_{t-1} = Bu_t$ and (5.63) becomes

$$(1 - \phi B)u_t = \varepsilon_t \tag{5.64}$$

or

$$u_t = \frac{1}{1 - \phi B} \varepsilon_t. \tag{5.65}$$

The term on the right in (5.65) may be expanded as a geometric series (Appendix 5A) to give

$$
\begin{aligned}
u_t &= (1 + \phi B + \phi^2 B^2 + \cdots)\varepsilon_t \\
&= \varepsilon_t + \phi\varepsilon_{t-1} + \phi^2\varepsilon_{t-2} + \cdots;
\end{aligned} \tag{5.66}
$$

that is, the $AR(1)$ scheme may be represented as an *MA* scheme of infinite order.

Generally, an $AR(p)$ scheme may be written as

$$\phi(B)u_t = \varepsilon_t, \tag{5.67}$$

where

$$\phi(B) = 1 - \phi_1 B - \phi_2 B^2 - \cdots - \phi_p B^p \tag{5.68}$$

and, given stationarity, this becomes

$$
\begin{aligned}
u_t &= \frac{1}{\phi(B)} \varepsilon_t = \psi(B)\varepsilon_t, \\
&= \varepsilon_t + \psi_1\varepsilon_{t-1} + \psi_2\varepsilon_{t-2} + \cdots,
\end{aligned} \tag{5.69}
$$

where

$$\psi(B) = 1 + \psi_1 B + \psi_2 B^2 + \cdots. \tag{5.70}$$

Expression (5.69) is known as the *random-shock* form of the model and its coefficients are known, unsurprisingly, as psi-weights.

5.27 In the same way, any finite-order *MA* scheme may be represented as an *AR* scheme of infinite extent; however, there is one problem to be overcome. In (5.65), the geometric series expansion is valid because stationarity ensures that $|\phi| < 1$. However, in the $MA(1)$ scheme

$$u_t = \varepsilon_t - \theta\varepsilon_{t-1} \tag{5.71}$$

or

$$\varepsilon_t = \frac{1}{1 - \theta B} u_t \tag{5.72}$$

no restrictions have been placed on θ. Indeed, we can show that the two $MA(1)$

schemes (5.71) and

$$u_t = \varepsilon_t - \theta^{-1}\varepsilon_{t-1} \tag{5.73}$$

both yield $\rho_1 = -\theta/(1 + \theta^2)$. To represent the $MA(1)$ scheme in AR form, we must impose the *invertibility* condition, $|\theta| < 1$, to allow the use of the geometric series expansion. Then (5.72) becomes

$$\varepsilon_t = (1 + \theta B + \theta^2 B^2 + \cdots)u_t$$

or

$$u_t = -\theta u_{t-1} - \theta^2 u_{t-2} - \cdots + \varepsilon_t. \tag{5.74}$$

In general, the $MA(q)$ scheme

$$u_t = \theta(B)\varepsilon_t,$$

where $\theta(B) = 1 - \theta_1 B - \cdots - \theta_q B^q$, may be written as

$$\varepsilon_t = \frac{1}{\theta(B)} u_t = \pi(B)u_t$$

$$= u_t - \pi_1 u_{t-1} - \pi_2 u_{t-2} - \cdots, \tag{5.75}$$

provided all the roots of the auxiliary equation

$$\theta(x) = 1 - \theta_1 x - \cdots - \theta_q x^q = 0 \tag{5.76}$$

are greater than one in absolute value.

Autocorrelation generating function

5.28   An additional benefit derived from the random shock form of the model is that we can generate the autocorrelations by a different method. Starting with

$$u_t = \psi_0 \varepsilon_t + \psi_1 \varepsilon_{t-1} + \psi_2 \varepsilon_{t-2} + \cdots,$$

where $\psi_0 = 1$, we have

$$E(u_t u_{t-k}) = \gamma_k = \sigma^2 \sum_{j=0}^{\infty} \psi_j \psi_{j+k}. \tag{5.77}$$

Define the *autocovariance generating function* as

$$G(z) = \sum_{k=-\infty}^{\infty} \gamma_k z^k, \tag{5.78}$$

where z is a 'dummy variable' and $G(z)$ is of interest solely because the coefficient of z^k is the kth autocovariance, γ_k. From (5.77) and (5.78)

$$G(z) = \sigma^2 \sum_{k=-\infty}^{\infty} \sum_{j=0}^{\infty} z^k \psi_j \psi_{j+k}. \tag{5.79}$$

Putting $s = j + k$ this gives

$$G(z) = \sigma^2 \sum_{s=0}^{\infty} z^s \psi_s \sum_{j=0}^{\infty} z^{-j} \psi_j \tag{5.80}$$

since $s - j = k$ and $\psi_k = 0$, $k < 0$. In turn, (5.80) becomes

$$G(z) = \sigma^2 \psi(z)\psi(z^{-1}), \tag{5.81}$$

where $\psi(z)$ is given by (5.70) with z replacing B. Finally, the autocorrelations may be generated from

$$G_\rho(z) = \frac{\sigma^2}{\gamma_0} G(z). \tag{5.82}$$

In using (5.81), $\psi(z) = \theta(z)$ for MA schemes and $\psi(z) = [\phi(z)]^{-1}$ for AR schemes.

Example 5.7   The $MA(1)$ scheme has

$$\psi(z) = \theta(z) = (1 - \theta z)$$

so that

$$\begin{aligned} G(z) &= \sigma^2(1 - \theta z)(1 - \theta z^{-1}) \\ &= \sigma^2(1 + \theta^2 - \theta z - \theta z^{-1}). \end{aligned}$$

Hence $\gamma_0 = \sigma^2(1 + \theta^2)$ and $\gamma_1 = \gamma_{-1} = -\theta\sigma^2$, as before.

Example 5.8   The $AR(1)$ scheme has

$$\phi(z) = (1 - \phi z)$$

leading to

$$G(z) = \frac{\sigma^2}{(1 - \phi z)(1 - \phi z^{-1})}.$$

Using (5A.7) in Appendix 5A, this becomes

$$G(z) = \frac{\sigma^2}{(1 - \phi^2)} \left\{ \frac{1}{1 - \phi z} + \frac{\phi z^{-1}}{1 - \phi z^{-1}} \right\}.$$

Since $\gamma = \sigma^2/(1 - \phi^2)$, we have

$$\begin{aligned} G_\rho(z) &= \frac{1}{1 - \phi z} + \frac{\phi z^{-1}}{1 - \phi z^{-1}} \\ &= (1 + \phi z + \phi_2 z^2 + \cdots) + (\phi z^{-1} + \phi^2 z^{-2} + \cdots). \end{aligned}$$

Hence $\rho_k = \rho_{-k} = \phi^k$, as before.

As we shall see in Chapter 10, these generating functions provide an immediate way of developing the spectrum for AR and MA schemes. Indeed, the purist may have already noticed that z should be a complex number if $G(z)$ is to exist as a doubly infinite expansion; this does not cause any problems in practice, but complex z plays a natural role in spectrum analysis; see Section 10.8.

Mixed ARMA schemes

5.29   Given the developments of the last four sections, it is natural to think in terms of combining the AR and MA components to produce a mixed model such as

$$y_t = \delta + \phi_1 y_{t-1} + \cdots + \phi_p y_{t-p} + \varepsilon_t - \theta_1 \varepsilon_{t-1} - \cdots - \theta_q \varepsilon_{t-q}, \tag{5.83}$$

or, using the *B*-operator

$$\phi(B)y_t = \delta + \theta(B)\varepsilon_t. \tag{5.84}$$

We shall refer to this as an $ARMA(p, q)$ scheme. If differencing is necessary to produce a stationary series, so that

$$w_t = \nabla^d y_t \tag{5.85}$$

and

$$\phi(B)w_t = \delta + \theta(B)\varepsilon_t, \tag{5.86}$$

we refer to this as an autoregressive-integrated-moving average scheme of orders p, d and q, or $ARIMA(p, d, q)$. Note that when $d > 0$, many time series analysts would set $\delta = 0$ rather than retain a constant drift element corresponding to a dth-order polynomial. The term *integrated* refers to the fact that the model for y_t must be 'undifferenced' using (5.85).

5.30 The conditions for stationarity and invertibility are precisely those for the *AR* and *MA* components taken separately. Further, the autocovariance generating function is given by

$$G(z) = \frac{\sigma^2 \theta(z)\theta(z^{-1})}{\phi(z)\phi(z^{-1})}. \tag{5.87}$$

Example 5.9 The $ARMA(1, 1)$ scheme is

$$y_t - \phi y_{t-1} = \delta + \varepsilon_t - \theta \varepsilon_{t-1}. \tag{5.88}$$

It follows that

$$E(y_t) = \mu = \delta/(1 - \phi),$$

$$\text{var}(y_t) = \gamma_0 = \frac{(1 + \theta^2 - 2\theta\phi)}{1 - \phi^2}\sigma^2$$

$$\gamma_k = \frac{(1 - \phi\theta)(\phi - \theta)}{1 - \phi^2}\sigma^2$$

and

$$\gamma_k = \phi\gamma_{k-1}, \qquad k \geqslant 2.$$

Thus the *ACF* of the $ARMA(1, 1)$ scheme behaves like that of an $AR(1)$ scheme *after* lag 1. Likewise, the *PACF* behaves like that of an $MA(1)$ scheme after the first lag.

5.31 We now have a choice of *AR*, *MA* and *ARMA* schemes with relationships between each. When we come to examine the analysis of sample data in the next two chapters, we shall look for the simplest representation that is consistent with the observed series. Indeed, Box and Jenkins (1976) in their ground-breaking book on *ARIMA* model-building, refer to this as the *principle of parsimony* and use it as a major touchstone in model development.

Seasonal series

5.32 Thus far we have not considered the possibility that seasonal components may be present in a series. To go to the other extreme for a moment, suppose that a series consisted of *only* a seasonal component. In effect, we could consider a separate submodel for each month or whatever. Thus, if there are s 'months' in a 'year', we might consider a seasonal $AR(1)$ scheme of the form

$$y_t = \delta + \Phi y_{t-s} + \varepsilon_t, \tag{5.89}$$

where we shall follow the convention that capital Greek letters are reserved for seasonal parameters. If it is reasonable to assume that (5.89) holds for each month of the year, we find immediately from (5.89) that

$$\rho_1 = \rho_2 = \cdots = \rho_{s-1} = 0$$
$$\rho_s = \Phi$$

and all other ρ_j are zero except

$$\rho_{ks} = \Phi^k, \qquad k \geqslant 1. \tag{5.90}$$

Further, the *PACF* would comprise the single non-zero element $\Phi_{11} = \Phi$. In like manner, a seasonal $MA(1)$ scheme could be defined as

$$y_t = \mu + \varepsilon_t - \Theta \varepsilon_{t-s} \tag{5.91}$$

which has all autocorrelations zero except

$$\rho_s = -\frac{\Theta}{1+\Theta^2}. \tag{5.92}$$

Generally, a seasonal $ARMA(P, Q)$ model may be defined as

$$y_t - \Phi_1 y_{t-s} - \cdots - \Phi_P y_{t-P_s} = \delta + \varepsilon_t - \Theta_1 \varepsilon_{t-s} - \cdots - \Theta_Q \varepsilon_{t-Qs} \tag{5.93}$$

or

$$\phi(B^s) y_t = \delta + \Theta(B^s) \varepsilon_t, \tag{5.94}$$

where $B^s y_t = y_{t-s}$ and $\Phi(B^s)$, $\Theta(B^s)$ are appropriate polynomials in B^s. All our earlier results carry over with B^s replacing B throughout.

The problem is that with the exception of occasional items such as the sales of Christmas cards, few series are purely seasonal. We need to combine seasonal and regular effects into a single model.

Mixed seasonal models

5.33 There are two ways of combining seasonal and regular effects. One is to combine both effects into a single polynomial containing terms in both B and B^s; the other is to take the product of two polynomials. We pursue the second path for the following reason. Suppose that the data are first deseasonalized; if y_t denotes the original series and x_t the deseasonalized series, we might have

$$x_t = \pi(B^s) y_t. \tag{5.95}$$

A regular *ARIMA* model for x_t might then take the form

$$x_t = \psi(B)\varepsilon_t. \tag{5.96}$$

Combining (5.95) and (5.96) suggests the multiplicative scheme

$$y_t = \frac{\psi(B)}{\pi(B^s)}\varepsilon_t$$

or, more generally

$$\phi(B)\Phi(B^s)y_t = \theta(B)\Theta(B^s)\varepsilon_t. \tag{5.97}$$

The ordering of the operators is immaterial. In keeping with our earlier discussion of the components of a time series, (5.97) allows a more ready interpretation of the coefficients than does a single polynomial form involving both regular and seasonal terms.

5.34 We have used regular differences to eliminate trends in the series; should we use seasonal trends such as

$$\nabla_s y_t = y_t - y_{t-s} \tag{5.98}$$

to remove trends in the seasonal components? The answer is yes, but in a way that may seem rather surprising. Suppose we start with the monthly series y_t and produce a deseasonalised series using a moving average over all twelve months:

$$\begin{aligned} x_t &= y_t + y_{t-1} + \cdots + y_{t-11} \\ &= (1 + B + \cdots + B^{11})y_t. \end{aligned} \tag{5.99}$$

If we then difference the x_t series to remove trends, we obtain

$$\begin{aligned} \nabla x_t &= (1 - B)x_t \\ &= (1 - B)(1 + B + \cdots + B^{11})y_t \\ &= (1 - B^{12})y_t = \nabla_{12}y_t, \end{aligned} \tag{5.100}$$

so that seasonal differencing accomplishes an element of both deseasonalisation and trend removal. Comparing (5.97) and (5.100), we see that multiplicative models are still preferable, although individual terms may not be interpreted unambiguously.

5.35 The models of form (5.97) used in practice usually contain relatively few seasonal terms; we now proceed to give a few examples. We shall use the notation $ARIMA(p, d, q)(P, D, Q)_s$ to refer to a mixed seasonal model of period s with regular and seasonal components of order p, P for AR, q, Q for MA and d, D for differencing.

Example 5.10 The $ARIMA(1, 0, 0)(1, 0, 0)_s$ model is

$$(1 - \phi B)(1 - \Phi B^s)y_t = \delta + \varepsilon_t \tag{5.101}$$

or

$$y_t = \delta + \phi y_{t-1} + \Phi y_{t-s} - \phi\Phi y_{t-s-1} + \varepsilon_t.$$

It follows that

$$E(y_t) = \delta/(1-\phi)(1-\Phi)$$

$$\text{var}(y_t) = \frac{\sigma^2(1+\phi^s\Phi)}{(1-\phi^s\Phi)(1-\phi^2)(1-\Phi^2)}$$

$$\rho_k = \frac{\phi^k + \phi^{s-k}\Phi}{1+\phi^s\Phi}, \qquad k = 1, \ldots, s$$

$$\rho_{k+s} = \frac{\phi^{k+s} + \Phi^2\phi^{s-k}}{1+\phi^s\Phi}, \qquad k = 1, \ldots, s \quad \text{and so on.}$$

The autocorrelation function tends to have extrema at lags 1, s, 2s and so on.

Example 5.11   The $ARIMA(0,0,1)(0,0,1)_s$ model is

$$y_t = \mu + (1 - \theta B)(1 - \Theta B^s)\varepsilon_t \tag{5.102}$$

or

$$y_t = \mu + \varepsilon_t - \theta\varepsilon_{t-1} - \Theta\varepsilon_{t-s} + \theta\Theta\varepsilon_{t-s-1}.$$

It follows that

$$E(y_t) = \mu$$
$$\text{var}(y_t) = \sigma^2(1+\theta^2)(1+\Theta^2)$$
$$\rho_1 = -\theta/(1+\theta^2)$$
$$\rho_s = -\Theta/(1+\Theta^2)$$
$$\rho_{s+1} = \rho_{s-1} = \rho_1\rho_s$$
$$\rho_k = 0 \text{ otherwise.}$$

The simple structure for the *ACF* is a major attraction of the mixed-seasonal models. When (5.102) is used for $w_t = \nabla\nabla_s y_t$, the resulting *ARIMA* $(0,1,1)(0,1,1)_s$ scheme is often known as the 'airline' model, following its use by Box and Jenkins (1976) on data similar to those of Table 1.3. This scheme often works surprisingly well for seasonal series, as we shall see in Section 6.24.

Exercises

5.1   Plot the *ACF* and the *PACF* for the Markov scheme with (i) $\phi = 0.8$, (ii) $\phi = -0.8$.

5.2   For the Yule scheme (5.38), verify equation (5.46):

$$\rho_k = \phi_1\rho_{k-1} + \phi_2\rho_{k-2}.$$

Use (5.50) and (5.51) to show that $\phi_{kk} = 0$, $k \geqslant 3$.

5.3   Plot the form of the *ACF* and the *PACF* for the Yule scheme (5.38) with (i) $\phi_1 = 1.1$, $\phi_2 = -0.3$, (ii) $\phi_1 = -0.2$, $\phi_2 = -0.24$, (iii) $\phi_1 = 0.8$, $\phi_2 = -0.8$.

5.4   By flipping a coin, generate a sequence of 20 independent observations from the distribution

$$P(\varepsilon_t = +1) = P(\varepsilon_t = -1) = 0.5.$$

Starting with $u_0 = 0$, generate time series corresponding to the $AR(1)$ model $u_t = \phi u_{t-1} + \varepsilon_t$ when (i) $\phi = 0.8$, (ii) $\phi = -0.8$. Plot both series and compare the results. (*Hint*: Consider some of the simple tests described in Chapter 2.)

5.5   Using your data from Exercise 5.4 and starting with $\varepsilon_0 = 0$, generate time series corresponding to the $MA(1)$ model $u_t = \varepsilon_t - \theta\varepsilon_{t-1}$ when (i) $\theta = 1$, (ii) $\theta = -1$. Plot both series and compare the results.

5.6   Derive the partial autocorrelations ($k = 1, 2, 3$) for the $MA(1)$ scheme given in (5.60) using (5.50–51).

5.7   Find the ACF of the $MA(2)$ scheme.

5.8   Express the $MA(2)$ scheme in AR form, going as far as terms in y_{t-4}.

5.9   Show that these four $MA(2)$ schemes all have the same ACF, but that only one is invertible.
   (i) $\theta_1 = 1.3, \theta_2 = -0.40$
   (ii) $\theta_1 = 1.75, \theta_2 = -0.625$
   (iii) $\theta_1 = 2.80, \theta_2 = -1.60$
   (iv) $\theta_1 = 3.25, \theta_2 = -2.50$

5.10 Derive the autocovariances for the $ARMA(1, 1)$ scheme given in Example 5.9.

5.11 For *any ARMA* scheme, show that

$$E(y_t\varepsilon_t) = \sigma^2.$$

5.12 Plot the ACF of the $(1, 0, 0)(1, 0, 0)_{12}$ model when $\phi = \Phi = 0.6$.

5.13 Plot the ACF of the $(0, 0, 1)(0, 0, 1)_{12}$ model when (i) $\theta = \Theta = 0.6$, (ii) $\theta = \Theta = -0.6$.

5.14 Demonstrate that the autocorrelations listed in Example 5.11 are correct.

Appendix 5A: The geometric series

The geometric series may be written as

$$S = 1 + \alpha + \alpha^2 + \cdots + \alpha^n + \cdots, \tag{5A.1}$$

where α is a constant. Clearly,

$$\alpha S = \alpha + \alpha^2 + \cdots + \alpha^{n+1} + \cdots$$

so that

$$S - \alpha S = (1 - \alpha)S = 1,$$

or

$$S = 1/(1 - \alpha), \tag{5A.2}$$

provided that $|\alpha| < 1$ so the series is convergent. To see this more clearly, let

$$S_n = 1 + \alpha + \cdots + \alpha^n,$$

then $S_n - \alpha S_n = 1 - \alpha^{n+1}$ or

$$S_n = (1 - \alpha^{n+1})/(1 - \alpha). \tag{5A.3}$$

Provided $|\alpha| < 1$, $\lim_{n \to \infty} \alpha^{n+1} = 0$ and so

$$\lim_{n \to \infty} S_n = \frac{1}{1 - \alpha} \tag{5A.4}$$

The geometric series expansion (5A.1) is also useful when dealing with expressions involving the backshift operator, B. Thus, provided $|\alpha| < 1$,

$$\frac{1}{1 - \alpha B} = 1 + \alpha B + \alpha^2 B^2 + \cdots \tag{5A.5}$$

so that

$$\frac{1}{1 - \alpha B} y_t = y_t + \alpha y_{t-1} + \alpha^2 y_{t-2} + \cdots. \tag{5A.6}$$

A further useful property is that if

$$g(B) = \frac{1}{(1 - \alpha B)(1 - \alpha B^{-1})},$$

then

$$g(B) = \left\{ \frac{1}{1 - \alpha B} + \frac{\alpha B^{-1}}{1 - \alpha B^{-1}} \right\} / (1 - \alpha^2). \tag{5A.7}$$

Also, when $\gamma \neq \alpha$, we have

$$\frac{1}{(1 - \alpha B)(1 - \gamma B)} = \left\{ \frac{\alpha}{(1 - \alpha B)} - \frac{\gamma}{1 - \gamma B} \right\} / (\alpha - \gamma). \tag{5A.8}$$

Appendix 5B: Solution of difference equations

First order. The first-order difference equation may be written as

$$z_t - a z_{t-1} = c. \tag{5B.1}$$

On setting $u_t = z_t - d$, where $c = d(1 - a)$, we obtain the homogeneous form

$$u_t - a u_{t-1} = 0. \tag{5B.2}$$

This has the general solution

$$u_t = A a^t,$$

where $u^0 = A$; hence

$$z_t = d + A a^t. \tag{5B.3}$$

If $|a| < 1$, $z_t \to d$ as $t \to \infty$, but if $|a| \geqslant 1$, z_t does not converge to a limit (the series is explosive if $a \geqslant 1$ or $a < -1$ and oscillatory if $a = -1$).
Second order. The second-order difference equation is

$$z_t - a z_{t-1} - b z_{t-2} = c \tag{5B.4}$$

or, in homogeneous form

$$u_t - a u_{t-1} - b u_{t-2} = 0, \tag{5B.5}$$

where $d = c(1 - a - b)$. The solution of (5B.5) depends upon the roots of the auxiliary equation

$$1 - ax - bx^2 = 0; \tag{5B.6}$$

three different cases are possible.

Case I: roots real and different, α_1 and α_2, say. The solution is

$$u_t = A_1 \alpha_1^{-t} + A_2 \alpha_2^{-t}. \tag{5B.7}$$

Case II: roots real and equal, to α say. The solution is

$$u_t = (A_1 + A_2 t)\alpha^{-t}. \tag{5B.8}$$

Case III: roots complex, $\alpha_1 = \beta^{-1}e^{i\theta}$ and $\alpha_2 = \beta^{-1}e^{-i\theta}$ say, where $e^{i\theta} = \cos\theta + i\sin\theta$, $i^2 = -1$. The solution is

$$u_t = A\beta^{-t}\sin(\theta t + \omega), \tag{5B.9}$$

where

$$|\alpha_1| = |\alpha_2| = \beta = (-b).^{-\frac{1}{2}} \tag{5B.10}$$

In all cases, the constants may be determined from the initial values, u_0 and u_1. The series converges to a limit only if α_1, α_2, α or β, respectively, are greater than one in absolute value. The form of (5B.6) has been chosen for consistency with (5.45) and (5.46).

6

Serial correlations and model identification

6.1 In Chapter 5, we developed a class of stochastic models for time series which could be characterized by their autocorrelations. We now develop inferential tools which enable us to describe the behaviour of an observed series and, later, to fit a model to the observations. We begin by defining the *serial correlations*, or *sample autocorrelations*; both terms are common in the literature and will be used interchangeably in the remainder of the book. We shall then examine the sampling distributions of the serial correlations. Generally, only asymptotic results are available, and the mathematical development of these is beyond the scope of this volume; the interested reader is referred to Kendall *et al.* (1983, Chapter 48) and Fuller (1976, Chapter 6).

Definitions

6.2 We may define the sample variance as

$$\hat{\gamma}_0 = \frac{1}{n} \sum_{i=1}^{n} (y_i - \bar{y})^2 \tag{6.1}$$

and the kth sample covariance as

$$\hat{\gamma}_k = \frac{1}{n} \sum_{i=1}^{n-k} (y_i - \bar{y})(y_{i+k} - \bar{y}), \tag{6.2}$$

where $n\bar{y} = \sum_{i=1}^{n} y_i$. Then the kth serial correlation is

$$r_k = \hat{\gamma}_k / \hat{\gamma}_0, \qquad k = 1, 2, \ldots; \tag{6.3}$$

by symmetry, $r_k = r_{-k}$. Inspection of (6.1)–(6.3) reveals that we have made several rather arbitrary decisions concerning these definitions. For example, we could modify these in any of the following ways:

(1) Use $(n - 1)$ and $(n - k - 1)$ in place of n, in (6.1) and (6.2), respectively. This would take account of the degree of freedom lost in estimating μ by $\bar{y}$ and the fewer number of terms in (6.2).

(2) Calculate r_k as one would a regular correlation coefficient between the pairs of values (y_1, y_{k+1}), (y_2, y_{k+2}), ..., (y_{n-k}, y_n), or

$$r_k = \frac{\sum_{i=1}^{n-k} (y_i - \bar{y}_{1k})(y_{i+k} - \bar{y}_{2k})}{\{\sum_{i=1}^{n-k} (y_i - \bar{y}_{1k})^2 \sum_{i=1}^{n-k} (y_{i+k} - \bar{y}_{2k})^2\}^{1/2}}, \qquad (6.4)$$

where

$$(n - k)\bar{y}_{1k} = \sum_{i=1}^{n-k} y_i \text{ and } (n - k)\bar{y}_{2k} = \sum_{i=1}^{n-k} y_{i+k}.$$

6.3   Version (6.3) is often preferred on the grounds that it corresponds to the maximum likelihood estimator when the data are normally distributed. The estimator tends to be biased towards zero as k/n increases, but this damping of the estimates by the factor $(n - k)/n$ is often considered desirable. This is so because model selection is often based upon visual inspection of the *correlogram*, or sample autocorrelation function (*SACF*) and the damping helps the analyst to avoid giving undue weight to large but possibly spurious r_k values based upon a small number of observations. Sargan (1953) demonstrates theoretically why the correlogram fails to damp for shorter series, although its general oscillatory shape for *AR* series may be preserved.

At one time, version (6.4) was discounted because of the computational effort involved, but this is no longer a substantive objection. When dealing with series that appear to be close to non-stationarity, (6.4) may be preferable to (6.3). Both (6.3) and (6.4) produce sample covariance matrices which are positive semidefinite, that is,

$$\sum_{i=1}^{m} \sum_{j=1}^{m} a_i a_j r_{|i-j|} \geq 0, \qquad (6.5)$$

for all $m \geq 1$ and all possible choices of $\{a_1, a_2, ..., a_m\}$, not all zero. Property (6.5) is clearly desirable for inferential purposes since it ensures that the estimated variance of any function $\sum a_i y_i$ is non-negative. It should be noted that (6.5) is not satisfied by modification (1) suggested above.

6.4   Once the serial correlations have been computed, the sample partial autocorrelations may be found using the sample analogue of the Durbin–Levinson algorithm described in Section 5.19. This procedure is not always stable numerically, and the method developed by Pagano (1972) using the Cholesky decomposition is preferable for general computational purposes.

Central limit theorem

6.5   Given a random sample of n observations $(y_1, ..., y_n)$ from a population with mean μ and variance σ^2, the central limit theorem states that the distribution of

$$z_n = (\bar{y}_n - \mu)\sqrt{n}/\sigma \qquad (6.6)$$

approaches the standard normal, $N(0, 1)$, as $n \to \infty$. The conditions under which this result holds are generally mild, but do require independence of the observations, clearly an unrealistic assumption for time-series data. To see how we can accommodate dependence, consider the usual $AR(1)$ scheme:

$$y_t = \mu(1 - \phi) + \phi y_{t-1} + \varepsilon_t. \tag{6.7}$$

We may consider

$$\sum_1^n \varepsilon_t = \sum_1^n \{y_t - \phi y_{t-1} - \mu(1 - \phi)\}$$

$$= n\mu(1 - \phi) + n\bar{y}_n(1 - \phi) + \phi(y_n - y_0). \tag{6.8}$$

Since the ε_t are independent and identically distributed $(0, \sigma^2)$, the central limit theorem holds and the distribution of

$$n^{-1/2} \sum \varepsilon_t / \sigma \text{ approaches } N(0, 1). \tag{6.9}$$

Clearly the distribution of the right-hand side of (6.8) must be of the same form; the first term is a constant and the last term becomes negligible as n increases, so it follows that z_n is asymptotically $N(0, 1)$, where

$$E(\bar{y}_n) = \mu$$

and

$$\text{var}(\bar{y}_n) = \frac{1}{n} \left\{ \gamma_0 + \frac{2}{n} \sum_{k=1}^n (n - k)\gamma_k \right\}, \tag{6.10}$$

where $\gamma_k = \text{cov}(y_t, y_{t+k})$. It may also be shown that $\bar{y}_n$ is asymptotically efficient in the class of linear unbiased estimators; for details, see Fuller (1976, pp. 230–6, 244–5). In general, stationarity plus a finite variance for the ε_t is sufficient for the central limit theorem to hold. Expression (6.10) is completely general but simplifies in particular cases. For example, when the process is $AR(1)$,

$$\gamma_0 = \sigma^2 / (1 - \phi^2), \qquad \gamma_k = \phi^k \gamma_0$$

and, for large n, (6.10) reduces to

$$\text{var}(\bar{y}_n) = \frac{\sigma^2}{n(1 - \phi)^2} = \frac{\gamma_0(1 + \phi)}{n(1 - \phi)} \tag{6.11}$$

as may be verified from (6.8) directly if we ignore the last term on the right.
6.6 Central limit theorems may also be proved for the serial correlations; see Fuller (1976, pp. 245–57) for details. Using the approach described in Kendall *et al.* (1983, pp. 547–52), it may be shown that

$$E(r_j) = \rho_j + O(n^{-1}), \tag{6.12}$$

$$\text{var}(r_j) = \frac{1}{n} \sum_{-\infty}^\infty [\rho_i^2 + \rho_{i-j}\rho_{i+j} - 4\rho_i\rho_j\rho_{i+j} + 2\rho_i^2\rho_j^2] + O(n^{-2}), \tag{6.13}$$

where $O(n^{-j})$ means that the next terms in the series are some fixed number multiplied by n^{-j} and can be ignored in longer series. These results were derived originally by Bartlett (1946) under the assumption that the sixth moment of the errors exists. Recent work by Burn (1985) shows that this condition can be relaxed to the requirement that the variance of the errors should exist.
6.7 The results we have given are asymptotic and we may expect them to be

rather inaccurate for short series. For example, the $AR(1)$ series with r_1 given by (6.3) has

$$E(r_1) = \phi - (1 + 4\phi)/n + O(n^{-2})$$

$$E(r_j) = \phi^j\left(1 - \frac{j}{n}\right) - \left\{\frac{(1+\phi)}{(1-\phi)}(1 - \phi^j) + 2j\phi^j\right\}/n + O(n^{-2}), \qquad j > 1.$$

For example, when $\phi = 0.5$ and $n = 25$, the expected value of r_1 given by the first two terms is 0.38 not 0.5, a sizeable bias towards zero.

6.8 For short series, therefore, it may be desirable to remove the bias of order $1/n$ by a device due to Quenouille (1956). If we split the series into two halves, so that r is the serial correlation for the whole series and $r_{(1)}$, $r_{(2)}$ those for the halves, then

$$R = 2r - \tfrac{1}{2}\{r_{(1)} + r_{(2)}\} \tag{6.14}$$

will be unbiased to order n^{-2}. For if

$$E(r) = \rho + \frac{k}{n} + O(n^{-2}),$$

where k is a constant independent of n,

$$E\{r_{(1)}\} = E\{r_{(2)}\} = \rho + \frac{2k}{n} + O(n^{-2}),$$

and on substitution in (6.14) we find that

$$E(R) = \rho + O(n^{-2}). \tag{6.15}$$

Unfortunately, the variance of R is generally higher than that of the original r.

6.9 When the emphasis is on model-building, a major issue is whether the correlogram suggests non-randomness in the series; recall our discussion of tests for trend in Section 2.7. In this case, it is important to ensure that we have a valid estimate of var(r_j). In particular, (6.13) may not be very accurate for shorter series, particularly when the ρ_j are replaced by their estimates. This has led to the development of a computationally intensive re-sampling scheme known as the *bootstrap* (Efron, 1982). For many problems, this appears to give better estimates of true sampling variances than do the asymptotic formulae, although the method has had only limited application in time series analysis (cf. Findley, 1986; Holbert and Son, 1986).

Notwithstanding the reservations we have just expressed, we believe that approximate tests based upon (6.12) and (6.13) will usually be sufficiently accurate for model-building purposes.

6.10 We now give some examples using (6.13) which illustrate the effects of autocorrelation in the series upon the sampling variances of the serial correlations.

Example 6.1 For the random series $\gamma_0 = \sigma^2$, $\rho_k = 0$, $k > 0$ and we obtain

$$\mathrm{var}(r_j) \doteq 1/n. \tag{6.16}$$

For the $MA(q)$ scheme, when $j > q$, we obtain

$$\text{var}(r_j) = \frac{1}{n} \left\{ 1 + 2 \sum_{i=1}^{q} \rho_i^2 \right\}. \tag{6.17}$$

Example 6.2 For the $AR(1)$ scheme (6.7), when j is large, we find

$$\text{var}(r_j) = \frac{1}{n} \left(1 + 2 \sum_{1}^{j} \phi^{2i} \right), \tag{6.18}$$

which fairly rapidly approaches the limiting value

$$\frac{1}{n} (1 + \phi^2)/(1 - \phi^2),$$

unless $|\phi|$ is close to 1. Also, it may be shown that the correlation between r_j and r_{j+k} is approximately

$$\phi^k \{ (k + 1) - (k - 1)\phi^2 \}/(1 + \phi^2). \tag{6.19}$$

For example, when $k = 1$ and $\phi = 0.5$, the correlation is 0.80; when ϕ increases to 0.9, it becomes 0.994! This very high correlation between successive terms must be borne in mind when interpreting the correlogram.

Exact results

6.11 As might be expected, exact results on the sampling distributions of the serial correlations are available only under rather restrictive conditions. Nevertheless, they serve as a useful check upon the asymptotic results given earlier. Detailed derivations are not given here but may be found in Kendall *et al.* (1983, pp. 553–70). In all cases, it is assumed that the errors are normally distributed; for comparative purposes, we use $r_1^* = nr_1/(n - 1)$ rather than r_1 in (6.3).

(a) Moran (1947/48; 1967) showed that, when the observations are independent,

$$E(r_1^*) = -\frac{1}{n - 1} \tag{6.20}$$

and

$$\text{var}(r_1^*) = \frac{(n - 2)^2}{(n - 1)^3} = \frac{1}{n + 1} + O(n^{-2}). \tag{6.21}$$

In fact, for *any* error distribution, (6.20) holds and

$$\text{var}(r_1^*) \leqslant \frac{1}{n - 1} + O(n^{-2}). \tag{6.22}$$

(b) By defining the serial correlation coefficient to be circular (i.e. by taking y_{n+1} to be y_1), we have

$$r_{1c} = \frac{y_1 y_2 + y_2 y_3 + \cdots + y_n y_1 - n\bar{y}^2}{\sum_{i=1}^{n} (y_i - \bar{y})^2}. \tag{6.23}$$

R. L. Anderson (1942) was able to obtain an explicit form for the distribution of r_{1c} in the case of a random normal series. The distribution is cumbrous but tends to normality quickly and yields

$$E(r_{1c}) = -\frac{1}{n-1} \qquad (6.24)$$

$$\text{var } r_{1c} = \frac{n(n-3)}{(n+1)(n-1)^2}.$$

About 30 years later it was discovered that some of Anderson's work had been anticipated in a most remarkable way by Ernst Abbe in 1854 – see Kendall (1971).

(c) Dixon (1944) showed that a very close approximation to the Anderson distribution is given by the Beta distribution

$$\frac{\Gamma(\tfrac{1}{2}n + \tfrac{1}{2})}{\Gamma(\tfrac{1}{2}n)\Gamma(\tfrac{1}{2})}(1 - r^2)^{(n-2)/2} \qquad (6.26)$$

for which

$$E(r_{1c}) = -\frac{1}{n-1}$$

$$\text{var } r_{1c} = \frac{n(n-2)}{(n+1)(n-1)^2}. \qquad (6.27)$$

(d) Madow (1945) and Leipnik (1947) extended the results to the Markov, or $AR(1)$, scheme, obtaining a distribution of type (6.26) with

$$E(r_{1c}|\rho) = \frac{n\rho}{n+2} \qquad (6.28)$$

$$\text{var}(r_{1c}|\rho) = \frac{1-\rho^2}{n} \qquad (6.29)$$

(e) Daniels (1956) developed expressions for the sampling distribution of r_1 for the non-circular Markov scheme using a saddlepoint approximation. This is extremely accurate for small values of $|\rho|$, but less so for large values.

(f) Phillips (1978) gives an Edgeworth expansion which is particularly useful in the tails.

Partial serial correlations

6.12 The partial serial correlations may be computed using (5.49) and (5.50), as observed in Section 6.4. Denoting these by $\hat{\phi}_{jj}$, it may be shown that

$$E(\hat{\phi}_{jj}) = \phi_{jj} + O(n^{-1}) \qquad (6.30)$$

and that for an $AR(p)$ scheme

$$\text{var}(\hat{\phi}_{jj}) \simeq 1/n, \qquad j > p; \qquad (6.31)$$

these results were originally established by Quenouille (1947).

Durbin (1980a,b) has developed a general Edgeworth expansion approach which allows the determination of sampling distributions in a wide range of circumstances, including those of r_1 and higher-order partial serial correlation coefficients.

Model identification

6.13   Now that we have established the approximate sampling distributions of the serial and partial serial correlations, we may use these sample statistics to help us select, or identify, an appropriate *ARMA* model. The general approach is straightforward. We plot the correlogram (or sample autocorrelation function, *SACF*) and the partial correlogram (or sample partial autocorrelation function, *SPACF*) and examine these to determine one or more possible models. These models may then be fitted and examined using the methods described in the next chapter. The reason for this initial model selection phase is that if an inappropriate model is used, the estimation procedure may fail to converge and the researcher is left empty handed! Model identification procedures allow the investigator to build up a model systematically, taking account of the main features exhibited in the *SACF* and *SPACF*. These diagnostics may be computed from the residuals to see whether any systematic variation remains.

6.14   Initially our discussion concentrates on non-seasonal models. The extension to seasonal models is covered in Sections 6.23–24. Table 6.1 summarises the expected behaviour of the *ACF* and *PACF* for different *ARMA* schemes and Fig. 6.1 shows typical patterns for $ARMA(p, q)$ schemes, $p + q \leqslant 2$.

6.15   The next question that arises is how well do the correlogram and partial correlogram conform to their expected patterns? We have seen already in Section 6.10 that successive terms in the sample *ACF* may be highly correlated. Further, as noted by O. D. Anderson (1980), for *any* observed series with the

Table 6.1 Behaviour of ACF and PACF for selected ARMA schemes

MODEL	ACF	PACF
$AR(1)$	decays exponentially, (alternating signs if $\phi < 0$)	single spike
$MA(1)$	single spike	decays exponentially (alternating signs if $\theta < 0$)
$AR(p)$	decays exponentially, may contain damped oscillations	p spikes
$MA(q)$	q spikes	decays exponentially, may contain damped oscillations
$ARMA(p, q)$	both decay exponentially and may contain damped oscillations	

r_k defined by (6.3)

$$\sum_{k=1}^{n} r_k = -\tfrac{1}{2}; \qquad (6.32)$$

which means that the average r is negative even for schemes like $AR(1)$ with $\phi > 0$, where $\rho_k > 0$ for all k.

To illustrate the behaviour of the sample functions, several examples are given in Fig. 6.2. In each case, a sample of size $n = 100$ was used and the errors were normally distributed. Figure 6.2(a) shows the sample ACF for a purely random series or $y_t = \varepsilon_t$. From (6.16) individual coefficients have standard errors of $1/\sqrt{n} = 0.1$ and the coefficients lie well inside the interval ± 2 standard errors, or ± 0.2. This $SACF$ confirms the lack of any apparent structure in the series.

Figure 6.2(b) shows the first 20 terms of the $SACF$ of an $AR(1)$ scheme with $\phi = 0.75$. The exponential decay in the r_k is apparent although it is somewhat more rapid than the expected pattern in Fig. 6.1(a). The secondary peak around lag 15 is unexpected and has no parallel in the original ACF. Figure 6.2(c) shows the sample ACF for lags up to 50, rather than 20. We now see a long string of negative values, as is required to satisfy (6.32), even though the theoretical ACF would have $\rho_k > 0$ for all k and $\rho_k \doteq 0$ for all $k > 20$. As noted earlier, reasons for the slow damping of the sample ACF have been given by Sargan (1953); if version (1) of r_k is used rather than (6.3), the lack of damping becomes even more pronounced. The sample $PACF$ for the $AR(1)$ scheme is given in Fig. 6.2(d). This gives a much clearer picture and suggests an $AR(1)$ process, with $r_1 = 0.77$.

6.16 From these preliminary explorations, we may draw several tentative conclusions:

(a) The high-order r_k may be unreliable as a guide to model identification; most time series analysts favour taking the maximum lag to be at most $n/3$ or $n/4$, although a somewhat higher value may be useful for seasonal (especially monthly) series. Of course, greater reliability is to be expected as n increases. Kendall (1946) illustrated this by considering samples ranging from $n = 60$ to $n = 480$.

(b) We should be careful to allow for sampling variation in the sample plots. Many computer programs now routinely plot the limits $\pm 2SE(r_k)$ on the sample ACF, where the SE is calculated iteratively using (6.17) with $q = k - 1$ at each stage and ρ_j is estimated by r_j. The sample $PACF$ has $SE(\hat{\phi}_{kk}) \doteq 1/\sqrt{n}$ as in (6.31) and $\pm 2SE$ limits may be plotted in this diagram also. Even with this precaution, it is possible to be misled by spurious 'spikes' on the plots since the performance of 20 tests at the $\alpha = 0.05$ level would lead to an average of one significant result just by chance! A conservative correction procedure is to use a Bonferroni correction and test each coefficient at the level $\alpha_K = \alpha/K$ when K serial correlations are tested. For example, with $K = 20$, $\alpha = 0.05$, we have $\alpha_K = 0.0025$ and the cutoff value from normal tables is 3.03 rather than 1.96.

(c) The sample $PACF$ is more useful for detecting AR schemes; conversely, the sample ACF does a better job for MA schemes.

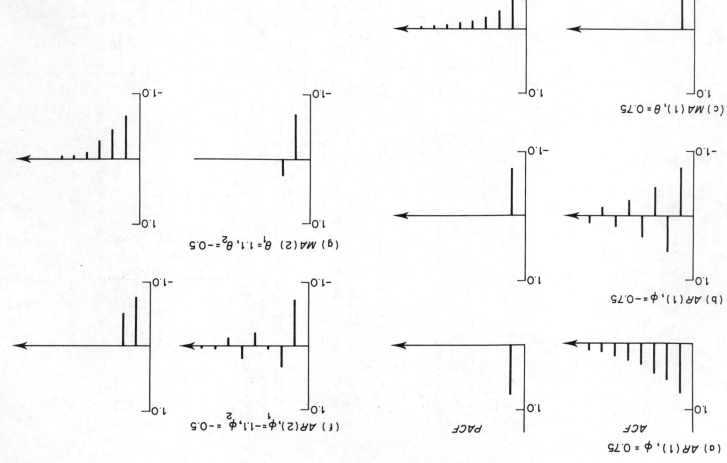

(d) $MA(1)$, $\theta = -0.75$

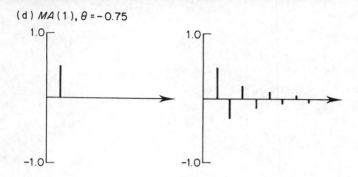

(h) $MA(2)$ $\theta_1 = -1.1$, $\theta_2 = -0.5$

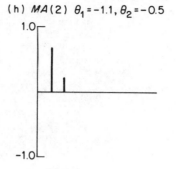
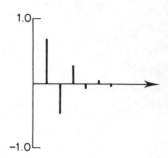

(e) $AR(2)$, $\phi_1 = 1.1$, $\phi_2 = -0.5$

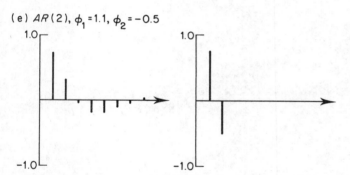

(i) $ARMA(1,1)$, $\phi = 0.9$, $\theta = 0.5$

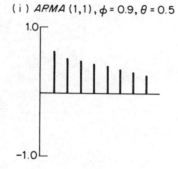
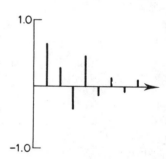

Fig. 6.1 Typical ACF and PACF plots for ARMA schemes

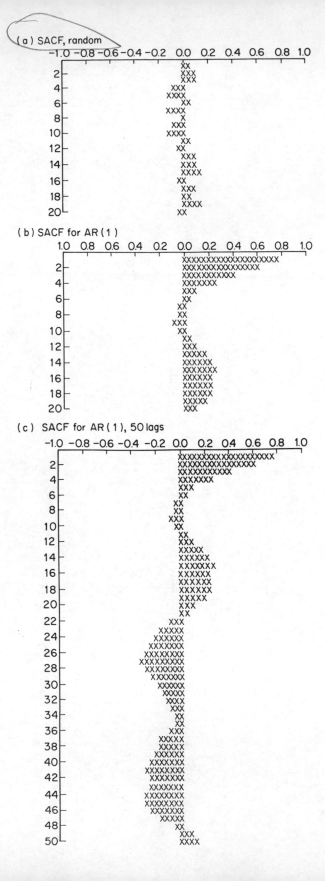

(d) SPACF for AR (1), 50 lags

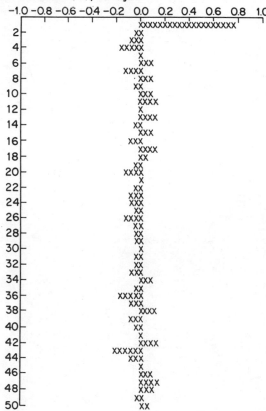

Fig. 6.2 Correlograms for simulated series containing $n = 100$ observations (a) sample ACF for a purely random series; (b) and (c) sample ACF for AR(1) with $\phi = 0.75$, showing 20 and 50 terms respectively; (d) sample PACF for AR(1) with $\phi = 0.75$

Although one would hesitate to draw these conclusions solely on the evidence of Fig. 6.2, there is a general consensus among time series analysts that they are valid.

6.17 As further examples, Fig. 6.3 shows the sample *ACF* and *PACF* for the $MA(2)$ scheme with $\theta_1 = 1.1$, $\theta_2 = -0.5$; $n = 100$ and normal errors as before. The reader may wish to draw in the $\pm 2SE$ limits to assist in model selection. Figure 6.3 is consistent with an $MA(2)$ scheme rather than $MA(1)$ since $r_1 = -0.66$ well outside the range $|r_1| \leqslant 0.5$ for an $MA(1)$ process. Note that the sample *ACF* is more useful this time. Figure 6.4 shows the correlograms for an $AR(2)$ scheme with $\phi_1 = 1.1$, $\phi_2 = -0.5$. The *PACF* clearly suggests an $AR(2)$ process. The reader may find it useful to compare these sample results with Figs 6.1(g, e), respectively.

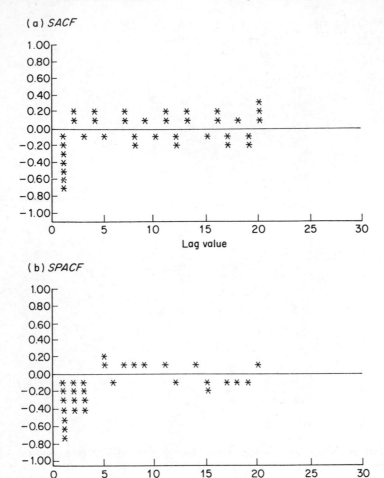

Fig. 6.3   SACF and SPACF for MA(2) with $\theta = 1.1$, $\theta_2 = -0.5$

Detecting non-stationarity

6.18   Thus far, we have assumed that the data were generated by a stationary process, but clearly this may not be true in practice. How is non-stationarity to be detected? The rank tests described in Sections 2.9–10 represent an effective way of recognizing the presence of a trend, but may not detect a random walk or 'stochastic' trend.

The correlograms will often suggest the possibility of non-stationarity in the mean. The typical shape of the sample *ACF* in such cases is a very regular *linear* decay; indeed for the random walk Wichern (1973) showed that, for k/n

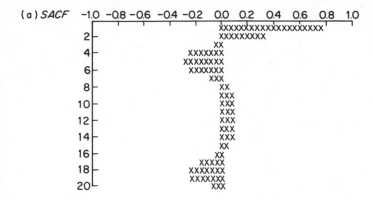

(a) *SACF*

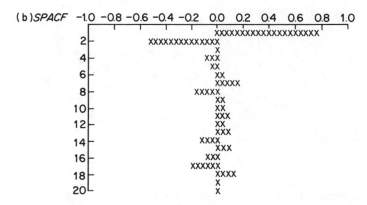

(b) *SPACF*

Fig. 6.4 SACF and SPACF for AR(2) scheme with $\phi_1 = 1.1$, $\phi_2 = -0.5$ and $n = 100$

small,

$$r_k \doteq 1 - \frac{5k}{n}. \tag{6.33}$$

Figure 6.5 shows the correlograms for the $ARIMA(0, 1, 1)$ process

$$\nabla y_t = \varepsilon_t - 0.5\varepsilon_{t-1}; \tag{6.34}$$

the sample $PACF$ is less useful, but will often appear to correspond to an $AR(1)$ scheme. This is not altogether surprising, since model (6.34) and an $AR(1)$ scheme with $\phi = 1 - \theta$, θ small, will have similar-looking correlograms despite the different theoretical underpinnings.

Taking first differences of the data yields the correlograms in Fig. 6.6. These suggest an $MA(1)$ scheme, as they should. Differencing again yields the plots shown in Fig. 6.7. The large negative value of r_1 and the string of negative partial serial correlations suggest that the series has been over-differenced. Figures 6.5–7 taken together indicate that plotting the correlograms for the

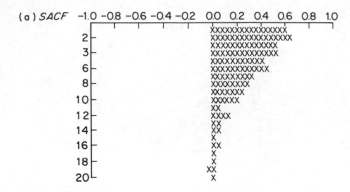

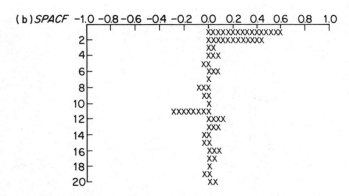

Fig. 6.5 SACF and SPACF for IMA(1), $\theta = 0.5$, $n = 100$

original series and suitable differences will often guide the investigator to an appropriate degree of differencing.

6.19 More formal tests are available for testing for non-stationarity, developed by Dickey and Fuller (1976); see Fuller (1976, pp. 366–82; 1985), for details.

In particular, suppose that the process can be represented by an $AR(p)$ scheme and that we wish to test whether there is a single unit root; that is, whether we need a first-order difference. We may consider the regression model for

$$y_t \text{ on } (y_{t-1}, y_{t-1} - y_{t-2}, ..., y_{t-p+1} - y_{t-p}), \qquad (6.35)$$

where the coefficient for y_{t-1} is β_1. A reasonable choice for p may be achieved by using stepwise regression, forcing y_{t-1} into the model and allowing later lags in if they are significant.

Given the estimate $\hat{\beta}_1$ and its standard error, $SE(\hat{\beta}_1)$ we may employ the usual t-statistic

$$\hat{\tau} = (\hat{\beta}_1 - 1)/SE(\hat{\beta}_1). \qquad (6.36)$$

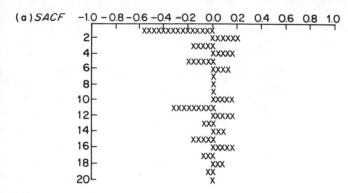

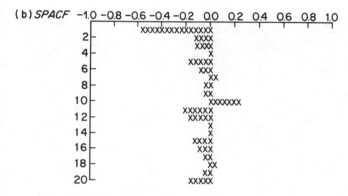

Fig. 6.6 SACF and SPACF for IMA(1), $\theta = 0.5$, $n = 100$, differenced once

However, the distribution of τ is not that of the Student's t-statistic. Tables of $\hat{\tau}$, developed by Dickey (1975) are given in Fuller (1976, p. 373). To a close approximation, the percentage points are

> lower 1%: $-3.43 - 0.08c$; upper 1%: $0.60 + 0.03c$
> lower 5%: $-2.86 - 0.03c$; upper 5%: $-0.07 - 0.02c$,

where $c = 100n^{-1}$.

Example 6.3 The IMA model with $\theta = 0.5$ may be written in AR form as

$$y_t = y_{t-1} - 0.5\nabla y_{t-1} - 0.25\nabla y_{t-2} - \cdots + \varepsilon_t. \tag{6.37}$$

For the simulated series of length $n = 100$ we obtained the fitted values:

$$\hat{y}_t = 0.976y_{t-1} - 0.487\nabla y_{t-1} - 0.209\nabla y_{t-2} - 0.214\nabla y_{t-3}, \tag{6.38}$$

where the first coefficient had $SE = 0.0464$. Thus,

$$\hat{\tau} = (0.976 - 1)/0.0464 = -0.517,$$

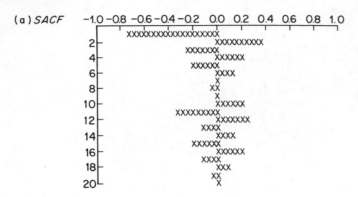

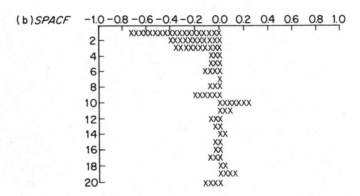

Fig. 6.7 SACF and SPACF for IMA(1), $\theta = 0.5$, $n = 100$, differenced twice

well above the lower 5% value of -2.89 so that we would accept $H_0: \beta_1 = 1$ and proceed to difference.

A more accurate test for general *ARIMA* schemes is given by Dickey *et al.* (1984).

Data analysis

6.20 Now that our tools for model selection have been identified, we proceed to examine some of the series introduced in Chapter 1 of this book. The correlograms for the barley yields data (Table 1.1) are given in Fig. 6.8. These suggest that the series is random, confirming our earlier analyses in Sections 2.4 and 2.6.

6.21 The plot of the sheep population series (Table 1.2) clearly suggests a series with trend. The sample *ACF* in Fig. 6.9 also suggests the need for trend removal. Finally, we fitted the regression model

$$\hat{y}_t = 0.935\,y_{t-1} + 0.466\nabla y_{t-1} - 0.215\nabla y_{t-2} - 0.216\nabla y_{t-3}$$

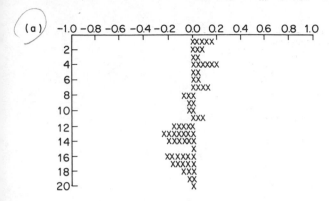

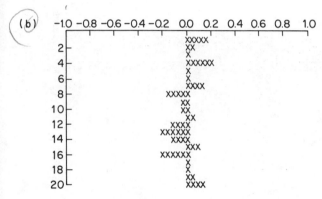

Fig. 6.8 SACF and SPACF of barley series (Table 1.1)

yielding

$$\hat{\tau} = (0.935 - 1)/0.0416 = -1.56$$

confirming the need for differencing (recall that $n = 73$).

The correlograms for the differenced series, Fig. 6.9(b), do not give clear indications for a particular model, but we might consider

$$\text{(i) } ARMA(1, 1), \qquad \text{(ii) } AR(3), \qquad \text{or even} \qquad \text{(iii) } MA(4).$$

Further refinement of the choice must await the development of estimation procedures in Chapter 7.

6.22 The correlograms for the FT index series (Table 1.6) are given in Fig. 6.10(a). Again, the combination of slow decay in the sample *ACF* and $AR(1)$-like behaviour in the sample *PACF* suggest possible differencing. Further, the stepwise model reduced to y_t on y_{t-1}, yielding

$$\hat{\tau} = (0.911 - 1)/0.0678 = -1.31,$$

confirming the diagnosis. The correlograms for the differenced series in Fig. 6.10(b) suggest either $AR(1)$ or $MA(1)$ with further activity at a lag of 8 periods (or 2 years). We take this analysis further in Section 7.10.

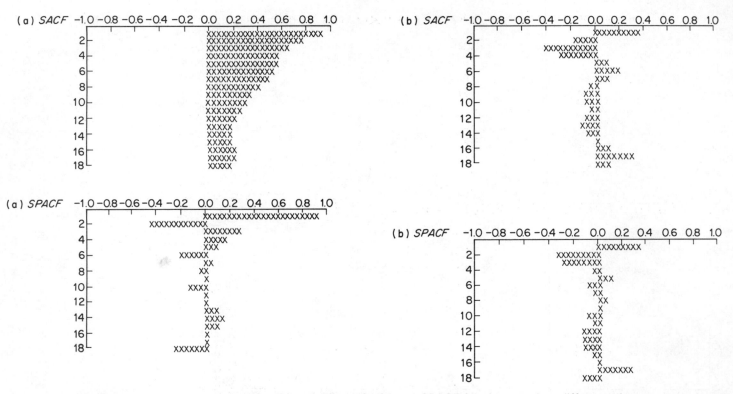

Fig. 6.9 (a) SACF and SPACF of sheep series (Table 1.2); (b) SACF and SPACF for sheep series, differenced once

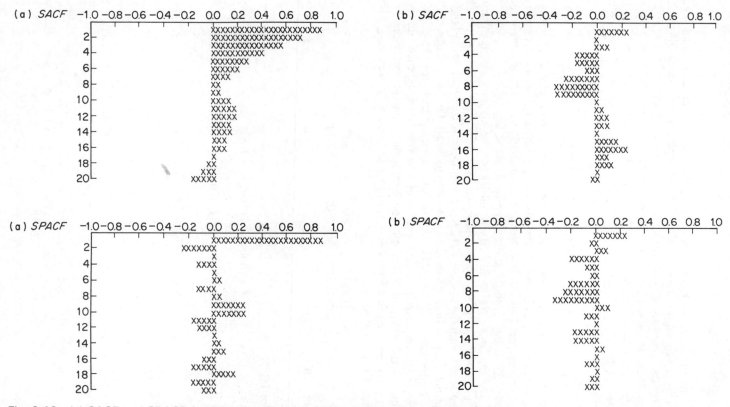

Fig. 6.10 (a) SACF and SPACF for FT index (Table 1.6); (b) SACF and SPACF for FT index, differenced once

Seasonal series

6.23   The methods we have developed for series without a seasonal component apply to seasonal series also. For example, if we are dealing with quarterly data, we should look for spikes in the correlograms at lags 4, 8, 12, ..., and so on. Particularly when dealing with monthly data, we may need to generate more than $n/3$ serial correlations in order to get a better feel for the seasonal component. Figure 6.11 shows the correlograms for a simulated series of $n = 100$ observations generated from the model

$$(1 - 0.6B)(1 - 0.8B^4)y_t = \varepsilon_t, \qquad (6.39)$$

the errors being normally distributed. Again, the sample *PACF* provides a clearer picture for autoregressive processes and clearly suggests an *AR* scheme with components at lags 1 and 4. Note that model (6.39) yields positive values in the *PACF* at lags 1 and 4 and a negative value at lag 5.

6.24   We shall now try to identify a model for the airline passenger series listed in Table 1.3. The sample *ACF* shows a slow rate of decay in the seasonals, although the sample *PACF* shows a spike at lag 1 but not at lag 12. Our earlier analysis in Sections 2.12–13 strongly suggested both seasonal and trend components. Figure 6.12 shows the correlograms for ∇y_t, $\nabla_{12} y_t$ and $\nabla \nabla_{12} y_t$, respectively. It appears that the regular difference alone is inadequate; either ∇_{12} or $\nabla \nabla_{12}$ may be appropriate. An *approximate* procedure corresponding to Section 6.19 is to regress y_t on y_{t-12} and y_{t-13}. This produces the estimates

$$y_t = 0.453 y_{t-1} + 0.886 y_{t-12} - 0.302 y_{t-13}.$$
$$\quad (0.100) \qquad (0.070) \qquad (0.116)$$

For lag 1 this produces

$$\hat{\tau} = (0.453 - 1)/0.10 = -5.47$$

and for lag 12

$$\hat{\tau} = (0.886 - 1)/0.07 = -1.63$$

suggesting seasonal differencing only. Thus, provisional models may be identified as

$$(1 - \phi B)\nabla_{12} y_t = (1 - \Theta B^{12})\varepsilon_t \qquad (6.40)$$

and, possibly,

$$\nabla \nabla_{12} y_t = (1 - \theta B)(1 - \Theta B^{12})\varepsilon_t. \qquad (6.41)$$

Indeed, model (6.41) is sometimes known as the 'airline' model since it was proposed by Box and Jenkins (1976) for a similar and oft-analysed set of data on airline activity.

6.25   Given the importance of selecting a suitable model for a series, it is not surprising that there have been several other attempts to develop identification procedures. One such procedure is the use of the *inverse* autocorrelation function, *IACF*, proposed by W. S. Cleveland (1972) and further developed by

(a) *SACF*

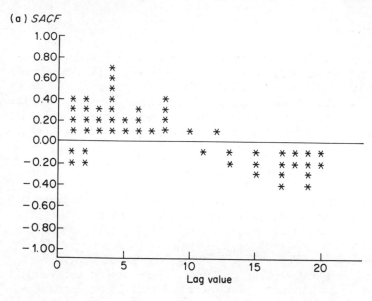

(b) *SPACF*

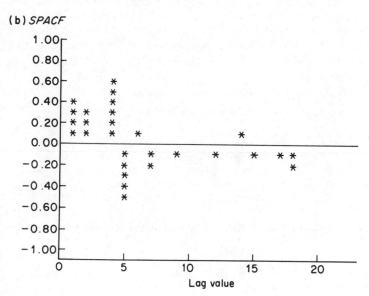

Fig. 6.11   SACF and SPACF for simulated ARIMA $(1,0,0)(1,0,0)_4$ scheme, $n = 100$

Chatfield (1979). If we start with the model

$$\phi(B)y_t = \theta(B)\varepsilon_t, \qquad (6.42)$$

it is possible to think of an 'inverse' model of the form

$$\theta(B)y_t^* = \phi(B)\varepsilon_t^*, \qquad (6.43)$$

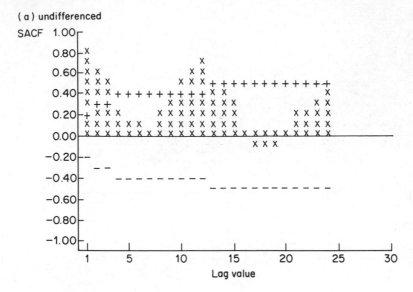

(a) undifferenced

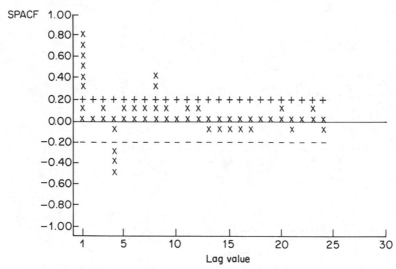

Fig. 6.12   SACF and SPACF for airline data (Table 1.3), (a) undifferenced (b) seasonal difference only; (c) seasonal and regular differences

where the ε^* denote a different random error process and y^* is not observable but, as we shall see in Section 10.37, it is possible to estimate the 'sample ACF' of y^*, using the spectrum. This is the sample inverse autocorrelation function. Because of the duality between (6.42) and (6.43), the $IACF$ behaves similarly to the $PACF$, apart from a change of sign; that is, autoregressive terms will generate spikes in the $IACF$. Some computer packages (e.g. SAS) routinely generate the $IACF$ to assist in model identification. The sample $IACF$ for the first differences of the sheep data of Table 1.2 is shown in Fig. 6.13.

(b) seasonal difference only

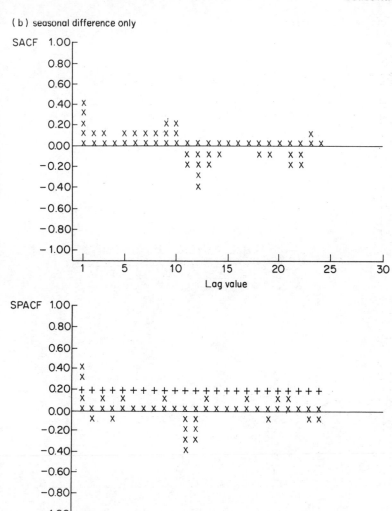

Fig. 6.12 *(Continued)*

6.26 One of the difficulties we have encountered in selecting a model is whether or not to difference. Tsay and Tiao (1984) have suggested an ingenious way around this problem by means of what they call the *extended* sample autocorrelation function (*ESACF*). This is constructed as a two-way table as shown in Table 6.2.

The serial correlations in the first row ($AR = 0$) correspond to the regular correlogram. Subsequent rows are defined in the following way. First we define the *extended autoregression* of orders k and j, $EAR(k, j)$ as the regression of

$$y_t \text{ on } (y_{t-1}, y_{t-2}, \ldots, y_{t-k}; \ e_{k,t-1}(j-1), \ e_{k,t-2}(j-2), \ldots, e_{k,t-j}(0)); \qquad (6.44)$$

(c) seasonal and regular differences

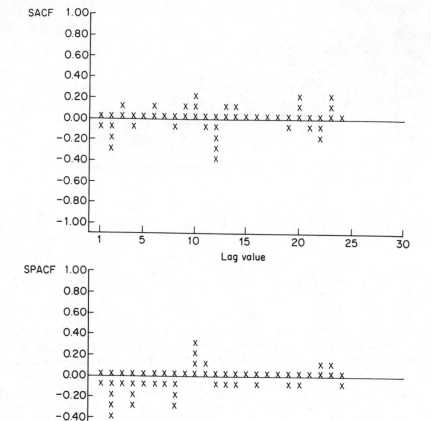

Fig. 6.12 (*Continued*)

this produces estimates $\hat{\phi}_{ik}(j)$ for the coefficient of y_{t-i} and the residuals for the regression are $e_{k,t}(j)$. Thus, the residuals are defined recursively with $e_{k,t}(0)$ denoting the regular residuals for an $AR(k)$ scheme. We then set

$$w_{k,t}(j) = y_t - \sum_{i=1}^{k} \hat{\phi}_{ik}(j)y_{t-i}. \tag{6.45}$$

Finally, the (j, k)th extended sample autocorrelation, $r_{j(k)}$, is the jth-order autocorrelation of $w_{k,t}(j)$. The computations may be performed by an extension of the Durbin–Levinson algorithm described in Section 5.19; see Tsay and Tiao (1984) for details.

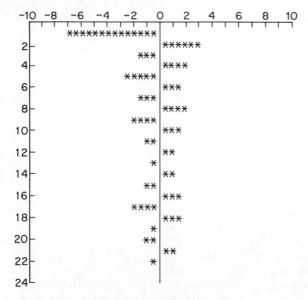

Fig. 6.13 The sample inverse IACF for the first differences of the sheep data (Table 1.2).

A pure $MA(q)$ process cuts off the $SACF$ after q lags. Similarly, an $ARMA(p, q)$ scheme cuts off after the qth lag for the pth $ESACF$. That, is, after q entries on the row corresponding to $AR = p$. This result is unaffected by non-stationarity so that an $ARIMA(p - d, d, q)$ scheme behaves in the same way as an $ARMA(p, q)$ for $ESACF$ entries in row p and beyond. In general, if the underlying process is $ARIMA(p - d, d, q)$, this will generate a theoretical extended ACF with

$$\rho_{j(k)} = 0 \quad \text{if } j > q \text{ and } k = p$$
$$\neq 0 \quad \text{if } j = q \text{ and } k = p. \tag{6.46}$$

This produces a wedge of near-zero values in the $ESACF$. For example, an $ARIMA(1, 1, 0)$ scheme should produce a pattern such as that shown in Table 6.3 ($X =$ non-zero, $0 =$ zero, $* =$ no set pattern).

Table 6.2

		MA			
		0	1	2	3
AR	0	$r_{1(0)}$	$r_{2(0)}$	$r_{3(0)}$	$r_{4(0)}$	...
	1	$r_{1(1)}$	$r_{2(1)}$	$r_{3(1)}$	$r_{4(1)}$	...
	2	$r_{1(2)}$	$r_{2(2)}$	$r_{3(2)}$	$r_{4(2)}$	...
	3	$r_{1(3)}$	$r_{2(3)}$	$r_{3(3)}$	$r_{4(3)}$	...
	⋮	⋮	⋮	⋮	⋮	⋰

Table 6.3 Pattern of zero terms in ESACF.

		MA				
		0	1	2	3	4
AR	0	*	X	X	X	X
	1	*	0	0	0	0
	2	*	X	0	0	0
	3	*	X	X	0	0
	4	*	X	X	X	0

6.27   As an example, we computed the *ESACF* for the sheep data of Table 1.2. The numerical values are ($\times 100$) as shown in Table 6.4.

Scoring values within $\pm n^{-1/2} = \pm 0.234$ as 0, and everything else as X, we obtain the pattern shown in Table 6.5.

This suggests an $AR(4)$, an $ARMA(3, 1)$ or an $ARMA(1, 4)$ scheme, where the AR coefficient row includes possible differencing.

6.28   The *ESACF* is clearly a method with some potential, although it has yet to be extended to seasonal series. Other alternative procedures have been suggested by Gray *et al.* (1978), Beguin *et al.* (1981), and Hannan and Rissanen (1982). In addition, several 'automatic' selection procedures have been proposed. We shall examine these in the next chapter after we have discussed estimation procedures.

Table 6.4 ESACF for sheep series (Table 1.2).

		MA						
		0	1	2	3	4	5	6
AR	0	91	76	63	57	56	54	48
	1	42	−3	−38	−27	7	17	11
	2	51	−4	−29	−3	16	15	10
	3	−47	15	18	−17	14	−3	6
	4	−11	11	19	−6	16	9	5

Table 6.5 Pattern of zeros in ESACF for sheep series.

		MA						
		0	1	2	3	4	5	6
AR	0	X	X	X	X	X	X	X
	1	X	0	X	X	0	0	0
	2	X	0	X	X	0	0	0
	3	X	0	0	0	0	0	0
	4	0	0	0	0	0	0	0

Exercises

6.1 Suppose that

$$y_t = 6 - t, \qquad t = 1, 2, ..., 11.$$

Compute the first four sample autocorrelations using (6.3) and the two modified versions suggested in Section 6.2.

6.2 Verify (6.10) using (6.8), but ignoring the last term.

6.3 Suppose y_t follows an $AR(1)$ scheme with $\phi = 0.8$. Use (6.18) to compute the standard error of r_j for $j = 1, 2, ..., 5, 10$, and compare these with the limiting values.

6.4 Estimate the standard errors of r_j, $j = 1, 2, ..., 5$ using (6.17) given

j	1	2	3	4
r_j	0.6	0.2	-0.3	0.7

6.5 Verify that (6.32) holds. (*Hint:* Expand $\mathrm{var}\{\Sigma \ (y_i - \bar{y})\}$, which is identically zero.)

6.6 Generate simulated data from an $AR(1)$ scheme with $n = 100$ and $\phi = 0.7$. Compute the sample correlograms and see how well these conform to expectation. Repeat the process for $n = 50$ and $n = 25$.

6.7 Repeat Exercise 6.6 for an $MA(2)$ scheme with $\theta_1 = 0.9$ and $\theta_2 = -0.4$.

6.8 Generate simulated data from an $ARIMA(1, 1, 0)$ scheme with $\phi = 0.6$ and $n = 100$. Examine the correlograms and perform the Dickey–Fuller text. Decide whether differencing is needed, and determine a provisional model for the data. (Clearly, there is no limit to the number of simulation games one can play. The reader is urged to try a variety of combinations to gain familiarity with examining the sample ACF and $PACF$ and, if available, the sample $IACF$ and the $ESACF$.)

6.9 Generate simulated data from an $ARIMA(1, 0, 1)$ scheme with $\phi = 0.6$ and $\theta = 0.65$. What form of the model do the correlograms suggest? Why? (*Hint*: What happens to this scheme when $\phi = \theta$?)

6.10 Select provisonal models for some or all of the following non-scasonal series (tables marked A are in Appendix A):
   (a) Kendall's simulated $AR(2)$ series (Table A1);
   (b) UK gross domestic product (Table A2);
   (c) Wölfer's sunspots data (Table A3);
   (d) Canadian lynx trappings (Table A4);
   (e) US interest rates (Table A5);
   (f) US immigration levels (Table 1.4).

6.11 Select provisional models for some or all of the following seasonal series given in Appendix A:
   (a) Chatfield and Prothero's sales data (Table A6);
   (b) UK unemployment (Table A7);
   (c) UK whisky production (Table A8).

7

Estimation and model checking

7.1 Up to this point we have taken a somewhat intuitive approach to estimation problems, assuming that sample quantities provide reasonable estimates for their population counterparts. However, in order to estimate the parameters of models such as those identified in the previous chapter, we must consider a more formal approach to model fitting. Some of the statistical issues which arise rapidly become quite complex and lie beyond the scope of this book. Therefore, we shall content ourselves with a rather brief discussion and refer the interested reader to Kendall *et al*, (1983, Chapter 50) and Fuller (1976, Chapter 8). The first part of this chapter considers estimation problems and the central part addresses model checking; that is, to decide whether the selected model is appropriate and, if not, how it should be modified. Finally, we consider the extent to which the modelling process can be 'automated'. Can the decisions regarding model selection be formulated in such a way that the whole operation can be performed by a computer program, without the investigator being directly involved?

Fitting autoregressions

7.2 We begin by considering purely autoregressive, or $AR(p)$ models, where the order p is assumed to be known:

$$y_t = \delta + \phi_1 y_{t-1} + \cdots + \phi_p y_{t-p} + \varepsilon_t, \tag{7.1}$$

where the ε_t are independent and identically distributed with mean 0 and variance σ^2. The model formulation is similar to that of ordinary least-squares (OLS) regression save that the 'independent' variables are now lagged dependent variables. Mann and Wald (1943, reproduced in Wald's *Collected Papers*, 1955) demonstrated that the OLS method is indeed applicable to (7.1) and that the estimators are both asymptotically unbiased, consistent and efficient. Straightforward application of the OLS procedure yields the $(p + 1)$

estimating equations

$$\begin{aligned}
\Sigma\ y_t &= (n - p)\delta + \Sigma\ y_{t-1} + \cdots + \Sigma\ y_{t-p} \\
\Sigma\ y_t y_{t-j} &= \delta\ \Sigma\ y_{t-j} + \Sigma\ y_{t-1}y_{t-j} + \cdots + \Sigma\ y_{t-p}y_{t-j}
\end{aligned}\bigg\}, \qquad (7.2)$$

where the summations are all over the range $t = p + 1$ to $t = n$. Equations (7.2) are similar to the sample version of the Yule–Walker equations (5.49) apart from end-effects. When the process lies well inside the stationary region, the difference between the estimates from (5.49) and (7.2) are slight for moderate or large n. However, the Yule–Walker equations assume stationarity, whereas equations (7.2) do not. Indeed, Tiao and Tsay (1983) show that these estimators remain consistent even when the process is non-stationary. In general, therefore, we shall use the OLS estimates. Then, it follows from Mann and Wald (1943) that the vector of estimators

$$\hat{\boldsymbol{\beta}} = n^{1/2}\{\hat{\delta} - \delta, \hat{\phi}_1 - \phi_1, ..., \hat{\phi}_p - \phi_p\} \qquad (7.3)$$

is asymptotically multivariate normal with mean zero and a finite covariance matrix. Further, the mean square error

$$s^2 = \Sigma\ (y_t - \hat{y}_t)^2/(n - p - 1) \qquad (7.4)$$

is a consistent estimator for σ^2 and the covariance matrix of $\hat{\boldsymbol{\beta}}$ is validly estimated by $s^2 H^{-1}$, where

$$H = \begin{bmatrix} n & \mathbf{m}' \\ \mathbf{m} & C \end{bmatrix}, \qquad (7.5)$$

$$\mathbf{m}' = (\Sigma\ y_{t-1}, ..., \Sigma\ y_{t-p})$$

and the (j, k)th element of C is $\Sigma\ y_{t-j}y_{t-k}$.

7.3 The next question of interest is whether specific distributional assumptions about the error process lead to improved estimation procedures. Specifically, we assume that the ε_t are $N(0, \sigma^2)$. For example, when $p = 1$ and the mean is known (set $\delta = 0$ with no loss of generality), it follows that the distribution of y_1 is $N\{0, \sigma^2/(1 - \phi^2)\}$. Thus, the log-likelihood function for (ϕ, σ^2) given $(y_1, y_2, ..., y_n)$ becomes

$$\begin{aligned}
l &= \text{const} - \tfrac{1}{2}n \ln \sigma^2 - \frac{1}{2}\sum_{t=1}^{n} \varepsilon_t^2/\sigma^2 \\
&= \text{const} - \tfrac{1}{2}n \ln \sigma^2 - \frac{1}{2}\sum_{t=2}^{n} (y_t - \phi y_{t-1})^2/\sigma^2 \\
&\quad + \tfrac{1}{2}\ln(1 - \phi^2) - \tfrac{1}{2}y_1^2(1 - \phi^2)/\sigma^2. \qquad (7.6)
\end{aligned}$$

If we examine the conditional likelihood, taking y_1 as given, the last two terms in (7.6) disappear and the maximum likelihood solution corresponds to the second equation in (7.2) with $j = p = 1$. However, the complete likelihood solution is derived from (7.6). As might be expected, differences between the two solutions become important when $|\phi|$ is near 1 and n is small, see Poirer (1978). For general autoregressive schemes, the conditional likelihood is based upon $(\varepsilon_{p+1}, ..., \varepsilon_n)$ and reduces to (7.2), whereas the complete solution also includes the joint distribution of $(y_1, ..., y_p)$.

7.4   Most of the work done on estimation in time series involves the normality assumption or the use of least squares. A notable exception is the work of R. D. Martin, who has developed robust estimators using generalized M procedures; see, for example, Martin and Yohai (1985). Robust procedures will be considered in Chapter 15.

Fitting MA schemes

7.5   The least squares equations provide effective estimators for AR processes, so it might be expected that the first q serial correlations would provide good estimators for the $MA(q)$ scheme. Unfortunately, as Whittle (1953) showed, the estimator $\hat{\theta}$ derived from

$$r_1 = \hat{\theta}/(1 + \hat{\theta}^2) \tag{7.7}$$

is very inefficient and similar problems arise for $q > 1$.

7.6   One method of developing improved estimators is due to Durbin (1959) and will be illustrated for $q = 1$. From (5.74) we may write the $MA(1)$ scheme

$$y_t = \varepsilon_t - \theta\varepsilon_{t-1}$$

as

$$y_t + \theta y_{t-1} + \theta^2 y_{t-2} + \cdots = \varepsilon_t. \tag{7.8}$$

If (7.8) is treated as an $AR(k)$ scheme for a suitably large choice of k, this gives rise to the estimates $\hat{\phi}_1, \hat{\phi}_2, ..., \hat{\phi}_k$. By an appeal to the Mann–Wald theorem Durbin showed that the estimator

$$\hat{\theta} = \sum_{j=0}^{k-1} \hat{\phi}_j\hat{\phi}_{j+1} \Bigg/ \sum_{j=0}^{k} \hat{\phi}_j^2, \tag{7.9}$$

where $\phi_0 \equiv 1$, is asymptotically efficient if $k \to \infty$ in such a way that $(k/n) \to 0$ as $n \to \infty$.

Fitting ARIMA schemes

7.7   There have been several other proposals for fitting pure MA schemes, but we shall now focus attention upon the general problem. Box and Jenkins (1976) developed a non-linear least-squares procedure and this led to a variety of least squares and approximate likelihood solutions. Later, Newbold (1974) developed an exact likelihood procedure which, with improvements in computational efficiency rendered by Ansley's (1979) use of the Cholesky decomposition, has been widely accepted. Other procedures, based on the Kalman filter, are discussed briefly in Section 9.13. The reader willing to accept the technical details may move on directly to Section 7.10.

7.8   To follow the basic steps of the Newbold–Ansley procedure, consider the $ARMA(1, 1)$ scheme

$$y_t = \phi y_{t-1} + \varepsilon_t - \theta\varepsilon_{t-1}, \qquad t = 1, 2, ..., n. \tag{7.10}$$

We may write this in matrix form as

$$\begin{bmatrix} \boldsymbol{\varepsilon}^* \\ \boldsymbol{\varepsilon} \end{bmatrix} = \begin{bmatrix} 0 \\ L \end{bmatrix} \mathbf{y} + \begin{bmatrix} I_2 \\ X \end{bmatrix} \boldsymbol{\varepsilon}^*, \tag{7.11}$$

where $\boldsymbol{\varepsilon}' = (\varepsilon_1, \ldots, \varepsilon_n)$, $\mathbf{y}' = (y_1, \ldots, y_n)$, $\boldsymbol{\varepsilon}^* = (\varepsilon_0, y_0)'$, I_2 is the identity matrix,

$$L = \begin{bmatrix} 1 & & & & \\ (\theta - \phi) & 1 & & & \\ \theta(\theta - \phi) & (\theta - \phi) & 1 & & \\ \vdots & \vdots & \vdots & \ddots & \\ \theta^{n-2}(\theta - \phi) & . & . & (\theta - \phi) & 1 \end{bmatrix}, \tag{7.12}$$

and

$$X' = \begin{bmatrix} \theta & \theta^2 \ldots \theta^n \\ -\phi & -\theta\phi \ldots -\theta^{n-1}\phi \end{bmatrix}. \tag{7.13}$$

Noting that $E(\varepsilon_t^2) = E(\varepsilon_t y_t) = \sigma^2$ and using Example 5.9, the covariance matrix of $\boldsymbol{\varepsilon}^*$ is

$$\sigma^2 \Lambda = \sigma^2 \begin{bmatrix} 1 & 1 \\ 1 & \dfrac{1 + \theta^2 - 2\theta\phi}{1 - \phi^2} \end{bmatrix}. \tag{7.14}$$

We may choose an orthogonal matrix T such that $T\Lambda T' = I_2$; here

$$T^{-1} = \begin{bmatrix} 1 & 0 \\ 1 & (\theta - \phi)(1 - \phi^2)^{-1/2} \end{bmatrix}. \tag{7.15}$$

Multiplying (7.11) through by

$$\begin{bmatrix} T & 0 \\ 0 & I \end{bmatrix},$$

we obtain expressions for $\boldsymbol{\varepsilon}$ and $\mathbf{u} = T\boldsymbol{\varepsilon}^*$. The random vectors $\boldsymbol{\varepsilon}$ and u^* are independent by construction, which enables us to transform from $(\boldsymbol{\varepsilon}^*, \boldsymbol{\varepsilon})$ to $(\mathbf{u}, \mathbf{y})$; the details are given in Newbold (1974). Finally, we may integrate out $\boldsymbol{\varepsilon}^*$ to obtain the marginal density for $\mathbf{y}$ which provides the exact log-likelihood:

$$l = \text{const} - \tfrac{1}{2} \ln \sigma^2 - \tfrac{1}{2} \ln |Z'Z| - \tfrac{1}{2} S(\theta, \phi)/\sigma^2, \tag{7.16}$$

where

$$Z = [I_2, (T^{-1})' X'] \tag{7.17}$$

and

$$S(\theta, \phi) = (L\mathbf{y} + Z\hat{\mathbf{u}})'(L\mathbf{y} + Z\hat{\mathbf{u}}), \tag{7.18}$$
$$\hat{\mathbf{u}} = -(Z'Z)^{-1}Z'L\mathbf{y}. \tag{7.19}$$

Equations (7.16)–(7.19) provide the basis for an iterative search procedure to derive the estimates. Ansley (1979) shows how to speed up the calculations using the Cholesky decomposition.

7.9   The method outlined here can be applied to general $ARMA(p, q)$ models by setting $\boldsymbol{\varepsilon}^* = (\varepsilon_0, ..., \varepsilon_{1-q}, y_0, ..., y_{1-p})'$ and redefining (7.12)–(7.15) appropriately. For example, re-working these expressions with $p = 1$, $q = 0$ yields (7.6).

Some examples

7.10   Our analysis in Section 6.21 suggested that an appropriate model for the FT Index of Table 1.6 is $ARIMA(0, 1, 1)$. Fitting by conditional least squares

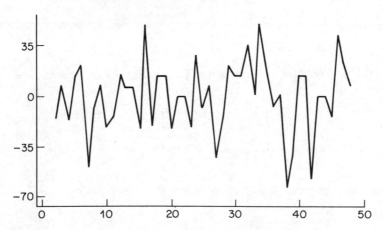

Fig. 7.1   Plot of residuals from model (7.20) for FT index

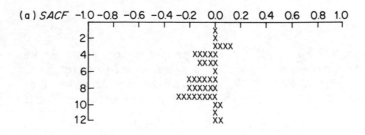

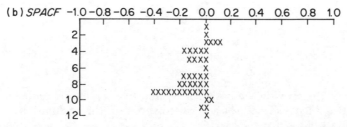

Fig. 7.2   SACF and SPACF of residuals from model (7.20) for FT index

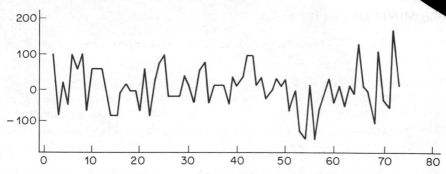

Fig. 7.3 Plot of residuals from model (7.21) for sheep series

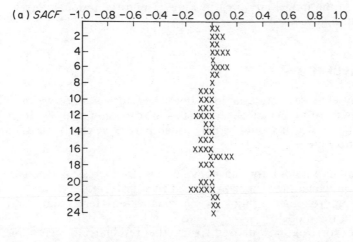

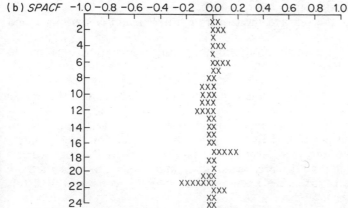

Fig. 7.4 SACF and SPACF of residuals from model (7.21) for sheep series

ve the model

$$\nabla y_t = 2.186 + \varepsilon_t + 0.256\varepsilon_{t-1}, \qquad \hat{\sigma} = 24.6. \qquad (7.20)$$
$$(0.49) \qquad (-1.77)$$

rackets under the estimated coefficients represent approximate
.s, t = coefficient/standard error. The sign of the t-ratio for θ_1
~egative sign in the original model. The residuals from the model
are giv. Fig. 7.1 and the correlograms for the residuals are given in
Fig. 7.2.

7.11 In Section 6.21, we suggested several models for the sheep series of
Table 1.2. One possibility was an $ARIMA\,(0, 1, 4)$ scheme; fitting using
MINITAB gave

$$\nabla y_t = -4.566 + \varepsilon_t + 0.252\varepsilon_{t-1} - 0.302\varepsilon_{t-2} - 0.544\varepsilon_{t-3} - 0.620\varepsilon_{t-4}, \ \hat{\sigma} = 64.8.$$
$$(-1.71) \qquad (-2.38) \qquad (2.77) \qquad (4.44) \qquad (4.87)$$

$$(7.21)$$

The residuals from the model are given in Fig. 7.3 and the residual correlograms are given in Fig. 7.4.

Modeling checking

7.12 These analyses raise more questions than they answer since we must
now address the issue of model adequacy. Just as in regression analysis, there
are a variety of checks which should be performed before we can be satisfied
that a model is adequate:

(1) plot the residuals and look for unusual values, or for increasing (decreasing) dispersion which may suggest the need to transform the data;
(2) examine the approximate t-ratios to see whether any terms should be
   dropped from the model;
(3) examine the correlograms derived from the residuals to determine
   whether additional terms are required;
(4) check whether the selected model (after differencing) is stationary and
   invertible;
(5) check the overall fit of the model (although this is less often done than in
   regression analysis).

We now examine each of these criteria in turn, developing them further where
necessary.

7.13 Plotting residuals is a familiar enough procedure, except that we now
plot against time rather than some independent variable. For the FT index,
Fig. 7.1 reveals nothing particularly unusual, although there is some evidence
of increasing variability in the later years. Stock market price models often use
'return' = log(price ratio) as the basic variable, which is consistent with this
observation.

   The plot for the sheep series in Fig. 7.3 also reveals somewhat greater
variation towards the end of the series when the long period of decline is

reversed in the early 1920s. This is a case where the inclusion of explanatory variables would seem more appropriate than a transformation.

7.14 Checking approximate *t*-ratios is also straightforward. From (7.20) it is evident that the constant term is not significant and the coefficient of ε_{t-1} only marginally so. In (7.21), all the *MA* coefficients are significant but the constant term is marginal. In each case, we shall suspend judgement until the diagnostic process is complete.

7.15 The correlogram plots for the FT index in Fig. 7.2 appear to show spikes at lags of about two years (eight and nine quarters). With $n = 48$, the S.E. of these coefficients is about 0.14 so there may be something of importance. This could reflect the 'General Election cycle' of which we spoke earlier, since there were elections in 1959, 1964, 1966, and 1970. Very tentatively, therefore, we might consider adding an *MA* or *AR* term at lag 9.

The correlograms for the sheep data appear very well behaved and do not suggest the inclusion of any additional terms.

The Box–Pierce–Ljung statistic

7.16 Simultaneous evaluation of all the terms in the *SACF* and *SPACF* can be somewhat unreliable since single coefficients may appear 'significant' due to the accumulation of type I errors or a pattern of non-significant coefficients may nevertheless indicate some lack of fit. Quenouille (1947) developed a test for the residuals of *AR* schemes, later extended to *MA* schemes also. However, the most widely used test is now that of Box and Pierce (1970), with modifications by Ljung and Box (1978). They suggested a 'portmanteau' test to contrast the null hypothesis:

$$H_0: \rho_1 = \rho_2 = \cdots = \rho_K = 0 \tag{7.22}$$

against the general alternative H_1: not all $\rho_j = 0$. Based on the residual

Table 7.1 Box–Pierce–Ljung statistics for the FT index and sheep series, using models (7.20) and (7.21)

	FT index			
K	12	24	36	
$Q(K)$	12.6	19.7	31.8	
D.F.	11	23	35	
$Q(K_2) - Q(K_1)$	—	7.1	12.1	
D.F.	—	12	12	

	Sheep			
K	12	24	36	48
$Q(K)$	5.8	14.6	18.8	33.3
D.F.	8	20	32	44
$Q(K_2) - Q(K_1)$	—	8.8	4.2	14.5
D.F.	—	12	12	12

correlogram, they suggested the statistic

$$Q \equiv Q(K) = n(n+2) \sum_{j=1}^{K} r_j^2/(n-j), \tag{7.23}$$

where n denotes the length of the series *after* any differencing. Box and Pierce (1970) showed that, under H_0, Q is asymptotically distributed as chi-squared with $(K - p - q)$ degrees of freedom. Typically, the statistic is evaluated for several choices of K. Under H_0, for large n,

$$E[Q(K_2) - Q(K_1)] = K_2 - K_1 \tag{7.24}$$

so that different sections of the correlogram can be checked for departures from H_0.

7.17   The values of Q for the FT index and sheep series are given in Table 7.1. Neither the original values nor the differences computed in accordance with (7.24) suggest any unusual behaviour.

Stationarity and invertibility check

7.18   In 5.20, we indicated that the conditions for stationarity are that the roots of the equation

$$\phi(x) = 0 \tag{7.25}$$

should be greater than one in absolute value; similarly, the conditions for invertibility are that the roots of

$$\theta(x) = 0 \tag{7.26}$$

should be greater than one in absolute value. When the roots are not available explicitly, a partial check is available by computing the values of (7.25) and (7.26) at $x = \pm 1$. The conditions can be satisfied only if

$$|\phi_p| < 1, \quad |\theta_q| < 1, \quad \phi(1) > 0, \quad \phi(-1) > 0, \quad \theta(1) > 0 \text{ and } \theta(-1) > 0. \tag{7.27}$$

The conditions in (7.27) are necessary but *not* sufficient.

7.19   From (7.20), the model for the FT index has $\phi(x) = 1$ and $\theta(x) = 1 + 0.256x$ so that it is clearly both stationary and invertible. The model for the sheep series has

$$\phi(x) = 1 \text{ and } \theta(x) = 1 + 0.252x - 0.302x^2 - 0.544x^3 - 0.620x^4,$$

where $\theta(1) = -0.234$, so the model is not invertible.

Goodness-of-fit

7.20   The natural measure of goodness-of-fit used in regression analysis is the coefficient of determination

$$R^2 = 1 - SSE/SSY, \tag{7.28}$$

where $SSE = \Sigma (y - \hat{y})^2$ and $SSY = \Sigma (y - \bar{y})^2$. This is sometimes used in adjusted form as

$$\bar{R}^2 = 1 - MSE/MSY, \tag{7.29}$$

where $MSE = SSE/(n - k - 1)$ and $MSY = SSY/(n - 1)$, when k additional parameters have been incorporated into the model. As noted by Harvey (1984) among others, this criterion may appear to be over-optimistic if the data are differenced before modelling yet (7.27) is still used. Harvey suggests using

$$R_D^2 = 1 - SSE/SS(\nabla Y), \tag{7.30}$$

where $SS(\nabla Y) = \Sigma \ (\nabla y - \overline{\nabla y})^2$. Again, this can be expressed in adjusted form by using MS in place of SS. Replacing $\overline{\nabla y}$ by zero yields $1 - R_D^2 = U^2$, a lack of fit measure proposed by Theil (1966). For s-period seasonal data, Harvey suggests first regressing the differenced series on s seasonal dummies

$$\nabla y_t = \sum_{j=1}^{s} \beta_j z_{tj} + \varepsilon_t, \tag{7.31}$$

where

$z_{tj} = 1$ if observation t occurs in season j
$\quad = 0$ otherwise

(Fitting model (7.31) is equivalent to performing a one-way analysis of variance on ∇y.)

The goodness-of-fit may then be measured by

$$R_{SD}^2 = 1 - SSE/SSE_0 \tag{7.32}$$

where SSE_0 denotes the sum of squared errors from (7.31). As before, an adjusted form may be defined by using the mean squares. Higher-order or seasonal differences may be dealt with in similar fashion. It should be noted that (7.30) and (7.32) are more stringent criteria than (7.28); indeed, R_{SD}^2 may well be negative, although the corresponding model could scarcely be considered acceptable.

7.21 Analysis of the FT index model produced

$$\bar{R}^2 = 0.797, \qquad \bar{R}_D^2 = 0.033$$

clearly demonstrating the ineffectiveness of the $ARIMA(0, 1, 1)$ model. For the sheep series, we find

$$\bar{R}^2 = 0.915, \qquad \bar{R}_D^2 = 0.401,$$

indicating that the inclusion of the MA terms is desirable. $\bar{R}^2$ shows that over 91 percent of the total variation is accounted for by the model, whereas $\bar{R}_D^2$ indicates that, after differencing, 40 percent of the remaining variation is explainable.

Updating a model

7.22 The information gleaned from the various diagnostics clearly indicates that the model suggested for the FT index is inadequate. The only potential improvements would be

(a)   include AR or MA terms at lags 8 or 9;
(b)   consider a transformation to logarithms.

We considered a model with an MA term at lag 9 and obtained

$$\nabla y_t = \varepsilon_t - 0.397\varepsilon_{t-9}, \qquad \hat{\sigma} = 23.4; \qquad (7.33)$$
$$(2.42)$$

the constant term again proving to be not significant. The residuals plot showed nothing unusual, and the residual serial correlations were all within $\pm 2SE$. The Box–Pierce–Ljung statistics had values

$Q(K)$	14.4	23.1	34.3
DF	11	23	35

which are eminently satisfactory. The model is invertible and

$$\bar{R}^2 = 0.815, \qquad \bar{R}_D^2 = 0.121.$$

Evidently the relationship is rather weak, but no major improvements are possible. This is hardly surprising since an effective model would enable the statistician to make money on the Stock Exchange! The weak nine-quarters relationship probably reflects General Election factors, which are as unpredictable as the market itself.

7.23   The model for the sheep series appeared to perform well in most respects, save that it is not invertible. Other possible models suggested in Section 6.21 for the differenced series were $ARMA(1, 1)$ and $AR(3)$. The $ARMA(1, 1)$ model was estimated as

$$\nabla y_t = -4.37 + 0.062\nabla y_{t-1} + \varepsilon_t + 0.437\varepsilon_{t-1}, \qquad \hat{\sigma} = 75.9$$
$$(-0.34) \quad (0.24) \qquad\qquad (-1.85)$$

and is unacceptable on several counts. The $AR(3)$ scheme yielded, writing w_t for ∇y_t

$$w_t = -6.59 + 0.420w_{t-1} - 0.211w_{t-2} - 0.327w_{t-3} + \varepsilon_t \qquad (7.34)$$
$$(-0.80) \quad (3.61) \qquad (-1.63) \qquad (-1.73)$$

with $\hat{\sigma} = 69.8$, $\bar{R}^2 = 0.902$, $\bar{R}_D^2 = 0.304$ and $\phi(1) = 1.118$, $\phi(-1) = 1.304$. The statistic Q is less than its expected value for $K = 12$, 24, 36 and 48. The constant term may be dropped, but this does not have a material effect upon the model. Although $\hat{\sigma}$ is slightly higher in (7.34) than in (7.21), (7.34) is both stationary and invertible.

Seasonal models

7.24   To demonstrate the approach for seasonal models, we again consider the airlines data from Table 1.3. Our preliminary investigations in Section 6.24

suggested two possible models:

$$(A): (1, 0, 0)(0, 1, 1)_{12} \qquad \text{or} \qquad (B): (0, 1, 1)(0, 1, 1)_{12},$$

which we examine in turn.

Model (A) yields the estimates

$$(1 - 0.343B)\nabla_{12}y_t = 424.8 + (1 - 0.611B^{12})\varepsilon_t \qquad (7.35)$$
$$\quad(3.24) \qquad\qquad (12.28) \qquad (5.75)$$

with $\hat{\sigma} = 740$, $\bar{R}^2 = 0.887$, and $\bar{R}^2_{SD} = 0.300$. However, the Q statistics indicate some problems with

K	12	24	36	48
$Q(K)$	30.2	44.6	53.6	64.6
DF	10	22	34	46

and spikes on the residual correlogram at lags 3, 6 and 9. Analysis of the residuals plot in Fig. 7.5 indicates problems with the observations for June and July 1968 and April 1969, as previously noted in Section 2.14. Examination of back copies of *The Times* revealed BOAC (now part of British Airways) pilots' strikes in those periods. At this stage, we can recognise the need to adjust the data but we do not have an adequate basis for doing so. We return to this topic in Chapter 13.

Model (B) gave the results

$$\nabla\nabla_{12}y_t = 7.08 + (1 - 0.854B)(1 - 0.733B^{12})\varepsilon_t \qquad (7.36)$$
$$\quad(1.63) \qquad (14.86) \qquad\quad (6.96)$$

with $\hat{\sigma} = 734$, $\bar{R}^2 = 0.889$, and $\bar{R}^2_{SD} = 0.312$. The Q-statistics are

K	12	24	36	48
$Q(K)$	27.2	43.2	49.5	57.8
DF	10	22	34	46

Dropping the constant yields $\theta_1 = 0.816$, $\Theta_1 = 0.712$ and virtually identical

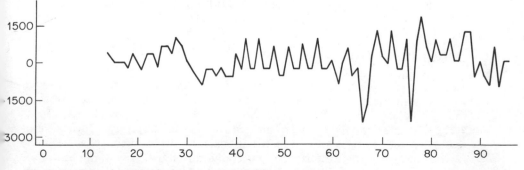

Fig. 7.5 Plot of residuals from model (7.36) for airlines data

diagnostics. The same three outlying observations appear on the residuals' plot in all cases; see Fig. 7.5 for the plot based on (7.36).

7.25   Further examination of Fig 7.5 reveals that, even when the outliers are ignored, the scatter of the residuals increases over time. Therefore, we re-analysed the data after transforming to logarithms, using model (B). The revised model is

$$\nabla \nabla_{12} \ln y_t = (1 - 0.790B)(1 - 0.661B^{12})\varepsilon_t \qquad (7.37)$$
$$(11.63) \qquad (6.61)$$

with $\hat{\sigma} = 0.030$, $\bar{R}^2 = 0.895$, and $\bar{R}^2_{SD} = 0.235$. Transforming back to the original units give $\{\Sigma(y - \hat{y})^2/DF\}^{1/2} = 788$, somewhat higher than $\hat{\sigma}$ for (7.36), as is to be expected since model (7.37) is fitted by minimising the sum of squares for the transformed observations. However, the Q-statistics are now

K	12	24	36	48
$Q(K)$	19.3	31.3	36.5	44.2
DF	10	22	34	46

and the first value lies between the 5 and 2.5 percentage points of the chi-squared distribution. The residuals are plotted in Fig. 7.6. Model (7.37) seems to be about as good as we can get without making adjustments for the outliers.

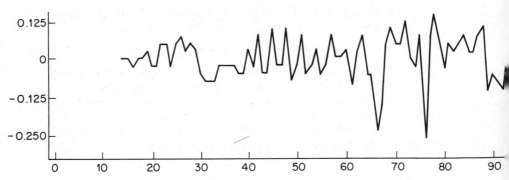

Fig. 7.6   Plot of residuals from model (7.37) for airlines series (logarithms)

Automatic model selection

7.26   The rather elaborate procedure necessary to identify and then to validate univariate ARIMA models has led, naturally, to attempts to automate the search.

7.27   The most straightforward procedure we can apply is stepwise autoregressive modelling. This proceeds in exactly the same way as stepwise regression, except that the regressor variables are not lagged y-values.

By way of example, we ran a stepwise autoregression for the sheep series, selecting from lags 1–8 to maximize $\bar{R}^2$. The selected model, based on $73 - 8 = 65$ observations, is

$$y_t = 171.4 + 1.374 y_{t-1} - 0.640 y_{t-2} + 0.165 y_{t-4} \qquad (7.38)$$
$$\quad\; (12.40) \quad\;\; (-4.84) \qquad\quad (2.48)$$

with $\hat{\sigma} = 65.8$ and $\bar{R}^2 = 0.886$. When (7.34) is expressed in terms of y_t rather than ∇y_t, the coefficients are

$$(1 - B)(1 - 0.420B + 0.211B^2 + 0.327B^3) = 1 - 1.420B + 0.631B^2$$
$$+ 0.166B^3 - 0.327B^4; \quad (7.39)$$

the leading terms are very similar between the two models.

Information criteria

7.28 When moving average terms also enter the model, a simple stepwise procedure is no longer available. A natural approach to the comparison of two models is to compare their likelihood functions (*LF*). If model 1, with k_1 parameters and $LF = L_1$ is correct and is compared with model 2, which has an additional $(k_2 - k_1)$ parameters and $LF = L_2$, it is well known (Kendall and Stuart, 1979, Section 24.7) that minus twice the log-likelihood ratio asymptotically follows a chi-squared distribution with $(k_2 - k_1)DF$. Thus, for large samples

$$E[-2 \ln(L_1/L_2)] = k_2 - k_1. \qquad (7.40)$$

Comparisons between log-likelihoods must take account of (7.40), and this led Akaike (1974) to propose an information criterion (*AIC*) of the form

$$AIC = -2 \ln(L) + 2k, \qquad (7.41)$$

where k denotes the number of parameters in the model. The model with the smallest *AIC* is deemed best in the sense of minimizing the forecast mean square error (*FMSE*). However, it was pointed out by Schwartz (1978) and others that *AIC* is not a consistent criterion in that it does not select the true model with probability approaching 1 as $n \rightarrow \infty$. To overcome this problem, Schwartz proposed the Bayesian information criterion (*BIC*):

$$BIC = -2 \ln(L) + k \ln(n). \qquad (7.42)$$

The *BIC* tends to include fewer terms than the *AIC* since the penalty term is greater. Several other criteria have been proposed; see, for example, Bhansali (1986). In general, these criteria either minimise *FMSE* or achieve consistency but not both. In practice, they usually provide similar, or even identical, results.

7.29 Table 7.2 summarises the results obtained when applying these criteria to the sheep series, once differenced. The following conclusions may be drawn:

(a) *AIC* and *BIC* select $(0, 1, 4)$ and $(3, 1, 0)$ schemes, respectively, in agreement with our earlier analyses.

(b) The models with higher numbers of parameters tend to be ill-conditioned; the parameter values are not well defined. Usually such higher-order

Table 7.2 Values of the *AIC* and *BIC* criteria for different *ARMA* (p, q) models for the sheep series, differenced once. (Values are shown as (AIC-824.77) and BIC-833.88), these being the numerical values for the *ARMA* (3, 0) scheme.)

		AR					
(a) AIC		0	1	2	3	4	5
MA	0	18.02	10.29	4.55	0.00	1.84	3.21
	1	7.48	9.34	1.66	1.90	2.03	4.93
	2	8.79	5.54^*	12.11^*	17.25^*	4.03	5.45^*
	3	6.70	2.24^*	3.00	31.52^*	5.94^*	7.45^*
	4	-2.14^*	-1.26^*	1.00^*	2.99^*	4.07^*	6.98^*
	5	-1.21^*	0.75^*	2.75^*	3.68^*	5.78^*	9.03^*

		AR					
(b) BIC		0	1	2	3	4	5
MA	0	11.19	5.73	2.27	0.00	4.12	7.76
	1	2.93	7.06	1.65	4.17	6.58	11.76
	2	6.51	5.54^*	14.39^*	21.80^*	10.86	14.79^*
	3	6.70	4.51^*	7.55	38.34^*	15.04^*	18.84^*
	4	0.13^*	3.29^*	7.82^*	12.10^*	15.45^*	20.64^*
	5	3.34^*	7.57^*	11.85^*	15.04^*	19.43^*	24.96^*

*Indicates that the estimates were numerically unstable because the sum of squares was very flat in the neighbourhood of the optimal solution. Different estimation procedures (conditional and unconditional, LS, ML) produced similar results, as did attempts to vary the convergence criteria.

models would not be considered; we fitted these higher-order schemes to illustrate the changes in the level of the criteria. The numerical difficulties suggest a nearly equal root in $\phi(B) = 0$ and $\theta(B) = 0$.

(c)   Note that the addition of a parameter should not increase *AIC* by more than 2.0 or *BIC* by ln (72) = 4.28; see (7.41–2).

(d)   *BIC* penalizes models with large numbers of parameters more heavily than does *AIC*, as expected.

This study and others like it suggest that the information criteria perform reasonably, although the method of searching through the set of possible models is not well defined. This becomes awkward, especially when seasonal models are being considered or when terms may be included with non-consecutive lags. For example, if we consider the *AR* models with lags 1 and 3 only, the resulting model has

$$AIC = 824.66 \quad \text{and} \quad BIC = 831.49;$$

in both cases these values are lower than the minima found in Table 7.2.

Autobox

7.30 Perhaps the most ambitious attempt to automate the *ARI...*
identification procedure is the work of Reilly (1981; 1984) who has ...

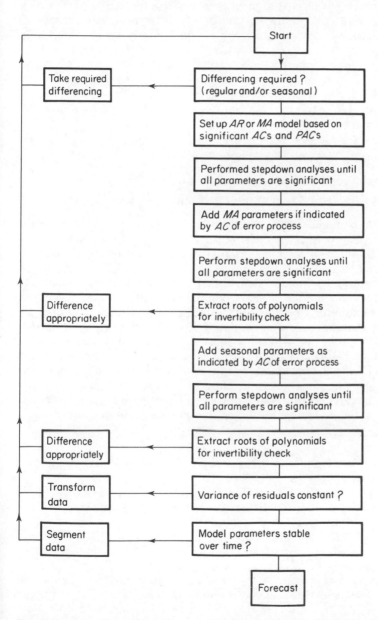

Fig. 7.7 Procedure flow chart to develop Box-Jenkins models (Reilly 1981)

computer program called AUTOBOX. Reilly's search procedure operates by performing series of tests to eliminate non-significant parameters and to include new terms where warranted. Additional terms are identified by finding significant spikes in the $SACF$ or $SPACF$. Once a model has been identified, the significance of each coefficient is examined and non-significant terms are discarded; this process is termed a 'stepdown' process by Reilly. AUTOBOX also includes criteria to determine whether or not differencing is necessary. A flow chart summarizing the procedure is given in Fig. 7.7, taken from Reilly (1981).

7.31   The performance of automatic selection procedures can only be evaluated empirically. By way of example, we applied the program to the sheep series. The model development process was, in this case, very brief. AUTOBOX identified an $AR(2)$ scheme. No additions or deletions were found to be necessary and the final model was

$$y_t = 231.6 + 1.277 y_{t-1} - 0.405 y_{t-2} + \varepsilon_t, \qquad \hat{\sigma} = 70.43. \qquad (7.43)$$
$$(12.17) \qquad (-3.92)$$

This is quite similar to (7.34), which may be written as

$$y_t = -6.59 + 1.420 y_{t-1} - 0.631 y_{t-2} - 0.116 y_{t-3} + 0.327 y_{t-4} + \varepsilon_t;$$

$\hat{\sigma} = 69.8$. Although model (7.43) is stationary whereas (7.34) is not, the short-term forecasts generated by the two models are likely to be quite similar; see Section 8.26. Interestingly, the original analysis of the sheep series in Kendall (1946) led to an $AR(2)$ scheme.

7.32   As occurred during the development of stepwise regression, automatic modelling methods in time-series analysis have met with a mixed reception. Those opposed to automatic selection argue that such procedures can produce poor models and that the context of the data is lost. Those in favour argue that automatic models often perform as well as those selected by 'experts' (cf. Texter and Ord, 1989) and that they are now much more economical to produce. Further, when a considerable number of series must be analysed, the time savings from automatic selection allow the 'expert' to focus attention upon the problem cases. In summary, our view is that automatic procedures, used with care, are a valuable addition to the time-series analyst's armoury.

Exercises

7.1   Given that $r_1 = 0.7$ and $r_2 = 0.2$, use the sample Yule–Walker equations to estimate ϕ_1 and ϕ_2.

7.2   Develop the exact likelihood function for the $AR(1)$ scheme with $E(y) = 0$, using the approach described in Section 7.8. Show that it is equivalent to (7.6).

7.3   (Follow-up to 6.6). Generate simulated data from an $AR(1)$ scheme with $n = 100$ and $\phi = 0.7$. Identify and then fit an appropriate model. Check the adequacy of your model using the diagnostic procedures.

7.4   (Follow-up to 7.3). Fit an $AR(2)$ scheme to your simulated series and then use the diagnostic checks to refine your model.

7.5   (Follow-up to 7.3). Fit an $MA(1)$ scheme to your simulated series and then use the diagnostic checks to refine your model. (*Note*: the process becomes more difficult because of the poor choice of initial model.)

7.6 A series consists of $n = 50$ observations. An $ARIMA(1, 1, 1)$ model is fitted and yields the following residual $SACF$:

lag	1	2	3	4	5	6
r_j	0.10	0.10	0.35	-0.05	0.10	0.15

Use the Q-statistic in (7.23) to test whether the residuals are random.

7.7 An $AR(3)$ model has estimated coefficients:

$$\hat{\phi}_1 = 0.20, \qquad \hat{\phi}_2 = 0.81 \quad \text{and} \quad \hat{\phi}_3 = 0.20.$$

Does the estimated model satisfy the stationarity conditions?

7.8 An $MA(3)$ model has estimated coefficients:

$$\hat{\theta}_1 = 1.70, \qquad \hat{\theta}_2 = -0.80 \quad \text{and} \quad \hat{\theta}_3 = -0.15.$$

Does this scheme appear to be invertible?

7.9 (Follow-up to 6.10). Develop appropriate models for some or all of the following non-seasonal series:

(a) Kendall's simulated $AR(2)$ series (Table A1);
(b) UK gross domestic product (Table A2);
(c) Wölfer's sunspots data (Table A3);
(d) Canadian lynx trappings (Table A4);
(e) US interest rates (Table A5);
(f) US immigration (Table 1.4).

7.10 (Follow-up to 6.11). Develop appropriate models for some or all of the seasonal series given in the data appendix:

(a) Chatfield and Prothero's sales data (Table A6);
(b) UK unemployment (Table A7);
(c) UK whisky production (Table A8).

7.11 If you have access to a package which performs automatic model selection or, at least, provides you with values of AIC and/or BIC, try automatic model selection for one of the series listed in Question 7.9, and 7.10. Compare the results with your own 'expert' model development.

8

Forecasting

Introduction

8.1 The English language is rich in words that describe attempts to see into the future: forecasting, foretelling, foreseeing, and even foretasting; prediction, prevision, and prognostication; as well as such phrases as 'gazing into a crystal ball'. Two of these are used to denote quantitative forecasting methods, namely forecasting and prediction: they are often used synonymously. If there is any difference, forecasting is perhaps used with reference to specific events of a quantifiable kind, prediction to more verbal descriptions relating to ambient circumstances. For example, we might speak of forecasting the results of a general election but of predicting its effect on foreign policy. For the purposes of this chapter, we shall not try to draw fine distinctions between the two terms.

8.2 The methods we used in forecasting depend on the purposes to which the forecasts are to be put. Broadly speaking, we may consider two cases: (1) we may require a forecast which is to be the basis of our own action, as in sales forecasting; (2) an agency may produce a forecast which is going to be used for all kinds of purposes by different people, as when someone forecasts the population, or a government agency forecasts growth rates in sectors of the economy. Prudent forecasters (and in forecasting one needs to be prudent, because mistakes are remembered much longer than successes) should regard their range of techniques much as mechanics regard their tool bags, using whichever instruments the circumstances require.

8.3 The time frame for forecasting may be divided into three levels: short term, medium term, and long term. These terms are relative to the subject under study. In meteorology, for example, 'short term' may mean only two or three days ahead, and 'medium term' refers to the next few months; whereas in economics 'short term' means a few months, perhaps as much as a year, 'medium term' refers up to the next five years, and everything after that is 'long term'. The time periods of interest may often be determined by external factors as for monthly sales, annual budgets, or longer-term strategic plans over several years.

We shall not examine long-term forecasting or most aspects of technological forecasting since these are too qualitative in nature; see Armstrong (1985) and Martino (1983), respectively, for full accounts of these topics. Forecasting using growth curves will be considered briefly in Section 15.19. Our emphasis is, therefore, on short-to-medium-term forecasting from a statistical viewpoint. It should be noted, however, that simple, direct methods are often very effective. For example, short-term weather forecasting is performed by observing approaching weather fronts, a simple enough method even though the observational equipment is technically sophisticated. Similarly, it is possible to forecast demands on our educational system over the next twenty years with some confidence because a good many of the young men and women who will be exposed to it are already born. Nor should one despise these simple-minded methods in the behavioural sciences – exports next month can be predicted from the order book for this and previous months; and forecasts of next year's consumption of consumer durables may be based upon surveys of buying intentions.

8.4 We begin by considering the criteria upon which forecasts may be judged, since this leads directly to the form of the forecast function. In particular, we consider the best linear predictor (BLP) in Sections 8.8–12. We then take the BLP as a standard by which to evaluate various more heuristic forecasting procedures that have been suggested; see Sections 8.13–24. Finally, in Sections 8.25–35, we review the literature on the relative performance of different forecasting procedures.

Forecast accuracy

8.5 If we have a known model for a process such as

$$y_t = \mu_t + \varepsilon_t, \tag{8.1}$$

where μ_t is a known function of past y- and ε-values, and ε_t is independent of μ_t, it is apparent that a suitable forecast for y_t is μ_t and that the error *after* y_t is observed is $e_t = y_t - \mu_t$. Note that e_t is a realization of the random variable ε_t. Clearly, the performance of a forecasting model at a single point in time does not tell us much about the overall quality of the method, but the long-run average properties of the e_t will be informative. That is, we may record the errors at time periods $t = n + 1, n + 2, \ldots, n + m$ and examine the corresponding observed forecast errors $e_{n+1}, e_{n+2}, \ldots, e_{n+m}$.

8.6 In standard statistical theory, we assume random sampling and develop point and interval estimators on that basis. We now continue to assume that the ε_t are independent and identically distributed (with mean zero and variance σ^2) which provides the necessary structure for statistical inference. Corresponding to the use of (squared) standard errors in random sampling, we consider the *forecast mean square error* or

$$FMSE = FMSE(\mu_t) = E\{(y_t - \mu_t)^2\} = E(\varepsilon_t^2) \tag{8.2}$$

which may be estimated by

$$\widehat{FMSE} = \sum_{i=1}^{m} e_{n+i}^2 / m. \tag{8.3}$$

Given our present assumptions, $FMSE = \sigma^2$, but $FMSE$ will be used under more general conditions, as we indicate below. Other measures of forecast performance include the forecast mean absolute error,

$$FMAE = E\{|\, y_t - \mu_t|\} = E(|\varepsilon_t|) \tag{8.4}$$

and the forecast mean absolute percentage error,

$$FMAPE = E\{|y_t - \mu_t|/y_t\} \tag{8.5}$$

provided y_t takes only positive values. We shall not examine the $FMAE$ and $FMAPE$ theoretically, but the corresponding sample versions are useful in evaluating forecasting procedures empirically.

8.7   The discussion so far is unrealistic in that only sampling errors are considered. In practice, we must consider two other sources of error:

(a)   *estimation* errors, when μ_t contains unknown parameters that have been estimated using past values of the series;
(b)   *specification* errors, when the model is incorrectly formulated or changes over time.

Estimation errors can be dealt with, at least in principle, by defining $FMSE$ as

$$
\begin{aligned}
FMSE(\hat{\mu}_t) &= E\{(y_t - \hat{\mu}_t)^2\} \\
&= E\{(y_t - \mu_t)^2\} + E\{(\mu_t - \hat{\mu}_t)^2\},
\end{aligned} \tag{8.6}
$$

where $\hat{\mu}_t$ is the estimator for μ_t based upon the estimated parameter values and the decomposition in (8.6) follows from the independence of the error terms. The other measures may be adapted similarly, but theoretical results are generally asymptotic in nature.

Specification errors are more difficult to handle, although some progress is possible, as we indicate in Section 9.14. The risks of incorrect specification based on past data can be reduced using the diagnostics described in Chapter 7, but the possibility of structural change can never be ruled out and may be almost impossible to foretell. For this reason, empirical evaluations of forecasting methods rely upon measures such as $FMSE$, $FMAE$, or $FMAPE$ rather than goodness-of-fit measures; see Section 8.25.

Best linear predictors

8.8   We now assume that the process has a known $ARIMA$ form and ask what is the best forecast we can make? For the purposes of this discussion, we assume that the model may be expressed in random-shock form (5.69) as

$$y_t = \sum_{j=0}^{\infty} \psi_j \varepsilon_{t-j}. \tag{8.7}$$

We now assume that observations are available up to and including time t and that we wish to predict

$$y_{t+k} = \sum_{j=0}^{\infty} \psi_j \varepsilon_{t+k-j}. \tag{8.8}$$

The information we have available is equivalent to knowing $\{\varepsilon_t, \varepsilon_{t-1}, \ldots\}$ so any linear predictor for k steps ahead, made at time t, is

$$m_t(k) = \sum_{j=0}^{\infty} \eta_{j+k}\varepsilon_{t-j}. \tag{8.9}$$

We now determine the best linear predictor, using *FMSE* as the criterion. The *FMSE* is

$$FMSE[m_t(k)] = E\{[y_{t+k} - m_t(k)]^2\}$$

$$= E\left\{\left[\sum_{j=0}^{k-1} \psi_j\varepsilon_{t+k-j} + \sum_{j=k}^{\infty} (\psi_j - \eta_j)\varepsilon_{t+k-j}\right]^2\right\}. \tag{8.10}$$

Since the ε_t are independent with zero means and variances σ^2, this reduces to

$$FMSE[m_t(k)] = \sigma^2\left[\sum_{j=0}^{k-1} \psi_j^2 + \sum_{j=k}^{\infty} (\psi_j - \eta_j)^2\right]. \tag{8.11}$$

Inspection of (8.11) reveals that the *FMSE* is minimized when $\psi_j = \eta_j$, $j \geq k$. That is, the *best linear predictor* is

$$y_t(k) = \sum_{j=k}^{\infty} \psi_j\varepsilon_{t+k-j} \tag{8.12}$$

with

$$FMSE(k) = \sigma^2 \sum_{j=0}^{k-1} \psi_j^2. \tag{8.13}$$

8.9 Given (8.12) and (8.13), we may set up a $100(1 - \alpha)$ percent *prediction* interval for y_{t+k}, assuming the errors to be normally distributed, as

$$y_t(k) \pm z_{\alpha/2}(FMSE)^{1/2}, \tag{8.14}$$

where $z_{\alpha/2}$ is the appropriate percentage point from normal tables. Note that this interval depends on the assumption of normality although (8.14) is approximately correct for many non-normal distributions when α is near 0.05. Even so, (8.14) does not allow for sampling errors so that the interval may be too narrow when the series is short. Fuller and Hasza (1980) show that the usual regression *FMSE* provides an excellent approximation when the parameters must be estimated. An empirical study by Gardner (1988) suggests that the combined effects of sampling variation and structural change are such that wider limits based on the Chebyshev inequality may give actual confidence levels closer to nominal levels. Clearly this is an area where further work is desirable.

8.10 An alternate determination of the best linear predictor is to consider the minimization of

$$E\{[y_{t+k} - \mu_{t+k}]^2\}| D_t\}, \tag{8.15}$$

where $D_t = \{y_t, y_{t-1}, \ldots, \varepsilon_t, \varepsilon_{t-1}, \ldots\}$ denotes the available information at time t. This leads directly to the result that

$$\mu_{t+k} = E(y_{t+k}| D_t) = y_t(k), \tag{8.16}$$

the conditional expectation of y_{t+k} at time t. Since the *ARIMA* models are linear in y and ε, it follows that for any *ARMA*(p, q) scheme

$$y_t(k) = \sum_{j=1}^{p} \phi_j E(y_{t+k-j} | D_t) - \sum_{i=0}^{q} \theta_i E(\varepsilon_{t+k-i} | D_t) \qquad (8.17)$$

taking $\theta_0 = -1$. For any past values with $s < 0$

$$E(y_{t+s} | D_t) = y_{t+s} \qquad (8.18)$$

and

$$E(\varepsilon_{t+s} | D_t) = \varepsilon_{t+s}, \qquad (8.19)$$

whereas for $s > 0$

$$E(y_{t+s} | D_t) = y_t(s) \qquad (8.20)$$

$$E(\varepsilon_{t+s} | D_t) = 0. \qquad (8.21)$$

Combining (8.17)–(8.21), we can generate $y_t(k)$ successively for $k = 1, 2, \dots$. Differenced schemes are readily handled by generating the forecasts for $w_{t+k} = \nabla^d \nabla_s^D y_{t+k}$ and then producing the 'integrated' predictions.

Example 8.1   Consider the model

$$y_t = \delta + \phi y_{t-1} + \varepsilon_t. \qquad (8.22)$$

From (8.17)–(8.21),

$$y_t(1) = \delta + \phi y_t$$

and

$$y_t(k) = \delta + \phi y_t(k-1), \qquad k \geqslant 2.$$

Hence,

$$y_t(k) = \delta(1 + \phi + \cdots + \phi^{k-1}) + \phi^k y_t$$
$$= \mu(1 - \phi^k) + \phi^k y_t \qquad (8.23)$$

since $E(y_t) = \mu = \delta/(1 - \phi)$. That is, the forecast is a weighted average of the mean, μ, and the latest observation y_t. As k increases, greater weight is attached to μ so that, as expected,

$$y_t(k) \to \mu \qquad \text{as } k \to \infty. \qquad (8.24)$$

Quite generally, the stationarity of a process implies that (8.24) holds for the forecasts.

Further, from (8.13) and (8.22)

$$FMSE(k) = \sigma^2 \sum_{j=0}^{k-1} \psi_j^2$$
$$= \sigma^2 (1 + \phi^2 + \cdots + \phi^{2k-2})$$
$$= \sigma^2 (1 - \phi^{2k})/(1 - \phi^2)$$

which approaches $\sigma^2/(1 - \phi^2)$ as $k \to \infty$, which is var(y_t), the unconditional variance. Again, the limiting *FMSE* is var(y_t) for any stationary process.

Example 8.2 Consider the model

$$y_t = \mu + \varepsilon_t - \theta\varepsilon_{t-1}. \tag{8.25}$$

Then

$$y_t(1) = \mu - \theta\varepsilon_t$$
$$y_t(k) = \mu, \qquad k \geqslant 2.$$

In general, the moving average processes have a finite memory property such that $y_t(k) = \mu$, $k \geqslant q + 1$.

From (8.13) and (8.25),

$$FMSE(k) = \sigma^2 \sum_{j=0}^{k-1} \psi_j^2$$

$$= \begin{cases} \sigma^2, & k = 1 \\ \sigma^2(1 + \theta^2), & k \geqslant 2. \end{cases}$$

Generally, for an $MA(q)$ scheme,

$$FMSE(k) = \sigma^2(1 + \theta_1^2 + \cdots + \theta_m^2), \tag{8.26}$$

where $m = \min(k - 1, q)$.

Non-stationary processes

8.11 When the process is non-stationary, the forecasts cannot revert back to a mean level as the mean is not defined. Similarly, the variance cannot converge to a stable value. The following example indicates the general behaviour.

Example 8.3 Consider the model

$$w_t = \nabla y_t = \varepsilon_t - \theta\varepsilon_{t-1}. \tag{8.27}$$

The forecasts for w_t follow directly from Example 8.2, so that those for y_t are given by

$$y_t(1) = y_t + w_t(1) = y_t - \theta\varepsilon_t = (1 - \theta)y_t + \theta y_{t-1}(1)$$
$$y_t(2) = y_t(1) + w_t(2) = y_t - \theta\varepsilon_t \tag{8.28}$$

and $y_t(k) = y_t(1)$ for all $k \geqslant 2$. The forecast does not and indeed cannot revert to a long-run mean level. It follows from (8.27) that

$$y_{t+k} - y_t = w_{t+k} + w_{t+k-1} + \cdots + w_{t+1}$$
$$= \varepsilon_{t+k} + (1 - \theta)\varepsilon_{t+k-1} + \cdots + (1 - \theta)\varepsilon_{t+1} - \theta\varepsilon_t$$

so that

$$\psi_0 = 1, \quad \psi_j = (1 - \theta), \qquad 1 \leqslant j \leqslant k - 1$$

and

$$FMSE(k) = \sigma^2\{1 + (k - 1)(1 - \theta)^2\}, \tag{8.29}$$

a linear function of k. The non-stationarity of the process means that the prediction intervals continue to widen as k increases. Indeed, if we difference d times, the *FMSE* is of order k^{2d-1} times σ^2; see Exercise 8.6.

From Examples 8.1 and 8.3, we see that the behaviour of forecasts for an $ARIMA(1, 0, 0)$ scheme with $\phi = 1 - \delta$, δ small and an $ARIMA(0, 1, 1)$ scheme with $\theta = \delta$ will be very similar for small k, but diverge with increasing k.

Seasonal models

8.12   The construction of forecasts for seasonal models proceeds in exactly the same way as for the non-seasonal case, but the pattern of the *FMSE* may be rather different.

Example 8.4   Consider the following model for monthly data:

$$\nabla_{12} y_t = (1 - \theta B)(1 - \Theta B^{12}) \varepsilon_t. \tag{8.30}$$

The forecasts are

$$y_t(1) = y_{t-11} - \theta \varepsilon_t - \Theta \, \varepsilon_{t-11} + \theta \Theta \, \varepsilon_{t-12}$$
$$y_t(k) = y_{t+k-12} - \Theta \, \varepsilon_{t+k-12} + \theta \Theta \, \varepsilon_{t+k-13}, \qquad 2 \leqslant k \leqslant 12$$
$$y_t(13) = y_t(1) + \theta \Theta \, \varepsilon_t$$
$$y_t(k) = y_t(k - 12), \qquad k \geqslant 14.$$

The random shock form of (8.30) is

$$y_t = \{1 - \theta B + (1 - \Theta)B^{12} - \theta(1 - \Theta)B^{13} + (1 - \Theta)B^{24} + \cdots\}\varepsilon_t \tag{8.31}$$

giving the forecast mean errors as

$$FMSE(1) = \sigma^2;$$
$$FMSE(k) = \sigma^2(1 + \theta^2), \qquad k = 2, \ldots, 12;$$
$$FMSE(13) = \sigma^2\{1 + \theta^2 + (1 - \Theta)^2\}, \qquad k = 13;$$
$$FMSE(k) = \sigma^2(1 + \theta^2)\{1 + (1 - \Theta)^2\}, \qquad k = 14, \ldots, 24;$$

and so on. The rate of increase of the *FMSE* is much slower than in Example 8.3 since the non-stationarity here arises only in the seasonal component.

Other forecasting procedures

8.13   Many of the forecasting procedures still used in practice have been developed on rather intuitive grounds. We shall now examine several of these methods and integrate them into our general framework.

Moving averages

8.14   We could just fit a polynomial to the entire series and use this for extrapolation. As indicated in Chapter 2, this is rarely a good idea, and it is usually better to give greater weight to the most recent observations. In Chapter 2, we discussed fitting local polynomials of order p to the last $(2m + 1)$ points of the series, and the tables given in Appendix C give the weights for forecasting one step ahead, under the column headed '0'. Alternatively, once the coefficients $\{a_j\}$ have been estimated, the k-step ahead forecasts are simply

$$y_m(k) = a_0 + a_1(m + k) + \cdots + a_p(m + k)^p, \tag{8.32}$$

where the time origin is set at the centre of the $(2m + 1)$ fitted values, as before.

Example 8.5 Suppose we fit a straight line by taking $p = 1$. It follows that

$$a_0 = \Sigma\, y_j/(2m + 1), \qquad a_1 = 3 \,\Sigma\, j y_j/m(m + 1)(2m + 1),$$

where the sums are over $j = -m$ to $j = +m$. Thus, the forecasts become

$$y_m(k) = a_0 + a_1(m + k)$$

$$= \frac{1}{m(m + 1)(2m + 1)} \Sigma\{m(m + 1) + 3j(m + k)\}y_j, \tag{8.33}$$

a linear function of the y_j. When the series is purely random with constant mean, the estimators a_0 and a_1 are uncorrelated so that

$$FMSE(k) = \frac{\sigma^2}{2m + 1}\left\{1 + \frac{3(m + k)^2}{m(m + 1)}\right\}, \tag{8.34}$$

which increases at the rate k^2. Of course, for a purely random process, better forecasts would be given by $y_m(k) = a_0 = \bar{y}$ with $FMSE = \sigma^2/(2m + 1)$.

8.15 One of the difficulties with the moving average model is that it is difficult to visualise a process which corresponds to (8.32); in turn, this makes it difficult to evaluate the *FMSE* for non-random schemes. Conversely, if we consider a process of the *ARIMA* class, we could indeed determine the *FMSE* corresponding to forecasts such as (8.32), but these forecasts would not be optimal. For these reasons, forecasting using local polynomials derived from moving averages is not recommended.

Exponential smoothing

8.16 An alternative way of giving most weight to the recent past is to consider a weighted average of the form

$$y_t(1) = \sum_{j=0}^{\infty} w_j y_{t-j}, \tag{8.35}$$

where $\Sigma\, w_j = 1$, and we might choose the weights to be monotonically decreasing. The simplest such choice is to set

$$w_j = (1 - \beta)\beta^j, \tag{8.36}$$

with $|\beta| < 1$ but, usually, $0 < \beta < 1$. Predictor (8.5) with weights (8.36) is known as an *exponentially weighted moving average* (EWMA) and the operation is also known as exponential smoothing. What makes the procedure particularly attractive is that (8.35) then satisfies a simple recurrence relation since

$$y_t(1) = (1 - \beta)\{y_t + \beta(y_{t-1} + \beta y_{t-2} + \cdots)\}$$
$$= (1 - \beta)y_t + \beta y_{t-1}(1). \tag{8.37}$$

Thus, to update the forecast, we need only the latest observation and the previous forecast. The k-step ahead forecast is also given by (8.37).

When (8.35) is based upon a finite data record of t-values, the weights in (8.36) become

$$w_j = c_t\beta^j, \quad j \leqslant t - 1$$
$$= 0, \quad j \geqslant t,$$

where $c_t = (1 - \beta)/(1 - \beta^t)$. We then have

$$y_t(1) = c_t y_t + (1 - c_t)y_{t-1}(1) \tag{8.38}$$

or

$$y_t(1) = c_t \sum_{j=0}^{t-1} \beta^j y_{t-j}. \tag{8.39}$$

8.17   Popular use of (8.37) often involves choosing a value for $\alpha = 1 - \beta$ and a start-up value $y_1(1)$. Recommended values for α are typically in the range of 0.05 to 0.30, depending on the volatility of the series, whereas a common choice for the start-up value is to set $y_1(1) = y_1$. Especially when the series is short or β is near one, we can see that this may be a poor procedure, giving undue weight to the first observation. It is better to use (8.39) in such circumstances. For a fuller discussion of start-up values for *EWMA* schemes, see Ledolter and Abraham (1984).

Returning to (8.37), we note that it is equivalent to (8.28) when $\beta = \theta$; that is, the *EWMA* corresponds to the forecast function of an *ARIMA*(0, 1, 1) scheme. We shall now explain why this is so.

8.18   Recall that, for any *ARIMA* process,

$$y_{t+1} = y_t(1) + \varepsilon_{t+1}. \tag{8.40}$$

From (8.37)

$$(1 - \beta B)y_t(1) = (1 - \beta)y_t;$$

substituting for $y_t(1)$ using (8.38) yields

$$(1 - \beta B)y_{t+1} = (1 - \beta)y_t + (1 - \beta B)\varepsilon_{t+1}$$

or

$$y_{t+1} - y_t = \varepsilon_{t+1} - \beta\varepsilon_t \tag{8.41}$$

which is exactly (8.27) when $\beta = \theta$. Thus, improved *EWMA* forecasts may be obtained by estimating β and the start-up value using the *ARIMA*(0, 1, 1) scheme. In addition, the diagnostics discussed earlier allow us to determine whether the *EWMA* is appropriate. This equivalence is discussed further in Section 9.5.

Discounted least squares

8.19 A different approach, suggested by Brown (1963), is to emphasise the most recent observations by use of discounted least squares. That is, we use a *discount factor*, $0 < \beta < 1$, and choose the estimator, a_t, to minimise

$$\sum_{j=0}^{\infty} (y_{t-j} - a_t)^2 \beta^j. \tag{8.42}$$

This leads to

$$a_t \sum \beta^j = \sum \beta^j y_{t-j}$$

or

$$\begin{aligned} a_t &= (1 - \beta) \sum \beta^j y_{t-j} \\ &= (1 - \beta) y_t + \beta a_{t-1}, \end{aligned} \tag{8.43}$$

exactly as in (8.37) when the same β is used. If we consider (8.42) for only t terms, forecast function (8.39) results. The advantage of this approach is that it may be extended to consider polynomial models. For example, when $p = 1$, we minimise

$$\sum_{j=0}^{\infty} (y_{t-j} - a_{0t} - a_{1t}j)^2 \beta^j. \tag{8.44}$$

This leads to the estimating equations

$$\sum y_{t-j}\beta^j - a_{0t} \sum \beta^j - a_{1t} \sum j\beta^j = 0$$
$$\sum jy_{t-j}\beta^j - a_{0t} \sum j\beta^j - a_{1t} \sum j^2\beta^j = 0.$$

Now $\sum \beta^j = 1/(1 - \beta)$, $\sum j\beta^j = \beta/(1 - \beta)^2$, $\sum j^2\beta^j = \beta(1 + \beta)/(1 - \beta)^3$, so the estimating equations become

$$(1 - \beta) \sum y_{t-j}\beta^j - a_{0t} - a_{1t}\frac{\beta}{1 - \beta} = 0$$

$$(1 - \beta)^2 \sum jy_{t-j}\beta^j - \beta a_{0t} - a_{1t}\frac{\beta(1 + \beta)}{(1 - \beta)} = 0.$$

Let

$$\begin{aligned} S_1(t) &= (1 - \beta) y_t + \beta S_1(t - 1) \\ &= (1 - \beta) \sum y_{t-j}\beta^j \end{aligned} \tag{8.45}$$

and define a second smoothing operation by

$$S_2(t) = (1 - \beta) S_1(t) + \beta S_2(t - 1)$$

or

$$(1 - \beta B) S_2(t) = (1 - \beta) S_1(t). \tag{8.46}$$

From (8.45) and (8.46),

$$(1 - \beta B)^2 S_2(t) = (1 - \beta)^2 y_t,$$

or

$$S_2(t) = \frac{(1 - \beta)^2}{(1 - \beta B)^2} \, y_t$$
$$= (1 - \beta)^2 \, \Sigma \, (j + 1) y_{t-j} \beta^j$$
$$= (1 - \beta) S_1(t) - (1 - \beta)^2 \, \Sigma \, j y_{t-j} \beta^j.$$

The estimating equations become, dropping the arguments in t, for convenience,

$$S_1 - a_0 - \frac{\beta a_1}{1 - \beta} = 0,$$

and

$$S_2 - (1 - \beta) S_1 - \beta a_0 - \frac{\beta(1 + \beta) a_1}{1 - \beta} = 0.$$

In turn, these expressions yield

$$a_0 = 2S_1 - S_2 \quad \text{and} \quad a_1 = (1 - \beta)(S_1 - S_2)/\beta. \qquad (8.47)$$

That is, the forecasting function is

$$y_t(k) = a_0 + a_1 k$$
$$= (2S_1 - S_2) + k(1 - \beta)(S_1 - S_2)/\beta, \qquad (8.48)$$

a locally linear function. Because of (8.45) and (8.46), this method is sometimes known as *double exponential smoothing*; form (8.48) gives rise to the alternative name *linear exponential smoothing*. Consideration of (8.48) with $k = 1$, together with (8.39), (8.45) and (8.46) leads to the conclusion that the underlying model may be written as

$$(1 - B)^2 y_t = (1 - \beta B)^2 \varepsilon_t; \qquad (8.49)$$

that is, an $ARIMA(0, 2, 2)$ scheme with a single parameter. For further discussion of such equivalences, see Abraham and Ledolter (1986).

Example 8.6   Suppose that $\beta = 0.8$, $y_1(1) = S_1(1) = S_2(1) = 20$ and $y_2 = 25$. Then, for single exponential smoothing

$$y_2(1) = (0.2)(25) + (0.8)(20) = 21$$

and $y_2(k) = 21$, $k \geqslant 2$. For double exponential smoothing,

$$S_1(2) = (0.2)(25) + (0.8)(20) = 21$$
$$S_2(2) = (0.2)(21) + (0.8)(20) = 20.2,$$

whence $a_0 = 21.8$, $a_1 = 0.2$, and $y_2(1) = 22$, $y_2(k) = 21.8 + 0.2k$.

8.20   As with single exponential smoothing, use of (8.48) historically required preselection of β, $S_1(1)$ and $S_2(1)$. Typical values would be $0.02 \leqslant \beta \leqslant 0.20$ and $S_1(1) = S_2(1) = y_1$, although this can be improved upon by using (8.49) and a least-squares procedure; also see Ledolter and Abraham (1984).

An excellent review of recent developments in exponential smoothing is provided by Gardner (1985).

8.21   The discounted least-squares procedure may be applied to higher-order polynomials; use of a quadratic in (8.44) produces *triple* exponential smoothing with an underlying model which is a special case of an $ARIMA(0, 3, 3)$ scheme of the form

$$(1 - B)^3 y_t = (1 - \beta B)^3 \varepsilon_t. \tag{8.50}$$

Use of (8.50) or higher orders is not recommended in general since the quadratic effects can produce rather wild oscillations in the forecast function.

Ameen and Harrison (1984) have developed a method of discount weighted estimation which enables different discount factors to be attached to different components, such as intercept, slope or seasonal effect. This considerably increases the flexibility of discounting methods while still allowing the forecast function to incorporate a start-up phase.

Holt's method

8.22 A key feature in fitting local polynomials and in using discounted least squares is the notion that the forecasts should be 'adaptive', in the sense that the low-order polynomials used for extrapolation have coefficients that are modified with each observation. Holt (1957) took this idea a step further by suggesting a forecast function of the form

$$y_t(k) = a_0(t) + k a_1(t), \tag{8.51}$$

where the $a_i(t)$ are updated according to the relations

$$a_0(t) = \alpha_1 y_t + (1 - \alpha_1)\{a_0(t - 1) + a_1(t - 1)\} \tag{8.52}$$

$$a_1(t) = \alpha_2\{a_0(t) - a_0(t - 1)\} + (1 - \alpha_2)a_1(t - 1). \tag{8.53}$$

Expression (8.52) is a weighted average of the new observation and its previous one-step ahead forecast, as for single exponential smoothing; (8.53) represents a similar average for the slope coefficient. When $\alpha_1 = \alpha(2 - \alpha)$ and $\alpha_2 = \alpha/(2 - \alpha)$, Holt's model is equivalent to double exponential smoothing. Generally (8.51)–(8.53) are equivalent to the forecast functions for an $ARIMA(0, 2, 2)$ scheme with two MA parameters; see Exercise 8.10.

As noted in Gardner (1985), (8.52)–(8.53) may be expressed in *error-correction* form as

$$a_0(t) = a_0(t - 1) + a_1(t - 1) + \alpha_1 e_t \tag{8.54}$$

$$a_1(t) = a_1(t - 1) + \alpha_1 \alpha_2 e_t, \tag{8.55}$$

where $e_t = y_t - y_{t-1}(1)$. This form is generally easier to use for updating purposes. Values of α_1, α_2, $a_0(1)$, and $a_1(1)$ are required to start; typical values would be $0.02 \leqslant \alpha_1$, $\alpha_2 \leqslant 0.20$ and $a_0(1) = y_1$, $a_1(1) = y_2 - y_1$. When the length of series allows, better values can be obtained by fitting the $ARIMA$ model; see also Ledolter and Abraham (1984).

The Holt–Winters seasonal model

8.23 When the data exhibit seasonal behaviour, several alternatives to $ARIMA$ models exist. A direct extension of Holt's method due to Winters

(1960) and often termed the Holt–Winters method, is to consider the forecast function

$$y_t(k) = \{a_0(t) + ka_1(t)\}c(t + k - s), \qquad (8.56)$$

where $c(t)$ is a *multiplicative* seasonal effect and s is the number of points in the year at which the series is observed (e.g. 12 for monthly data). The updating formulae are

$$a_0(t) = \alpha_1 \frac{y_t}{c(t - s)} + (1 - \alpha_1)\{a_0(t - 1) + a_1(t - 1)\} \qquad (8.57)$$

$$a_1(t) = \alpha_2\{a_0(t) - a_0(t - 1)\} + (1 - \alpha_2)a_1(t - 1) \qquad (8.58)$$

$$c(t) = \alpha_3 \frac{y_t}{a_0(t)} + (1 - \alpha_3)c(t - s). \qquad (8.59)$$

There is also an additive version of the Holt–Winters model where $c(t + k - s)$ is added, rather than multiplied, in (8.56). Expressions (8.57) and (8.59) then involve the differences $y_t - c(t - s)$ and $y_t - a_0(t)$ rather than the ratios. The multiplicative version is non-linear and clearly not an *ARIMA* scheme. The additive version of (8.56)–(8.59) was expressed in *ARIMA* form by McKenzie (1976), but in a form that would not be identified by any standard procedures. However, the additive model may be expressed in over-differenced form as

$$\nabla^2\nabla y_s = (1 - \theta_1 B - \theta_2 B^2 - \theta_s B - \theta_{s+1}B^{s+1} - \theta_{s+2}B^{s+2})\varepsilon_t$$
$$= \theta(B)\varepsilon_t, \qquad (8.60)$$

over-differenced in that $\theta(B)$ in (8.60) has a unit root. If a factor $(1 - B)$ is removed from both sides of (8.60), McKenzie's form results. One advantage of expression (8.60) is that the forecast mean square error can be developed directly using the approach of Section 8.8. The simpler Holt–Winters model, without the slope component, is considered explicitly in Exercises 8.9 and 8.11.

Chatfield (1978) considers parameter estimation and start-up values for the Holt–Winters multiplicative scheme, noting that the 'recommended' values of $0.02 \leqslant \alpha_i \leqslant 0.20$ are frequently inappropriate; also see Gardner (1985).

Harrison's seasonal model

8.24   One of the potential problems of the Holt–Winters procedures is the stability of the seasonal factors. Numerical studies by Sweet (1985) suggest that this is not a major problem when $s = 4$ but is much more troublesome when $s = 12$. One solution is to smooth the seasonal factors.

Harrison (1965) modified the Holt–Winters model by expressing the seasonal element in terms of harmonics, similar to Burman's treatment of seasonals described in Sections 10.33–34.

Let the forecast function be

$$y_t(k) = a_0(t) + ka_1(t) + c_{j+k}(t),$$

where the trend values are given by (8.54) and (8.55) with $\alpha_1 = \alpha(2 - \alpha)$, $\alpha_1\alpha_2 = \alpha^2$; that is, Brown's scheme. The seasonal index j is defined modulo

(s); that is, $j = j_2$ when $t = sj_1 + j_2$ and the seasonal elements are

$$c_j(t) = \sum_r \{g_r(t)\cos r\lambda_i + h_r(t)\sin r\lambda_i\}, \qquad (8.61)$$

with

$$\lambda_i = \frac{2\pi}{s}(i - 1) - \pi, \qquad i = (t + j) \text{ modulo } (s),$$

for $j = 1, 2, ..., s$. Recall that $\cos(2\pi + \lambda) = \cos \lambda$ and the same for the sine. The summation in (8.61) runs over $r = 1, 2, ..., M$, where $M = \frac{1}{2}s$ when s is even and $\frac{1}{2}(s - 1)$ when s is odd; $h_M = 0$ when s is even.

This construction ensures that $\Sigma_{j=1}^{s} c_j(t) = 0$. Further, the seasonals are smoothed by retaining only those harmonics found to be significant. The coefficients in (8.60) are updated by:

$$g_r(t) = g_r(t - 1) + \beta^* e_t' \cos r\lambda_j \qquad (8.62a)$$

$$h_r(t) = h_r(t - 1) + \beta^* e_t' \sin r\lambda_j, \qquad (8.62b)$$

where $\beta^* = 2\beta/s$, β being a smoothing constant and $e_t' = y_t - a_0(t) - c_j(t - 1)$.

The constraint $\Sigma c_j(t) = 0$ is not met by the Holt–Winters seasonals, which is sometimes seen as a weakness. Roberts (1982) gives a modified *HW* scheme that incorporates this constraint.

Evaluating forecasting methods

8.25 The best way to gauge the reliability of a forecasting method is to consider its performance over a period of time beyond that for which the model was fitted; residuals from fitted values have been found to be generally unreliable as a basis for evaluating methods since a particular model may fit historical data well but be unreliable for extrapolation purposes.

As an illustration, we used the sheep series (Table 1.2); the first 69 observations were used for model fitting and then the values of years 1936–39 were forecast using 1935 as the forecast origin. The results are given in Table 8.1 together with the prediction intervals. Similar sets of forecasts may

Table 8.1 Forecasts for 1936–39 for the sheep series, using the *ARIMA* (3, 1, 0) model fitted over 1867–1935 with 1935 as forecast origin

Year	Forecast	95 Percent prediction limits		Actual	Absolute error	Percentage error
		Lower	Upper			
1936	1 701	1 571	1 830	1 665	− 35	− 2.1
1937	1 777	1 544	2 011	1 627	− 150	− 9.2
1938	1 803	1 501	2 105	1 791	− 12	− 0.7
1939	1 779	1 449	2 108	1 797	18	1.0

be generated by other methods, and a summary of the errors for several different techniques appear in Table 8.2. These errors may be summarised using the mean square error (*FMSE*), mean absolute error (*FMAE*), and mean absolute percentage error (*FMAPE*) criteria introduced earlier; the results are given in Table 8.3. This procedure is followed by way of illustration only; in general, we should follow the steps outlined in Section 8.28 below. Finally, to assist comparisons, the coefficients for the four y_{t-k} values in $y_t(1)$

Table 8.2 Actual values and forecast errors for seven different forecasting procedures for the sheep series 1936–39; forecasts made using data for 1867–1935 with 1935 as forecast origin

	Year			
	1936	1937	1938	1939
Actual value	1 665	1 627	1 791	1 797
Forecast method		Errors		
ARIMA (3, 1, 0)	− 35	− 150	− 12	18
ARIMA (2, 0, 0)	− 2	− 69	63	40
SES				
Default ($\alpha = 0.200$)	− 22	− 60	104	110
Fitted ($\hat{\alpha} = -0.515$)	− 16	− 54	110	116
DES				
Default ($\alpha = 0.106$)	10	− 26	140	148
i Fitted ($\hat{\alpha} = 0.762$)	68	76	287	339
Holt				
Default ($\alpha_1 = \alpha_2 = 0.106$)	87	51	217	225
Fitted ($\hat{\alpha}_1 = 1.503$, $\hat{\alpha}_2 = 0.015$)	− 9	− 37	137	153

Notes:
(1) *SES* denotes single exponential smoothing and *DES* is double exponential smoothing.
(2) 'Default' means parameter values were set to default values; in the *SAS* procedure FORECAST the default values are $\alpha = 0.2$ for *SES* and α or $1 - \alpha_i = (0.8)^{0.05} = 0.894$ for *DES* and for Holt's method.
(3) The fitted values for *SES* and Holt's methods are given by fitting the corresponding *ARIMA* models; these estimates lie outside the usual ranges for these parameters.

Table 8.3 Comparison of forecast methods using different error criteria

Model	$(FMSE)^{1/2}$	FMAE	FMAPE
ARIMA (3, 1, 0)	78.1	54.1	3.3
ARIMA (2, 0, 0)	50.9	43.5	2.5
SES: Default	82.3	74.0	4.2
SES: Fitted	84.9	74.0	4.2
DES: Default	103.0	81.1	4.6
DES: Fitted	227.8	192.5	10.9
Holt: Default	164.2	145.0	8.3
Holt: Fitted	104.3	83.8	4.7

Table 8.4 The weights assigned to the first four *AR* terms by each method [in $y_t(1)$]

Method	Coefficient			
	y_t	y_{t-1}	y_{t-2}	y_{t-3}
ARIMA (3, 1, 0)	1.50	− 0.77	− 0.05	0.32
ARIMA (2, 0, 0)	1.40	− 0.50	0.00	0.00
SES: Default	0.20	0.16	0.13	0.10
SES: Fitted	1.51	− 0.78	0.40	− 0.21
DES: Default	0.21	0.18	0.15	0.12
DES: Fitted	0.76	− 0.22	− 0.19	− 0.08
Holt: Default	0.12	0.11	0.11	0.10
Holt: Fitted	1.53	− 0.78	0.40	− 0.21

are given in Table 8.4. We first compare results for the sheep series, then turn to more general issues.

8.26 The results in Table 8.1 are rather encouraging although the prediction intervals appear wide at first sight. However, the *ARIMA*(3, 1, 0) model yields $\hat{\sigma} = 66.1$, reflecting some rather extreme fluctuations in the past, whereas the years 1936−39 seem to be relatively stable. The errors in Table 8.2 show the general tendency of all the heuristic (default) methods to underestimate three and four years ahead, because of their failure to pick up the appropriate lag structure. The two *SES* methods perform vary similarly because $\hat{y}_t(k)$ has the same value for all k. On this occasion, the different criteria listed in Table 8.3 provide similar rank orderings for each method, although there is no guarantee of such consistency. It is noticeable from Table 8.4 that the fitted *SES* and Holt's model have coefficients very close to the *ARIMA* schemes for the first two lags, but diverge thereafter. Overall, conclusions must be very tentative for such a small study, but the *ARIMA* models appear to have performed somewhat better in this example.

8.27 A similar analysis is provided for the airlines data of Table 1.3 in Tables 8.5−8.7. The seasonal *ARIMA* model identified earlier is contrasted with two Holt−Winters models. In each case, the models were developed using the data for 1963−69 with December 1969 as forecast origin. Default values for the α_i were used in the Holt−Winters schemes. In this case, the *ARIMA* and *HW2* versions clearly outperform *HW3*. The *HW2* model is similar to the *ARIMA* model (cf. Exercise 8.11) save that the seasonal effects are multiplicative. Better parameter estimates may well have improved the performance of the *HW*(2) model even further. The actual values are generally well inside the prediction limits except for April, which is a result of the extreme value for April 1969.

8.28 From our limited examples, it is apparent that any major comparison of methods must involve making a considerable number of forecasts. This may involve a large number of series, although such series must be selected with a specific purpose in mind since there is no such thing as a random sample from the set of all possible time series! Further, each series should be fitted over a variety of time periods such as $[1, m]$, $[1, m + 1]$, $[1, m + 2]$, and so on, and then 1, 2, ..., k-step ahead forecasts generated for each forecast origin. At one time this was a very tedious endeavour, but with modern computing facilities it is no longer unduly onerous. Finally, several criteria should be used to evaluate performance at each step ahead as well as the overall averages which we used in Tables 8.3 and 8.7.

Table 8.5 Forecasts for 1970 for the airlines series, using the *ARIMA* $(0, 1, 1)(0, 1, 1)_{12}$ model fitted over 1963–69 with December 1969 as forecast origin

Month	Forecast	95 percent prediction limits		Actual
		Lower	Upper	
January	10 796	9 334	12 259	10 840
February	10 396	8 910	11 882	10 436
March	12 411	10 902	13 920	13 589
April	11 660	10 128	13 192	13 402
May	13 356	11 802	14 910	13 103
June	14 679	13 102	16 255	14 933
July	14 262	12 663	15 860	14 147
August	14 744	13 124	16 363	14 057
September	15 592	13 951	17 233	16 234
October	12 933	11 271	14 596	12 389
November	11 381	9 699	13 064	11 595
December	12 553	10 849	14 256	12 772

Table 8.6 Forecast errors for the airlines series for 1970; forecasts made using data for 1963–69, with December 1969 as forecast origin

Month	Errors			Percentage errors		
	ARIMA	*HW*(3)	*HW*(2)	*ARIMA*	*HW*(3)	*HW*(2)
January	44	881	114	0.4	8.1	1.1
February	40	851	34	0.4	8.1	0.3
March	1 178	2 172	1 146	8.7	16.0	8.4
April	1 742	2 352	1 237	13.0	17.5	9.2
May	− 253	616	− 657	− 1.9	4.7	− 5.0
June	254	1 215	− 213	1.7	8.1	− 1.4
July	− 115	764	− 659	− 0.8	5.4	4.7
August	− 687	293	− 1 184	− 4.9	2.1	− 8.4
September	642	1 890	375	4.0	11.6	2.3
October	− 544	540	− 729	− 4.4	4.4	− 5.9
November	214	1 228	83	1.8	10.6	0.7
December	219	1 816	373	1.7	14.2	2.9

Notes:
(1) The ARIMA model is $(0, 1, 1)(0, 1, 1)_{12}$.
(2) The Holt–Winters models $HW(3)$ is that of equations (8.56)–(8.59) with default values $\alpha_i = 1 - (0.8)^{1/3} = 0.072$. Model $HW(2)$ uses (8.56), (8.57) and (8.59) with $a_1(t) = 0$ and default values $\alpha_i = 1 - (0.8)^{1/2} = 0.106$.

Table 8.7 Comparison of forecasting methods using error criteria

	$(FMSE)^{1/2}$	*FMAE*	*FMAPE*
ARIMA	698	494	3.6
HW(3)	1 383	1 218	9.2
HW(2)	706	567	4.2

Empirical comparisons

8.29 A considerable variety of empirical studies has been undertaken, conforming more or less to these guidelines, and we shall now attempt to summarise this literature. A more comprehensive and detailed summary is provided by Armstrong (1985, Chapter 15).

8.30 One of the first major studies of this kind was undertaken by Reid (1971). He took 113 time series, including annual, quarterly, and monthly data, mostly on macro-economic variables in the UK, although some were from the USA. The series were in some cases short, the quarterly figures, for example, having fewer than 60 terms. Many comparisons are possible and not every method was applied to every series (because, *inter alia*, some were not seasonal). Table 8.8, however, presents a general picture of the results. The criterion of 'best' in this table is the minimization of forecast errors one step ahead. Reid also made comparisons for longer lead times. The results were similar, but Brown's method and Harrison's method improved somewhat, though not to overtake Box–Jenkins. On the other hand, there were certain types of series for which certain methods seem particularly appropriate; for instance, Harrison's method did particularly well on unemployment figures, which have a marked seasonal and irregular component.

8.31 In a further study of macro-economic series, Newbold and Granger (1974) also found that *ARIMA* models performed well, although the Holt –Winters scheme also performed well on the seasonal series. Another major study was the work of Makridakis and Hibon (1979) and the larger-scale

Table 8.8 Comparison of forecasting performance for 113 series (Reid, 1971)

	Number of series		
Method	Method used on	Method best on	Percentage
Box–Jenkins	113	76	67
Brown	113	2	2
Modified Brown	87	18	21
Winters	69	10	15
Harrison	47	7	15

Note:
The 'modified Brown' method is one in which Reid, finding serial correlation in the errors, fitted a first-order autoregressive series to those errors.

follow-up study reported in Makridakis *et al.* (1982); see also the discussion of this study in Armstrong and Lusk (1983). These studies examined a large number of series (1001) making forecasts up to 18 months/9 quarters/6 years ahead. Many of the series were very short and had undergone major structural changes; consider for example, the changes in the economy of any western country during 1900–50. The general conclusions from these investigations were that simpler methods, such as exponential smoothing, appeared to perform just as well as more complex methods. Another development, begun by Bates and Granger (1969) and continued by Makridakis *et al.* (1982) and Winkler and Makridakis (1983), is to take a (weighted) average of several forecasts and use this instead. This method was shown to be surprisingly effective in these studies. The reasons for this success are not yet fully understood, although it is evident that the probability of correct identification is low when the sample size is small (cf. Ord, 1988).

8.32   These studies, particularly the more recent ones, appear to challenge the conventional wisdom that *FMSE*-optimal forecasts from *ARIMA* models are bound to be superior. The reasons are not hard to find. Such supremacy rests upon the assumptions that (a) the selected model is the true representation of the process, and (b) the series does not undergo structural changes, particularly during the forecast period. When these assumptions are violated, the *ARIMA* models may not perform well. Unfortunately, some forecasting practitioners have rushed to judgement, taking the view that the 1001 series somehow 'represent the real world' and that the results of these studies may be extrapolated, without qualification, to every future forecasting task. Needless to say, we venture to disagree. A well-considered critical appraisal of this somewhat conflicting evidence is given in McGee (1986).

8.33   There was one other, less obvious, assumption that was made in the course of developing the forecast function in Sections 8.8–12; namely, that the one-step ahead forecast will, in turn, generate the optimal k-step ahead forecast function. This is true only if the model is *perfectly* specified, a rather unlikely occurrence. To improve k-step ahead forecasts, Findley (1984) suggests using a modified $AR(p)$ scheme fitted by minimising

$$\Sigma \{y_{t+k-1} - \delta - \phi_k y_{t-1} - \phi_{k+1} y_{t-2} - \cdots - \phi_{k+p-1} y_{t-p}\}^2. \qquad (8.63)$$

Makridakis (1988) investigates this general idea empirically by fitting models and generating forecasts in accordance with the steps given in Section 8.28. The best one-, two-, ..., m-step ahead forecast function is then selected by looking at the *FMAPE*. Finally, the forecast functions are used to produce a further round of forecasts to assess the effectiveness of the selection rules. The results show systematic reductions in the final *FMAPE* and indicate that the use of horizon-specific forecast functions deserves serious consideration in applied work.

8.34   It is clear from this discussion that there is some confusion in the realm of single-series forecasting, despite its apparently rather straightforward statistical basis. What conclusions can be drawn at this stage? Our own tentative assessment is given below:

(A)   When a series is of reasonable length and the environment is not expected to undergo major structural changes,

(i) use an *ARIMA* model for non-seasonal processes or those with additive seasonal structure;

(ii) use Holt–Winters or *ARIMA* after transformation (to logs) if the process has a multiplicative seasonal structure;

(iii) if there are outliers in the series, these should be adjusted or modelled (see Section 13.10) or robust estimation procedures should be employed (see Section 15.10).

(B) When the series is short and/or subject to structural changes, the choice of approach depends in part upon the importance of the study, as follows:

(i) use exponential smoothing or Holt–Winters with little or no attempt at model selection and then examine series that are behaving erratically. (This is appropriate when a large number of series must be considered with limited resources.);

(ii) attempt to model the structural changes or, at least, incorporate them into future forecasts. This approach is relevant when individual series are of considerable importance.

In B(i), there is only limited scope for statistical input, and factors such as the speed with which a method adapts to structural change should be considered in setting the parameter values. B(ii) is potentially more challenging to the statistician and must await our discussion of intervention analysis in Chapter 13.

8.35 A further aspect of model selection for forecasting is the extent to which the selection process may be automated, as discussed in Sections 7.26–32. An exploratory study by Texter and Ord (1989) suggests that the automatic selections produced by AUTOBOX perform comparably to the models selected by reasonably qualified forecasters. At very least, such automatic procedures provide a good starting point for the researcher seeking to develop an effective forecasting scheme.

Gardner and McKenzie (1989) demonstrate that automatic model selection among exponential smoothing schemes may be effected by considering the orders of differencing that minimize the variance; that is, choose (d, D) to minimize

$$\text{var}\{\nabla^d \nabla_s^D y(t)\}.$$

See Section 3.15 for earlier discussion of this criterion, which appears to work well for shorter series and is being implemented in Gardner's AUTOCAST; see Appendix D. Other automatic forecasting packages are described in Hill and Fildes (1984) and Libert (1984).

8.36 An excellent guide to developments in forecasting, whether using time series or other approaches, is the special issue of the *International Journal of Forecasting* (1988, No. 3). The special issue of *The Statistician* (1988, No. 2), on forecasting and decision making, is another valuable source of information.

Exercises

8.1 If
$$y_t = \mu_t + \varepsilon_t$$
and
$$\hat{\mu}_t = \mu_t + \delta_t$$

where $\hat{\mu}_t$ denotes a forecast based on past values of y, show that (8.6) holds when ε_t and δ_t are independent.

8.2   Develop the forecast function for the $AR(2)$ scheme

$$y_t = \delta + \phi_1 y_{t-1} + \phi_2 y_{t-2} + \varepsilon_t$$

Hence, find the k-step ahead *FMSE* for $k = 1, 2$ and $k \to \infty$.

8.3   Develop the forecast function for the $MA(2)$ scheme

$$y_t = \mu + \varepsilon_t - \theta_1 \varepsilon_{t-1} - \theta_2 \varepsilon_{t-2}$$

Hence, find the k-step ahead *FMSE* for $k = 1, 2$ and $\geqslant 3$.

8.4   Given $\delta = 2$, $\phi_1 = 1.0$, and $\phi_2 = -0.4$ for the model in Exercise 8.2, compute $y_t(k)$, $k = 1, 2$ and $k \to \infty$ given $y_t = 13$ and $y_{t-1} = 10$.

8.5   Show that (8.26) holds for any $MA(q)$ scheme.

8.6   Show that the k-step *FMSE* for

$$\nabla^2 y_t = \varepsilon_t$$

is $\sigma^2 k(k+1)(2k+1)/6$; that is, increases at the rate k^3.

8.7   Find the forecast function for the airline model

$$\nabla \nabla_s y_t = (1 - \theta B)(1 - \Theta B^s)\varepsilon_t$$

Hence, find the *FMSE* for $k = 1, 2, ..., s$.

8.8   Find the forecast function for the model

$$\nabla \nabla_s y_t = (1 - \theta_1 B - \theta B^s - \theta_{s+1} B^{s+1}) \varepsilon_t$$

and find its *FMSE* for $k = 1, 2, ..., s$.

8.9   Researcher A selects the model given in Exercise 8.7 and estimates the parameters as $\theta = \Theta = 0.80$. Researcher B chooses the model given in Exercise 8.8 with $\theta_1 = 0.80$, $\theta_s = 0.84$, and $\theta_{s+1} = -0.64$. Verify numerically that these two schemes will yield very similar results. Take $s = 4$. Show that S's model is invertible, but B's is not.

8.10   Rewrite the Holt forecast function (8.51)–(8.53) in terms of the B-operator. Using these expressions together with

$$y_{t+1} = y_t(1) + \varepsilon_t$$

verify that this procedure corresponds to an $ARIMA(0, 2, 2)$ model.

8.11   Consider the additive Holt–Winters forecasting scheme without the slope component:

$$y_t(1) = a(t) + c(t + 1 - s)$$
$$a(1) = \alpha_1 \{ y_t - c(t - s) \} + (1 - \alpha_1)a(t - 1)$$
$$s(t) = \alpha_3 \{ y_t - a(t) \} + (1 - \alpha_3)c(t - s).$$

Using the fact that

$$y_{t+1} = y_t(1) + \varepsilon_{t+1}$$

use the B operator to eliminate $a(t)$ and $c(t)$ from these equations and show that the underlying model is

$$(1 - B)(1 - B^s)y_{t+1} = \{1 - (1 - \alpha_1)B - (1 - \alpha_3 + \alpha_1\alpha_3)B^s$$
$$+ (1 - \alpha_1)(1 - \alpha_3)B^{s+1}\}\varepsilon_{t+1}$$

By substituting $B = 1$ in the polynomial on the right-hand side, show that this model is not invertible.

  Notes: This scheme is very close to the airline model (6.24), especially when it is borne in mind that α_1 and α_3 are generally both small. The forecast functions for the airline model and this scheme are discussed in Exercises 8.7–9.

8.12 For the series considered previously in Exercises 6.10, 6.11, 7.9 and 7.10 develop forecasts and prediction intervals for 1 to 6 periods ahead (non-seasonal) and 1 to s periods ahead (seasonal).

9

State-space models

Introduction

9.1 In early chapters of the book, we emphasised the notion that a time series is a set of components and attempted to decompose the series into its several parts. By contrast, the *ARIMA* models tend to treat a series as an integrated whole and do not filter out separate components. We now return, in a sense, to the original theme and develop the overall model as a set of distinct components. Thus, we may suppose that observations relate to the mean level of the process through an *observation* equation, whereas one or more *state* equations describe how the individual components change through time. A related advantage of this approach is that observations can be added one at a time, and the estimating equations are then updated to produce new estimates. This is in contrast to the fixed sample size approach hitherto adopted for *ARIMA* models.

The primary development of these components, or *state-space*, models was due to engineers interested in recursive methods for multi-sensor systems such as tracking mechanisms or servo-controls. The seminal work in this field is due to Kalman (cf. Kalman and Bucy, 1961) although the basic statistical ideas were formulated in a pioneering paper by Plackett (1950). The context in which these methods were developed inevitably affected the underlying assumptions that were made. In particular, a great deal of prior information is often available about the operating characteristics of engineering systems, which provides good initial estimates of some of the system parameters. Initially, we assume that such parameters are known; estimation procedures for these parameters are discussed at the end of the chapter. In Sections 9.3–6, we describe the formulation of state-space models and their relationships to *ARIMA* schemes; relationships for updating parameter estimates and forecasts are then examined in Sections 9.7–12. Finally, several extensions and modifications are reviewed.

9.2 The multidisciplinary background of state-space modelling has led to considerable variety in the terminology and notation used. The engineering literature typically refers to the *Kalman filter* whereas statisticians often prefer the terms *Bayesian forecasting* and the *dynamic linear model* (Harrison and Stevens, 1976; Fildes, 1983) or *linear dynamic regression* (Duncan and Horn, 1972). Inevitably different authors have approached the topic in different ways, but we shall use the term *state-space modelling* to cover any or all of these as may be convenient. Finally, as noted in Kendall *et al.* (1983, Section 50.41), we recall that the adjective 'Bayesian' refers primarily to the use of Bayes' theorem to update the probability distributions, and the results can be derived equally well by likelihood arguments.

The state-space formulation

9.3   The research of Harrison (especially Harrison and Stevens, 1976), and Harvey (1981, 1984 and references cited therein) has done much to make state-space modelling more accessible to statistical audiences. In this chapter, we shall rely heavily upon these contributions.

9.4   We begin with an *observation*, or *measurement*, equation of the form:

$$y_t = \mathbf{x}_t' \boldsymbol{\mu}_t + \delta_t, \tag{9.1}$$

where $\mathbf{x}_t$ is an $(r \times 1)$ vector of known values, $\boldsymbol{\mu}_t$ is an $(r \times 1)$ *state* vector of random variables describing the underlying process, and the δ_t are independent, normally distributed error terms with mean zero and variance $\sigma^2 h_t$, h_t known, or

$$\delta_t \sim NID(0, \sigma^2 h_t). \tag{9.2}$$

Further, we assume that current errors do not depend upon current or previous states of the process, so that

$$\text{cov}(\delta_t, \boldsymbol{\mu}_{t-k}) = 0 \qquad \text{for all } k \geqslant 0; \tag{9.3}$$

recalling that zero correlations plus normality imply independence. The variables $\boldsymbol{\mu}_t$ are not directly observable and are also known as *latent* variables.

9.5   The coefficients $\boldsymbol{\mu}_t$ are updated according to a set of *state* or *transition* equations:

$$\boldsymbol{\mu}_t = G_t \boldsymbol{\mu}_{t-1} + \boldsymbol{\eta}_t, \tag{9.4}$$

where

$$\boldsymbol{\eta}_t \sim NID(0, \sigma^2 Q_t); \tag{9.5}$$

the matrices G_t and Q_t are assumed to be known. Further,

$$\text{cov}(\delta_{t+k}, \boldsymbol{\eta}_t) = \mathbf{0}, \qquad k = 0, \pm 1, \pm 2, \ldots; \tag{9.6}$$

that is, there is no correlation between the disturbance terms in the state and observation equations.

The state equations are *Markovian*; the future values $\boldsymbol{\mu}_t$, $\boldsymbol{\mu}_{t+1}, \ldots$ are determined only by the current state $\boldsymbol{\mu}_{t-1}$. Some examples serve to illustrate the flexibility of this approach.

Example 9.1 When $\eta_t \equiv 0$ and $G = I$, the equation reduce to

$$y_t = \mathbf{x}_t' \boldsymbol{\mu} + \delta_t \tag{9.7}$$

since $\mu_t \equiv \mu$ for all t. Equation (9.7) is the standard multiple regression model.

Example 9.2 Consider the model

$$y_t = \mu_t + \delta_t, \tag{9.8}$$
$$\mu_t = \mu_{t-1} + \eta_t. \tag{9.9}$$

Equation (9.9) can be rewritten using the backshift operator B as

$$(1 - B)\mu_t = \eta_t;$$

multiplying (9.8) by $(1 - B)$ and then substituting yields

$$\begin{aligned}
(1 - B)y_t &= (1 - B)\delta_t + (1 - B)\mu_t \\
&= (1 - B)\delta_t + \eta_t \\
&= \delta_t + \eta_t - \delta_{t-1}.
\end{aligned} \tag{9.10}$$

We refer to (9.10) as the *random shock* form of the model. If we now suppose that the error processes are stationary so that $h_t = Q_t = 1$, we can employ a general aggregation result due to Granger and Morris (1976) which states that

> if u_1 is $ARMA(p, q)$ and u_2 is $ARMA(m, n)$, then $u = c_1 u_1 + c_2 u_2$ is $ARMA(r, s)$, where $r \leqslant p + m$ and $s \leqslant q + n$. In general, the equalities hold; for a proof, see Kendall *et al.* (1983, Section 47.29).

If we set

$$u_{1t} = \delta_t - \delta_{t-1}$$
$$u_{2t} = \eta_t,$$

it follows that $u_t = u_{1t} + u_{2t} = \delta_t + \eta_t - \delta_{t-1}$ is an $MA(1)$ scheme. We note that (9.10) has the form

$$(1 - B)y_t = u_t \tag{9.11}$$

so that the model for y_t is $ARIMA(0, 1, 1)$. That is, equations (9.8)–(9.9) are a state-space version of the $ARIMA(0, 1, 1)$ model. In particular, the discussion in Section 8.17 implies that the forecasts would be given by simple exponential smoothing. Also, it follows that

$$\text{var}(u_t) = 2\sigma^2 + \sigma_\eta^2 \tag{9.12}$$
$$\text{cov}(u_t, u_{t-1}) = -\sigma^2 \tag{9.13}$$
$$\text{cov}(u_t, u_{t-j}) = 0, \qquad j > 1 \tag{9.14}$$

so that the MA component is restricted to $0 < \theta < 1$. It is important to note that the stationarity assumptions corresponding to (9.12)–(9.14) were introduced at the end of the discussion to make explicit the equivalence to the $ARIMA$ structure. The original model (9.8)–(9.9) is equally valid in nonequilibrium conditions when the forecast function will have time-depedent coefficients; see Example 9.5. This property is not model specific and is a major feature of the state-space approach.

Example 9.3 Consider the model

$$y_t = \mu_t + \delta_t \qquad (9.15)$$

$$\mu_t = \mu_{t-1} + \beta_t + \eta_{1t} \qquad (9.16)$$

$$\beta_t = \beta_{t-1} + \eta_{2t} \qquad (9.17)$$

and let $h_t = 1$, $Q_t = I$. If we then eliminate μ_t and β_t between the equations, we finish up with an $ARIMA(0, 2, 2)$ scheme; that is, (9.15)–(9.17) generate Holt's forecasting equations (8.51)–(8.53). The additive Holt–Winters scheme is also expressible in a state-space form; see Exercises 9.3 and 9.4.

9.6 In addition to building models in state space, it is possible to write any $ARIMA$ process as a state-space scheme. The original development by Akaike (1974) expresses the model parameters in terms of the autoregressive parameters ϕ_i and the random shock coefficients ψ_i. This can be tedious to use if we wish to switch back and forth between the $ARIMA$ and state-space versions. Harvey (1981, 1984) has provided a more direct representation as follows:

$$y_t = \mathbf{x}_t' \, \boldsymbol{\mu}_t \qquad (9.18)$$

$$\boldsymbol{\mu}_t = G_t \boldsymbol{\mu}_{t-1} + \boldsymbol{\eta}_t, \qquad (9.19)$$

where $\boldsymbol{\mu}_t' = (\mu_{1t}, \mu_{2t}, \ldots, \mu_{rt})$, $r = \max(p, q + 1)$,

$$G_t \equiv G = \begin{bmatrix} \phi_1 & \vdots & \\ \phi_2 & \vdots & I_{r-1} \\ \vdots & \vdots & \\ \phi_{r-1} & \vdots & \\ \hline \phi_r & \vdots & \mathbf{0}' \end{bmatrix}, \qquad \mathbf{x} = \begin{bmatrix} 1 \\ \hline 0 \end{bmatrix}, \qquad \boldsymbol{\eta}_t = \begin{bmatrix} 1 \\ -\theta_1 \\ \vdots \\ -\theta_{r-1} \end{bmatrix} \varepsilon_t, \quad (9.20)$$

and

$$\mathrm{var}(\boldsymbol{\eta}_t) = \sigma^2 Q_t.$$

In (9.20) $\mathbf{0}$ is an $(r-1) \times 1$ vector of zeros and I_{r-1} is the identity matrix of order $(r-1)$. If $p < r$ or $q + 1 < r$, the corresponding ϕ or θ coefficients are set equal to zero.

Example 9.4 Consider the $ARIMA$ model with $p = 2$ and $q = 1$:

$$y_1 = \phi_1 y_{t-1} + \phi_2 y_{t-2} + \varepsilon_t - \theta_1 \varepsilon_{t-1}. \qquad (9.21)$$

Representation (9.18)–(9.20) has $r = 2$ and

$$\boldsymbol{\mu}_t = \begin{bmatrix} \mu_{1t} \\ \mu_{2t} \end{bmatrix}, \qquad \mathbf{x}_t = \begin{bmatrix} 1 \\ 0 \end{bmatrix}, \qquad \boldsymbol{\eta}_t = \begin{bmatrix} 1 \\ -\theta_1 \end{bmatrix} \varepsilon_t, \qquad (9.22)$$

$$G_t \equiv G = \begin{bmatrix} \phi_1 & 1 \\ \phi_2 & 0 \end{bmatrix} \qquad \text{and} \qquad Q_t = \begin{bmatrix} 1 & -\theta_1 \\ -\theta_1 & \theta_1^2 \end{bmatrix}. \qquad (9.23)$$

Several features are worth noting from this example. The matrices G_t and Q_t involve the AR and MA parameters, so that our present discussion involves updating the system given these parameters. We shall consider parameter estimation briefly in Section 9.13. Secondly, we note that the covariance matrix

$\sigma^2 Q_t$ is singular; this does not pose any problems since, as we shall note, no matrix inversions are necessary. Finally, we note that the representation for an $AR(2)$ scheme would follow by setting $\theta_1 = 0$ or an $ARMA(1,1)$ by setting $\phi_2 = 0$; in each case, the number of state variable is $r = \max(p, q+1)$ and the element of redundancy does not affect the basic approach.

Recursive updating

9.7   A major attraction of the state-space approach is the ease with which estimates and forecasts may be updated. We state the general results here and then illustrate them with an example. For more detailed discussions, see Harvey (1981, pp. 104–110). Consider the process specified by (9.1)–(9.6); that is,

$$y_t = \mathbf{x}_t' \boldsymbol{\mu}_t + \delta_t \tag{9.24}$$

$$\boldsymbol{\mu}_t = G_t \boldsymbol{\mu}_{t-1} + \boldsymbol{\eta}_t, \tag{9.25}$$

where $\delta_t \sim N(0, \sigma^2 h_t)$ and $\boldsymbol{\eta}_t \sim N(0, \sigma^2 Q_t)$ with all appropriate correlations zero and h_t, Q_t known.

Let $\mathbf{m}_{t-1}$ denote the estimator for $\boldsymbol{\mu}_{t-1}$ at time $t-1$ and

$$\mathbf{m}_{t-1} \sim N(\boldsymbol{\mu}_{t-1}, \sigma^2 P_{t-1}), \tag{9.26}$$

where P_{t-1} is assumed known. The key steps in the recursive scheme are as follows:

(1)   Calculate the one-step ahead forecast for the state variables from (9.25). We write this as

$$\mathbf{m}_{t/t-1} = G_t \mathbf{m}_{t-1}. \tag{9.27}$$

(2)   From (9.25) and (9.26), this has covariance matrix $\sigma^2 R_t$ where

$$R_t = G_t P_{t-1} G_t' + Q_t. \tag{9.28}$$

(3)   From (9.24) and (9.27), the one-step ahead forecast for the next observation is

$$y_{t-1}(1) = \mathbf{x}_t' \mathbf{m}_{t/t-1} = \mathbf{x}_t' G_t \mathbf{m}_{t-1}. \tag{9.29}$$

(4)   The one-step ahead forecast error is

$$e_t = y_t - y_{t-1}(1) = y_t - \mathbf{x}_t' G_t \mathbf{m}_{t-1}, \tag{9.30}$$

with *FMSE* $\sigma^2 f_t$, where

$$f_t = h_t + \mathbf{x}_t' R_t \mathbf{x}_t. \tag{9.31}$$

(5)   The updated estimator $\mathbf{m}_t$ for $\boldsymbol{\mu}_t$ is then given by

$$\mathbf{m}_t = \mathbf{m}_{t/t-1} + \mathbf{k}_t e_t, \tag{9.32}$$

where $\mathbf{k}_t$ is sometimes known as the (Kalman) gain vector (or matrix in the multivariate case):

$$\mathbf{k}_t = R_t \mathbf{x}_t / f_t. \tag{9.33}$$

It then follows that the distribution of $\mathbf{m}_t$ is

$$\mathbf{m}_t \sim N(\boldsymbol{\mu}_t, \sigma^2 P_t) \tag{9.34}$$

with

$$P_t = R_t - R_t \mathbf{x}_t \mathbf{x}_t' R_t / f_t. \tag{9.35}$$

Finally, the one-step ahead forecast for y_{t+1} is

$$y_t(1) = \mathbf{x}_{t+1}' G_{t+1} \mathbf{m}_t. \tag{9.36}$$

The forecasts are unbiased and the forecast mean square error is

$$\sigma^2(h_{t+1} + \mathbf{x}_{t+1}' R_{t+1} \mathbf{x}_{t+1}). \tag{9.37}$$

9.8   The effect of equations (9.26)–(9.37) is that given the starting values $\mathbf{m}_{t-1}$ and the new observation y_t, we can derive the new estimate $\mathbf{m}_t$ and hence the new forecast $y_t(1)$. The updating expressions (9.27)–(9.33) enable us to move from $(\mathbf{m}_{t-1}, P_{t-1}, y_{t-1}(1))$ to the updated values $(\mathbf{m}_t, P_t, y_t(1))$. Furthermore, this set of calculations may be executed without any matrix inversion operations, thereby enabling high speed and considerable numerical accuracy to be achieved. The overall framework seems cumbersome, but the set of updating equations is so flexible that a large number of different problems can be handled in this way. In order to see what is happening, let us consider a specific example.

Example 9.5   Consider the state-space scheme formulated in Example 9.2:

$$y_t = \mu_t + \delta_t$$
$$\mu_t = \mu_{t-1} + \eta_t$$

with $h_t = 1$, $Q_t = q = \sigma_\eta^2/\sigma^2$, $x_t = G_t = 1$, and scalars $P_t = p_t$, $R_t = r_t$. Then, from (9.27–9.33), we obtain

$$m_{t/t-1} = m_{t-1}$$
$$r_t = p_{t-1} + q$$
$$y_{t-1}(1) = m_{t-1}$$
$$e_t = y_t - m_{t-1}$$

and

$$f_t = 1 + r_t$$

leading to

$$m_t = m_{t-1} + k_t e_t,$$

where $k_t = r_t / f_t$, so that

$$m_t = m_{t-1} + r_t(y_t - m_{t-1})/f_t, \tag{9.38}$$

Further, from (9.35) $p_t = r_t - r_t^2/f_t$, and finally,

$$y_t(1) = m_t. \tag{9.39}$$

Thus, we may start from time zero with values m_0 and p_0 and generate updated estimates and new forecasts as each observation becomes available.

Equation (9.38) and (9.39) combine to give a forecast that is a weighted average of m_{t-1} and y_t, but the weights change over time. However, if the process approaches a steady state wherein $p_t \to p$, we arrive at the forecast function

$$y_t(1) = m_t = (1 - \alpha)m_{t-1} + \alpha y_{t-1}, \qquad (9.40)$$

where

$$\alpha = (p + q)/(1 + p + q). \qquad (9.41)$$

Expression (9.40) is the forecast function for the $ARIMA(0, 1, 1)$ scheme in accordance with Example 9.2.

9.9   A further benefit obtained from the state-space formulation is that the k-step ahead estimator is just

$$y_t(k) = \mathbf{x}'_{t+k} G_{t+k} y_t(k - 1) = \cdots = \mathbf{x}'_{t+k} G_{t+k} G_{t+k-1} \cdots G_{t+1} \mathbf{m}_t, \quad (9.42)$$

and the *FMSE* may be obtained in similar fashion.

9.10   In general, the forecasting system may be started up by obtaining the estimates from the first s ($\geqslant r$) observations when $\boldsymbol{\mu}$ is an $(r \times 1)$ vector. Alternatively, rather than obtain such estimates explicitly we may simply start off the iterative process with $P_0 = cI$, where c is a large, but finite number. This is similar to using a *diffuse* prior in a Bayesian setting. The reason for using a large c-value is to ensure that the effect of the initial conditions dies away rapidly, while maintaining the numerical stability of the updating process over early observations. It is, in effect, a painless way of developing a starting solution from the first r observations and solves the problem of start-up values considered earlier in Section 8.17.

Properties of the state-space approach

9.11   An immediate benefit of the state-space formulation is that we may generate one-step ahead *recursive* residuals:

$$e_{t+1} = y_{t+1} - y_t(1). \qquad (9.43)$$

These residuals may be used to check for changing structure in the process over time. This feature is exploited by Brown *et al.* (1975) in their development of cusum test statistics to check for the constancy of regression coefficients over time.

9.12   Another feature of the state-space formulation is that if we restrict attention to the special case given in Example 9.1, we arrive at updating relations for the multiple regression model, as developed originally by Plackett (1950). Advances in on-line digitial data recording make such updating features especially valuable, whether in aircraft guidance, medical monitoring (cf. Smith and West 1987), or other continuously operating systems.

Estimation of *ARIMA* models

9.13   Given that *ARIMA* models may be represented in state-space form, it is possible to develop estimators in recursive fashion once estimators for the unknown variance elements are included. Harvey and Phillips (1979) derived an exact maximum likelihood estimation procedure for *ARIMA* schemes based upon Harvey's state-space formulation (see Section 9.6); Gardner *et al.* (1980) provide a computer algorithm to implement this procedure. The procedure appears to be numerically stable and to be competitive with the standard *ARIMA* algorithms in terms of running time. One advantage of this approach is the ability to handle missing observations since we may simply 'update' the estimates; that is, we may replace the missing observation by its one-step ahead forecast. For details of the procedure, see Harvey and Pierse (1984) and Jones (1985).

Structural models

9.14 Thus far in this chapter, we have sought to link state-space models to *ARIMA* schemes. However, there are obvious attractions in developing such models directly. In particular, Harvey (1984, 1985) and Harvey and Durbin (1986) consider a class of *structural* models built up as follows. The *observation* equation is

$$y_t = \mu_t + \gamma_t + \varepsilon_t, \tag{9.44}$$

where μ, γ and ε represent trend, seasonal, and irregular components, in much the spirit of Chapters 3 and 4. The *state* equations for the trend are

$$\text{level: } \mu_t = \mu_{t-1} + \beta_{t-1} + \eta_{1t}, \tag{9.45}$$

$$\text{slope: } \beta_t = \beta_{t-1} \qquad + \eta_{2t}, \tag{9.46}$$

that is, the same as for Holt's method in Section 8.22. The preferred approach for the seasonal component (with s seasons) is the trigonometric version defined in a similar fashion to Harrison's (1965) scheme in Section 8.24. Let $s = 2M$ or $2M + 1$ accordingly, as s is even or odd, then write

$$\gamma_t = \sum_{j=1}^{M} \gamma_{jt}. \tag{9.47}$$

Given $\gamma_j = 2\pi j/s$, we have the updating scheme

$$\begin{pmatrix} \gamma_{jt} \\ \gamma_{jt}^* \end{pmatrix} = \begin{pmatrix} \cos \gamma_j & \sin \gamma_j \\ -\sin \gamma_j & \cos \gamma_j \end{pmatrix} \begin{pmatrix} \gamma_{j,t-1} \\ \gamma_{j,t-1}^* \end{pmatrix} + \begin{pmatrix} \omega_{jt} \\ \omega_{jt}^* \end{pmatrix} \tag{9.48}$$

for $j = 1, \ldots, M$, with $\gamma_{Mt}^* = 0$ for s even. As in our earlier development, the ηs are mutually independent white noise errors. In the basic structural model, it is assumed that all the errors are normally distributed with constant variances:

$$\eta_{it} \sim IN(0, \sigma_i^2), \quad i = 1, 2, \qquad \text{and} \qquad \omega_{jt} (\text{or } \omega_{jt}^*) \sim IN(0, \sigma_\omega^2). \tag{9.49}$$

The variances in (9.49) are known as the *hyperparameters* of the model.

9.15 The estimation of structural models follows the general ideas outlined in Section 9.13; for further details, see Harvey and Peters (1984). The structural model as described here has been implemented in the computer program STAMP (see Appendix D). This system allows the user to choose between seasonal models or, of course, to drop γ for non-seasonal series. Individual variances (σ_i^2) can be examined to determine whether the corresponding

Table 9.1 Structural model for airlines data

(a) Variance (or hyperparameter) estimates

Component	Stochastic slope		Fixed slope	
	Estimate	t-ratio	Estimate	t-ratio
level	2.24×10^4	1.25	1.69×10^4	1.55
trend	50.1	0.72	—	—
seasonal	1.89×10^3	2.32	3.05×10^3	2.68
irregular	1.80×10^5	2.74	1.41×10^5	2.18

(b) Component estimates at 12/69

Component	Estimate	t-ratio	Estimate	t-ratio
level	12 565	44.1	12 441	$(6.49)^*$
slope	75.3	2.14	53.5	3.58
harmonics (j, j^*)				
(1)	$-1\,431$	-7.97	$-1\,402$	-7.30
(1^*)	$-1\,289$	-6.80	$-1\,381$	-6.69
(2)	-151	-0.90	-87	-0.48
(2^*)	268	1.54	284	1.48
(3)	5	0.03	-49	-0.27
(3^*)	361	2.11	346	1.83
(4)	1 060	6.41	1 156	6.41
(4^*)	-86	-0.50	-130	-0.69
(5)	174	1.05	206	1.14
(5^*)	276	1.61	324	1.70
(6)	-181	-1.28	-200	-1.28

(c) Forecasts for 1970 from 12/69

Month	Actual	Forecast	Standardized residual
Jan.	10 840	10 838	0.00
Feb.	10 436	10 399	0.05
Mar.	13 589	12 808	0.98
Apr.	13 402	11 202	2.69
May	13 103	13 602	-0.59
June	14 933	14 997	-0.07
July	14 147	14 360	-0.24
Aug.	14 057	14 828	-0.85
Sept.	16 234	16 010	0.24
Oct.	12 389	13 098	-0.75
Nov.	11 595	11 567	0.03
Dec.	12 772	12 945	-0.18

$^*t$-statistic not comparable with stochastic trend case.

components should be stochastic or deterministic. Finally, we may use *t*-tests to examine the magnitudes of individual components.

9.16 As an illustration of this approach, we once again consider the airlines data given in Table 1.3. The full model (9.44)–(9.48) was fitted for the period 1963–69 and forecasts generated for 1970. The results are given in Table 9.1 and Figs 9.1–3. Part (a) of Table 9.1 indicates that the irregular (as expected) and seasonal components are clearly stochastic although the level and slope could conceivably be treated as deterministic. The results for a fixed slope component are also given in Table 9.1(a); the net effect is to increase the stochastic term for the seasonals, suggesting that some trend variation is now being treated as seasonal. Part (b) of Table 9.1 gives the values of the components at the end of the fitting period. The results for the two models are very similar. As expected, the level and slope terms are significant as are the harmonics for the yearly cycle ($j, j^* = 1$). The highly significant harmonic at $j = 4$ is unexpected and indicates a changing seasonal pattern, or, possibly, the effects of outliers. We return to the discussion of this point in Section 13.10.

9.17 An advantage of the structural approach is that, once the hyperparameters have been estimated, we can go back through the series and compute smoothed estimates for the individual components. We now present this analysis for the stochastic-slope model. Figure 9.1 shows the trend component superimposed on the original series, which shows a high degree of smoothness except in mid-1968. The seasonal component in Fig. 9.2 starts out smoothly but becomes increasingly erratic as time goes on. In Section 13.14, we examine how far this is due to the outliers and how far to a genuine shift in seasonal structure. Finally, Fig. 9.3 shows the irregular component, clearly highlighting the outliers for June and July 1968 and for April 1969. Returning to part (c) of Table 9.1, we see that the forecasts seem to perform quite well, save that the

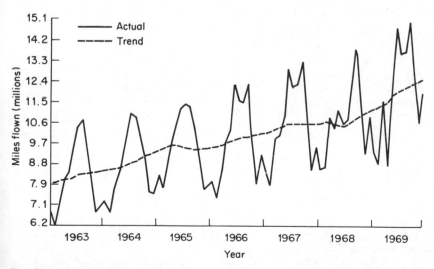

Fig. 9.1 Actual values and trends component for structural model of the airline data, 1963–69

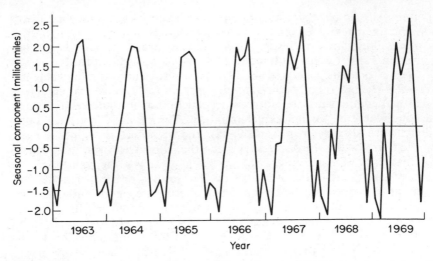

Fig. 9.2   Seasonal component for structural model of airline data, 1963–69

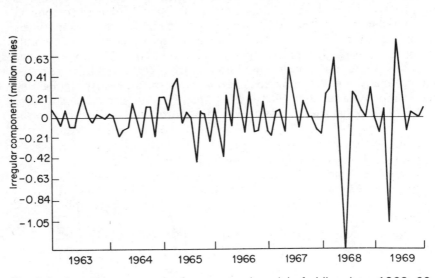

Fig. 9.3   Irregular component for structural model of airline data, 1963–69

April 1970 value is distorted by the outlier the year before. This is also reflected in the forecast seasonal pattern in Fig. 9.2.

Structural or *ARIMA* modelling?

9.18   *ARIMA* models offer the advantage of a stronger model-building paradigm whereas structural models offer clearer interpretations through the decomposition into components. Harvey and Todd (1983) and Harvey (1985) provide several applications of structural models and comparison with ARIMA schemes. The discussion following Harvey and Durbin (1986) provides an appraisal of the methods. What clearly can be said is that both offer major advances over the crude filtering of polynomial trends or regression models that ignore temporal structure.

9.19 In conclusion, the decomposition ability of structural models is a major attraction; however, it should be noted that *ARIMA* models may also be decomposed, albeit with somewhat greater difficulty, using signal extraction methods. For details, see Burman (1980). Interestingly enough, Burman found that the $ARIMA(0, 1, 1)(0, 1, 1)_s$ scheme often worked well in this context; this is very close to the structural model discussed in Exercise 9.3.

Exercises

9.1 If $u_t = \delta_t - \delta_{t-1} + \eta_t$, where the δ and η terms are independent normal variates with means zero and variances σ_δ^2 and σ_η^2, respectively, show that μ_t may be represented as an $MA(1)$ scheme with

$$\sigma_u^2 = 2\sigma_\delta^2 + \sigma_\eta^2 \quad \text{and} \quad \theta\sigma_u^2 = \sigma_\delta^2.$$

9.2 Verify that (9.15)–(9.17) reduce to an $ARIMA(0, 2, 2)$ process; that is, the forecast function corresponds to Holt's method.

9.3 Consider the seasonal state-space model

$$u_t = \mu_t + \gamma_t + \delta_t$$
$$\mu_t = \mu_{t-1} + \eta_{1t}$$
$$\gamma_t = \gamma_{t-s} + \eta_{2t},$$

where all errors have zero means and are uncorrelated, with variances σ_δ^2, σ_1^2, and σ_2^2, respectively. Show that $\nabla\nabla_s y_t$ is an MA scheme with non-zero autocorrelations at lags 1, $s - 1$, s, and $s + 1$. Compare this scheme with the models considered in Exercise 8.7 and 8.8.

9.4 Extend the model in Exercise 9.3 to include a slope term, as in (9.15)–(9.17). Derive the random shock form of this model.

9.5 Use (9.18)–(9.20) to develop a state-space representation for an $ARIMA(0, 2, 2)$ scheme.

9.6 Consider the damped trend structural model given by

$$\eta_t = \mu_t + \varepsilon_t$$
$$\mu_t = \mu_{t-1} + \beta_t + \eta_{1t}$$
$$\beta_t = \phi\beta_{t-1} + \eta_{2t}$$

with $|\phi| < 1$. If the error terms are stationary, show that this may be represented as an $ARIMA(1, 1, 2)$ scheme.

9.7 If you have access to a state-space package, develop models for one or more of the series in Appendix Tables A1–A8.

10

Spectrum analysis

Introduction

10.1 In earlier chapters, we have made periodic references to cyclical phenomena, which we defined as processes that repeat themselves in regular fashion. The simplest such example is the sine wave

$$y_t = a \cos \alpha t + b \sin \alpha t$$
$$= c \cos(\alpha t - \phi), \tag{10.1}$$

where $c^2 = a^2 + b^2$ and $\cos \phi = a/c$ since

$$\cos(\alpha + \beta) = \cos \alpha \cos \beta - \sin \alpha \sin \beta. \tag{10.2}$$

The constant c is the *amplitude* of the series, as y_t varies between the limits $-c$ and $+c$; see Fig. 10.1. The coefficient α is the (*angular*) *frequency* since the wave completes $\alpha/2\pi$ cycles in a unit time. It follows that y_t completes one cycle in time $\lambda = 2\pi/\alpha$; λ is known as the *wavelength*, or *period* of the sine wave. Finally, ϕ is the *phase* angle and indicates how far the wave has been shifted from the natural origin (corresponding to $\phi = 0$). For example, $\phi = -\pi/2$ would convert the cosine function into a sine function. It follows from (10.1) that a phase shift of ϕ corresponds to a time delay of $d = \phi/\alpha$ since

$$y_{t-d} = c \cos\{\alpha(t - d)\}$$
$$= c \cos(\alpha t - \phi).$$

It seems almost trivial to observe that c, α, and ϕ are fixed numbers, but it is important as the specification of these terms will change later.

10.2 In this chapter, we use trigonometric functions to describe the time series; this is termed the *frequency-domain* representation. In Sections 10.2–10, we consider a finite set of cosine waves, leading to the Fourier representation. As we shall show, this results in both inferential and structural problems which we resolve by considering all possible frequencies, $0 \leqslant \alpha \leqslant \pi$. This leads us to consider the *spectrum*; the main concept is developed in Sections 10.11–20 and then estimation procedures are discussed in Sections

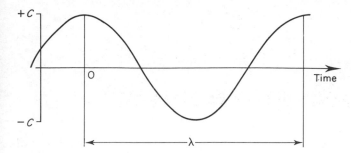

Fig. 10.1 A sinusoidal wave with wavelength λ

10.21–31. Seasonal components are examined from the frequency viewpoint in Sections 10.32–36 and several other topics are mentioned briefly in Sections 10.37–41.

Fourier representation

10.3 A natural extension of (10.1) is to consider the time series y_t $(t = 1, 2, ..., n)$ as a sum of sine waves:

$$y_t = \tfrac{1}{2} a_0 + \sum_{j=1}^{m} (a_j \cos \alpha_j t + b_j \sin \alpha_j t). \tag{10.3}$$

Again, the a, b, and α are fixed quantities, for the moment. We shall set $\alpha_j = 2\pi j/n$ so that the jth cycle has wavelength $\lambda_j = n/j$. Also, we take n to be odd (purely for convenience) and set $m = (n - 1)/2$, so that the frequencies lie in the range 0 to π. The factor $\tfrac{1}{2}$ in front of a_0 is also there for convenience and does not affect the basic argument. The selected frequencies, α_j, are known as the *harmonic* frequencies. The reason that the harmonic frequencies are of such interest derives from the work of Jean-Baptiste Joseph Fourier. Consider some function $f(x)$ defined for a finite range of values which we can reduce to $-\pi \leqslant x \leqslant \pi$ by a change of scale and origin. Then Fourier showed that $f(x)$ may be represented by the series expansion

$$f(x) = \tfrac{1}{2} a_0 + \sum_{j=1}^{\infty} (a_j \cos jx + b_j \sin jx) \tag{10.4}$$

provided only that $f(x)$ is single-valued, continuous except perhaps at a finite number of points, and possesses only a finite number of maxima and minima. Further, since

$$\int_{-\pi}^{\pi} \cos rx \sin sx \, dx = 0 \tag{10.5}$$

$$\int_{-\pi}^{\pi} \cos rx \cos sx \, dx = \int_{-\pi}^{\pi} \sin rx \sin sx \, dx$$

$$\left. \begin{aligned} &= 0, && r \neq s \\ &= \pi, && r = s \end{aligned} \right\}, \tag{10.6}$$

it follows that

$$a_j = 1/\pi \int_{-\pi}^{\pi} f(x)\cos jx \, dx \tag{10.7}$$

and

$$b_j = 1/\pi \int_{-\pi}^{\pi} f(x)\sin jx \, dx. \tag{10.8}$$

Note that (10.6) holds for $r = 0$ also.

10.4 The Fourier representation of a time series is thus seen to be a way of expressing the $n = 2m + 1$ values as a weighted average of sine waves. Mathematically, the result holds for any sequence of values; statistically, it is useful because the successive values belong to a time series. It follows that, in a formal sense, *any* time series may be decomposed into a set of cycles based on the harmonic frequencies. This does *not* mean that the phenomenon under study displays cyclical behaviour.

10.5 In order to develop the statistical framework for Fourier analysis, we return to (10.3). In discrete time (10.5) yields, for the harmonic frequencies:

$$\Sigma \cos \alpha_j t \sin \alpha_k t = 0 \tag{10.9}$$

$$\Sigma \cos \alpha_j t \cos \alpha_k t = \Sigma \sin \alpha_j t \sin \alpha_k t = 0, \qquad j \neq k$$
$$= n/2, \qquad j = k \tag{10.10}$$

all summations being over $t = 1, 2, \ldots, n$ and $j, k = 0, 1, \ldots, m$. Corresponding to (10.7)–(10.8), we find

$$a_0 = \bar{y}, \tag{10.11}$$

$$na_j = 2 \Sigma y_t \cos \alpha_j t, \qquad j = 1, \ldots, m; \tag{10.12}$$

$$nb_j = 2 \Sigma y_t \sin \alpha_j t, \qquad j = 1, \ldots, m. \tag{10.13}$$

When n is even, $n = 2m + 2$ say, these same results hold except that we add the term

$$na_{m+1} = 2 \Sigma y_t(-1)^t, \tag{10.14}$$

since $\alpha_{m+1} = \pi$; $b_{m+1} = 0$ since $\sin \pi t = 0$ for all integer t.

Finally, since we shall be interested in the variation within the series rather than its mean level, we may use the deviations from the mean

$$u_t = y_t - \bar{y}.$$

It is evident that the term a_0 drops out since $\bar{u} = 0$, whereas (10.12)–(10.14) continue to hold with u_t in place of y_t.

10.6 We saw from Fig. 10.1 that the coefficient c_j is of more natural interest than a_j or b_j. From (10.12) and (10.13), it follows that

$$c_j^2 = a_j^2 + b_j^2$$

$$= 4/n^2 \sum_{t=1}^{n} \sum_{s=1}^{n} u_t u_s(\cos \alpha_j t \cos \alpha_j s + \sin \alpha_j t \sin \alpha_j s). \tag{10.15}$$

From (10.2), the cosine terms reduce to $\cos \alpha_j(t - s)$. If we write $k = t - s$ and split (10.15) into three parts, for $k < 0$, $= 0$ or > 0, we obtain, after some reduction,

$$c_j^2 = 4/n^2 \left[\sum_{t=1}^{n} u_t^2 + 2 \sum_{t=1}^{n} \sum_{k=1}^{n-1} u_t u_{t+k} \cos \alpha_j k \right]. \tag{10.16}$$

From (6.1) and (6.2), the autocovariances are

$$\hat{\gamma}_0 = \Sigma \, u_t^2/n, \, \hat{\gamma}_k = \Sigma \, u_t u_{t+k}/n,$$

and the autocorrelations are $\hat{\rho}_k = \hat{\gamma}_k/\hat{\gamma}_0$ so that (10.16) may be rewritten as

$$nc_j^2 = 4 \left[\hat{\gamma}_0 + 2 \sum_{k=1}^{n-1} \hat{\gamma}_k \cos \alpha_j k \right] \tag{10.17}$$

or

$$nc_j^2/4\hat{\gamma}_0 = 1 + 2 \sum_{k=1}^{n-1} \hat{\rho}_k \cos \alpha_j k. \tag{10.18}$$

The quantities $\hat{I}(\alpha_j) = nc_j^2/4\pi$ define the *sample intensity function* and are considered further in Section 10.18.

10.7 Another way of interpreting our analysis is to think of (10.3) as a linear regression and to fit the $(2m + 1)$ parameters to the $(2m + 1)$ observations by least squares. The coefficients are then given by (10.11)–(10.13) and the sum of squares may be written as

$$\sum_{t=1}^{n} u_t^2 = \sum_{t=1}^{n} (y_t - \bar{y})^2 = \sum_{t=1}^{n} (y_t - a_0)^2$$

$$= \sum_{j-1}^{n} \sum_{j=1}^{m} (a_j \cos \alpha_j t + b_j \sin \alpha_j t)^2. \tag{10.19}$$

Using (10.9)–(10.11), this reduces to

$$\sum_{t=1}^{n} u_t^2 = (n/2) \sum_{j=1}^{m} (a_j^2 + b_j^2) = (n/2) \sum_{j=1}^{m} c_j^2,$$

or

$$\hat{\gamma}_0 = \frac{1}{2} \sum_{j=1}^{m} c_j^2. \tag{10.20}$$

Expression (10.20) reveals that the Fourier representation of the time series produces a decomposition of the total variance ($\hat{\gamma}_0$) into the variances accounted for by each harmonic component (c_j^2). Using the decomposition (10.20), we may plot

$$c_j^2/2\hat{\gamma}_0 \quad \text{against } j \tag{10.21}$$

which shows how the variance is apportioned to each frequency; (10.21) is known as the *sample line spectrum*. Alternatively, if we plot

$$\hat{W}(j) = \frac{1}{2} \sum_{i=1}^{j} c_i^2/\hat{\gamma}_0 \tag{10.22}$$

against j, we have a non-decreasing cumulative plot such that $\hat{W}(0) = 0$ and $\hat{W}(m) = 1$. $\hat{W}$ is termed the *sample spectral (distribution) function*.

10.8 Thus far, our development has been empirically based. There are two reasons for this. First, we feel that the notion of the frequency domain becomes accessible to a statistical audience if approached through the more familiar concepts of regression analysis. Secondly, the use of the spectrum in the social and economic sciences is primarily descriptive, and our discussion is consistent with that orientation. By contrast, in the physical sciences the spectrum is often the primary focus of attention when the *line* spectrum may correspond to strictly cyclical phenomena such as electric currents or sound waves. If we are to consider such interpretations, we must link our analysis in the frequency domain to that in the time domain. We shall do this in a rather heuristic fashion; a more formal treatment is available in Kendall *et al.* (1983, Chapter 49), Fuller (1976, Chapters 3 and 4) or Priestley (1981, Chapters 4–8).

10.9   In forging the link between frequency- and time-domain analyses, there are certain changes we must make in our underlying assumptions. First of all, if we use (10.3) as a regression model in the conventional sense, where the a_j and b_j are fixed parameters, we run into the difficulty that the model implies non-stationary behaviour, since $E(y_t)$ is a function of t through the sines and cosines; further, the dependence structure is not properly specified since y_t and y_s are uncorrelated, given their respective means. Both difficulties are resolved by using a *random effects* model. That is, we reformulate (10.3) as

$$y_t = \tfrac{1}{2} A_0 + \sum_{j=1}^{m} (A_j \cos \alpha_j t + B_j \sin \alpha_j t), \qquad (10.23)$$

where the As and Bs are independent, normally distributed random variables with zero means; that is,

$$A_j \sim N(0, \sigma_j^2) \quad \text{and} \quad B_j \sim N(0, \sigma_j^2), \qquad (10.24)$$

where A_j and B_j are uncorrelated with each other and with all other coefficients. The terms a_j and b_j in (10.11)–(10.13) may now be interpreted as realisations of A_j and B_j, respectively. Since a_j and b_j are linear functions of the observations, it follows that they too are normally distributed with zero means and variances that approach σ_j^2 in large samples. Further, a_j and b_j are asymptotically independent of each other and of all other (a_k, b_k). Also, since the sum of squares of normal variables follows a chi-squared distribution, we have, asymptotically, that

$$c_j^2/2\sigma_j^2 \quad \text{is} \quad \chi_j^2(2). \qquad (10.25)$$

This model forms the basis of our subsequent discussions of sampling properties.

10.10   There is one further matter we must consider, one that makes the case for a proper theoretical development even more compelling. If we re-examine (10.23), we observe that the 'regressor' variables – the sines and cosines – are defined in terms of the sample size through α_j. If one new observation is added to the series, an entirely new set of regressors needs to be defined. Since the choice of sample size is often arbitrary and is not usually linked to the choice of regressor variables, we appear to have hit a major stumbling block. As we shall see, this may be resolved, and the use of harmonic frequencies with their desirable properties retained, if we extend the definition of the spectrum to cover a continuous set of frequencies; that is, all values in the range $0 \leqslant \alpha \leqslant \pi$. The estimation of a continuous function from a finite number of observations then requires some form of smoothing, just as we use a histogram to provide estimates of the probability density function.

The spectrum

10.11 Consider a stationary time series with the theoretical autocorrelations ρ_k, $k = 1, 2, \ldots$. Then we define the *spectral density function* as

$$w(\alpha) = 1 + 2 \sum_{k=1}^{\infty} \rho_k \cos \alpha k, \qquad 0 \leqslant \alpha \leqslant \pi. \tag{10.26}$$

We note that (10.18) is of the same form as (10.26) if we replace $\hat{\rho}_k$ by ρ_k and then allow $n \to \infty$. However, $w(\alpha)$ is now defined for all frequencies in $[0, \pi]$ and not just at the harmonics.

Just as a probability density may be integrated to obtain the distribution function, we define the *spectral (distribution) function* as

$$W(\alpha) = \int_0^\alpha w(\alpha) \, d\alpha, \tag{10.27}$$

in the range $[0, \pi]$.

Some authors include the factor π^{-1} in (10.26) and (10.27) and others define W over $-\pi \leqslant \alpha \leqslant \pi$. These differences are relatively unimportant, provided consistency is maintained. Although $[-\pi, \pi]$ is more convenient for theoretical purposes, the range $[0, \pi]$ is more appealing intuitively. As now defined, $w(\alpha)$ exists only when there are no jumps in $W(\alpha)$; jumps would correspond to pure sine waves in the process. The form of (10.27) may be extended to include jumps, but we shall not do this as it leads to estimation problems later; see Section 10.26.

10.12 It may seem strange that we started with a *sample* line spectrum and have now ruled out the existence of line components in the theoretical spectrum. However, the situation is entirely analogous to sampling from a continuous population, where the density function is continuous although the sample is recorded as a set of unit frequencies. Indeed, the analogy becomes even more useful in understanding the sampling properties of the sample spectrum.

10.13 It follows from (10.26) that knowing the correlation structure in the time domain corresponds to knowing the form of the spectrum in the frequency domain. Indeed, the autocorrelation function and the spectrum are a *Fourier transform pair* and (10.26) may be inverted using (10.5) and (10.6) to give

$$\rho_k = 1/\pi \int_0^\pi w(\alpha) \cos \alpha k \, d\alpha. \tag{10.28}$$

Although the *ACF* and the spectrum determine each other uniquely, neither determines the original series. These descriptors are useful only when stationarity can be assumed.

10.14 The spectrum may be expressed in a different way which greatly simplifies its evaluation. Let

$$e^{i\alpha k} = \cos \alpha k + i \sin \alpha k, \tag{10.29}$$

where $i^2 = -1$, so that

$$\cos \alpha k = (e^{i\alpha k} + e^{-i\alpha k})/2. \tag{10.30}$$

Since $\rho_k = \rho_{-k}$, we may rewrite (10.26) as

$$w(\alpha) = \sum_{k=-\infty}^{\infty} \rho_k e^{i\alpha k}; \tag{10.31}$$

likewise, (10.28) yields

$$2\pi\rho_k = \int_{-\pi}^{\pi} w(\alpha) e^{-i\alpha k} \, d\alpha, \tag{10.32}$$

where $w(\alpha) = w(-\alpha)$ in (10.32). Writing $e^{-i\alpha} = z$, (10.31) becomes

$$w(\alpha) = \sum_{k=-\infty}^{\infty} \rho_k z^k, \tag{10.33}$$

so the spectrum of a process may be found from the autocorrelation generating function, introduced in Section 5.28. Incidentally, the use of the complex argument z justifies the doubly infinite expansion of $G_\rho(z) = w(\alpha)$ in (5.78). From (5.86), it follows that the stationary *ARMA* scheme

$$\phi(B)y_t = \theta(B)\varepsilon_t \tag{10.34}$$

has a spectrum of the form

$$w(\alpha) = \frac{\sigma^2}{\gamma_0} \frac{\theta(z)\theta(z^{-1})}{\phi(z)\phi(z^{-1})}, \tag{10.35}$$

where $z = e^{i\alpha}$. We now consider some examples.

Example 10.1   Suppose $y_t = \varepsilon_t$, so that

$$\gamma_0 = \sigma^2 \quad \text{and} \quad \gamma_k = 0, \quad k \neq 0.$$

Since $\phi(z) = \theta(z) = 1$, we have immediately from (10.35) that

$$w(\alpha) = 1, \quad 0 \leqslant \alpha \leqslant \pi.$$

Alternatively, the result follows directly from (10.26).

Thus, the spectrum of the purely random process is completely flat. We may say that all frequencies are 'equally represented'; this justifies use of the descriptive term *white noise* by analogy with white light, which is a combination of light of all available frequencies.

Example 10.2   Consider the Markov process with

$$y_t = \phi y_{t-1} + \varepsilon_t.$$

We know that $\gamma_0 = \sigma^2/(1 - \phi^2)$ and $\phi(z) = 1 - \phi z$ so that, from (10.35),

$$\begin{aligned} w(\alpha) &= \frac{\sigma^2}{\gamma_0} \cdot \frac{1}{(1 + \phi^2 - 2\phi \cos \alpha)} \\ &= (1 - \phi^2)/(1 + \phi^2 - 2\phi \cos \alpha). \end{aligned} \tag{10.36}$$

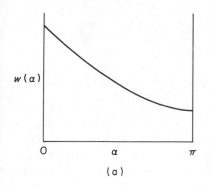 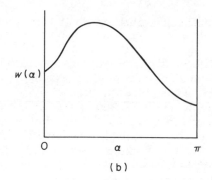

Fig. 10.2 Spectra of (a) the Markov scheme with $\phi > 0$; (b) the Yule scheme with $\phi_1 > 0$, $\phi_2 < 0$

An example of the shape of $w(\alpha)$ for $\phi > 0$ is given in Fig. 10.2(a). Result (10.36) could be derived from (10.26), but with much more effort.

Example 10.3 Consider the Yule process

$$y_t = \phi_1 y_{t-1} + \phi_2 y_{t-2} + \varepsilon_t.$$

From (10.35) it follows that

$$w(\alpha) = \frac{\sigma^2}{\gamma_0} \cdot \frac{1}{\{1 + \phi_1^2 + \phi_2^2 - 2\phi_1(1 - \phi_2)\cos\alpha - 2\phi_2 \cos 2\alpha\}}. \quad (10.37)$$

It may be shown that $w(\alpha)$ has a turning point at

$$\cos\alpha = -\phi_1(1 - \phi_2)/4\phi_2, \quad (10.38)$$

provided this value falls in the range $[0, \pi]$. This will be a maximum if $\phi_2 < 0$ or a minimum if $\phi_2 > 0$.

An example if given in Fig. 10.2(b) for $\phi_1 > 0$, $\phi_2 < 0$; this diagram illustrates that a low-frequency peak in the spectrum may result from a low-order *ARMA* scheme rather than a fixed cycle of long duration. This fundamental point was not grasped until Yule's pioneering work. For example, when $\phi_1 = 1.1$, $\phi_2 = -0.5$, the maximum occurs at $\alpha \doteq 0.6$ or about 0.1 cycles per unit time, just as our original discussion in Section 5.15 would suggest.

Example 10.4 The $MA(1)$ scheme

$$y_t = \varepsilon_t - \theta\varepsilon_{t-1}$$

has

$$w(\alpha) = (1 + \theta^2 - 2\theta \cos\alpha)/(1 + \theta^2). \quad (10.39)$$

The general shape of (10.39) is not dramatically different from that of the $AR(1)$ scheme in (10.36). Indeed, the spectrum is not particularly useful in selecting an *ARMA* model, although it is useful at the diagnostic stage to detect departures from white noise.

10.15   In early work with the spectrum, model (10.3) was used and then considerable effort was expended in trying to interpret the many resulting peaks in the line spectrum. Our discussion on sampling properties in Section 10.18 will reveal why this approach is unlikely to be successful, unless the phenomenon under study is exactly representable by a small number of sine waves. Thus, modern practice is to define the spectrum for all possible frequencies in $[0, \pi]$ and to regard the spectrum as an entity in itself. Although the spectrum and the *ACF* are a Fourier transform pair and, therefore, may be said to contain equivalent information, it is usually difficult to tell what either one would look like given the other. Therefore, the time series analyst often prefers to compute both and to draw conclusions about an observed process from two sources rather than one.

The Nyquist frequency

10.16   When we are dealing with a time series observed at equal time intervals, periodicities of less than one time unit may be overlooked. For example, it is clear that seasonal patterns cannot be detected if the variable is recorded only once a year. Generally, if the time between observations is t_0, we cannot detect periods smaller than $2t_0$ or angular frequencies higher than π/t_0. This limiting value is known as the *Nyquist frequency*. In our discussion, we take $t_0 = 1$, so that π becomes the Nyquist frequency.

Aliasing

10.17   A further problem of interpretation arises because $w(\alpha)$ is a periodic function; that is

$$w(\alpha + 2\pi) = w(\alpha), \qquad j = 1, 2, \dots.$$

Thus, all the frequencies $\alpha + 2\pi j$ are represented in the observed spectrum only at frequency α. Any spectral density at $\alpha + 2\pi j$ will be recorded at α; we say that these higher frequencies are *aliased*. For example, Granger and Hatanaka (1964) noted that when a series has a weekly fluctuation but is recorded monthly, the weekly harmonic occurs with frequency

$$\frac{365.25}{7 \times 12} = 4.348$$

per unit time period, which is aliased with 0.348. This is sufficiently close to 0.333 to make one wonder whether a peak in the spectrum corresponding to a three-monthly cycle is genuine or not. Whenever possible, the period between observations should be chosen short enough to ensure that no important frequencies are aliased. This is often feasible in experimental settings, but rarely so in the case of economic data.

Sampling properties

10.18 Since our definition of the spectral density in (10.26) followed from the sampling intensity in (10.18), it is natural to use

$$\hat{w}(\alpha) = \pi \hat{I}(\alpha)/\hat{\gamma}_0 \qquad (10.40)$$

to estimate $w(\alpha)$, at least at the harmonic frequencies α_j. From the assumptions underlying the random-effects model in Section 10.9, it follows that, asymptotically, for a white noise process.

$$E\{\hat{w}(\alpha)\} = 1 = \mathrm{var}\{\hat{w}(\alpha)\} \qquad (10.41)$$

and

$$\mathrm{cov}\{\hat{w}(\alpha), \hat{w}(\alpha')\} = 0, \qquad \alpha \neq \alpha', \qquad (10.42)$$

at $\alpha = \alpha_1, ..., \alpha_m$. When $\alpha = 0$ or π, the variance is doubled. Although these results were originally derived under the assumption of normality, they extend to the non-normal case (cf. Bartlett, 1978, although the result first appeared in the 1955 edition). Further, for general stationary processes, Bartlett showed that, asymptotically,

$$E\{\hat{w}(\alpha)\} = w(\alpha) \qquad (10.43)$$

$$\mathrm{var}\{\hat{w}(\alpha)\} = w^2(\alpha) \qquad (10.44)$$

and that $2\hat{w}(\alpha)$ is asymptotically distributed as $\chi^2(2)$. For further details, see Kendall *et al.* (1983, Sections 49.8–10). These rather remarkable results show that $\hat{w}(\alpha)$ is an asymptotically unbiased estimator for $w(\alpha)$, but that it is *not* consistent! As we increase n, we increase the number of harmonics at which w is estimated, but each estimator still has only two degrees of freedom. Since we have $n - 1$ asymptotically uncorrelated estimators, we are making full use of the available data; the problem arises because we are thinking in terms of the harmonic model (10.3) and allowing the number of parameters to increase directly with n. To overcome this, we must think in terms of the entire function $w(\alpha)$ and develop consistent estimators by some form of smoothing such as moving averages. This is similar to the use of histogram or kernel methods to produce estimates of the density function.

Examples of power spectra

10.19 Early users of spectrum analysis were unaware of the inconsistency of the estimators and this led to somewhat tortuous attempts to interpret the many (spurious) peaks that were observed. We now present several examples to illustrate the difficulties before going on to discuss smoothing procedures. The intensity function $\hat{I}(\alpha)$ is plotted in these examples together with a smoothed estimator; it is evident that plotting I rather than w involves only a change of scale.

Example 10.5 The results for the simulated Yule scheme, given in Fig. 5.3, are shown in Fig. 10.3. The unsmoothed intensity values show a

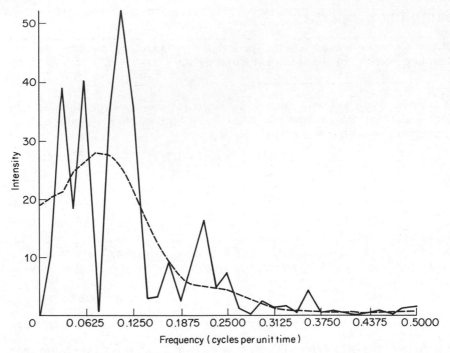

Fig. 10.3   The intensity and smoothed spectrum for the simulated Yule series $(\phi_1 = 1.1, \phi_2 = -0.5)$. A Parzen window − see (10.58) − was used for smoothing

very erratic pattern, whereas the smoothed estimator shows the general shape suggested in Fig. 10.2(b) with a maximum near $\alpha = 0.10$, as suggested by Example 10.3.

Example 10.6   Our second example is the time series on wheat prices developed by Lord Beveridge (1921) and reproduced in Table 10.1. An annual series of 370 observations is unprecedented in economic research and the series is deservedly recognised as one of the most famous in the time series literature. In his original analysis, Beveridge worked with what was then termed the *periodogram* − the plot of a multiple of the intensity function against the *wavelength* − although the term is now often used to refer to the (unsmoothed) intensity function plotted against *frequency*. Nevertheless, the two diagrams convey similar messages and Beveridge, working on the subject before the erratic nature of the estimates was understood, was led to identify 19 periodicities in the series, as illustrated in Fig. 10.4. By contrast, the smoothed estimate suggests that there are at most two. Indeed, the smoothed sample function in Fig. 10.4 is not so very different in its general shape from that in Fig. 10.3. It is interesting to note that both Kendall (1945) and Sargan (1953) suggested that the Yule scheme was appropriate for this series. Sargan gives the estimates

$$\hat{\phi}_1 = 0.73, \qquad \hat{\phi}_2 = -0.31$$

Table 10.1 Trend-free wheat-price index (European prices) compiled by Lord (then Sir William) Beveridge for the years 1500–1869

Year	Index	Year	Index	Year	Index	Year	Index	Year	Index	Year	Index	Year	Index
1500	106	1553	90	1606	81	1659	104	1712	115	1765	101	1818	94
01	118	54	100	07	98	60	120	13	134	66	106	19	86
02	124	55	123	08	115	61	167	14	108	67	113	20	84
03	94	56	156	09	94	62	126	15	90	68	108	21	76
04	82	57	71	10	93	63	108	16	89	69	108	22	77
05	88	58	71	11	100	64	91	17	89	70	131	23	71
06	87	59	81	12	99	65	85	18	94	71	136	24	71
07	88	60	84	13	100	66	73	19	89	72	119	25	69
08	88	61	97	14	94	67	74	20	107	73	106	26	82
09	68	62	105	15	88	68	80	21	79	74	105	27	93
10	98	63	90	16	92	69	74	22	75	75	88	28	114
11	115	64	78	17	99	70	78	23	91	76	84	29	103
12	135	65	112	18	82	71	83	24	94	77	94	30	110
13	104	66	100	19	73	72	84	25	76	78	87	31	105
14	96	67	86	20	81	73	106	26	84	79	79	32	82
15	110	68	77	21	97	74	134	27	94	80	87	33	80
16	107	69	69	22	124	75	122	28	101	81	88	34	78
17	97	70	93	23	106	76	102	29	90	82	94	35	82
18	75	71	71	24	106	77	107	30	96	83	94	36	88
19	86	72	72	25	121	78	115	31	83	84	92	37	102
20	111	73	73	26	105	79	113	32	76	85	85	38	117
21	125	74	74	27	84	80	104	33	84	86	84	39	107
22	78	75	75	28	97	81	92	34	91	87	93	40	95
23	86	76	76	29	109	82	84	35	94	88	108	41	101
24	102	77	77	30	148	83	86	36	101	89	108	42	92
25	71	78	78	31	114	84	101	37	93	90	86	43	88
26	81	79	79	32	108	85	74	38	91	91	78	44	92
27	129	80	90	33	97	86	75	39	122	92	87	45	115
28	130	81	87	34	92	87	66	40	159	93	85	46	139
29	129	82	83	35	97	88	62	41	110	94	103	47	90
30	125	83	85	36	98	89	76	42	90	95	130	48	80
31	139	84	76	37	105	90	79	43	81	96	95	49	74
32	97	85	110	38	97	91	97	44	84	97	84	50	78
33	90	86	161	39	93	92	134	45	102	98	87	51	86
34	76	87	97	40	99	93	169	46	102	99	120	52	105
35	102	88	84	41	99	94	111	47	100	1800	139	53	138
36	100	89	89	42	107	95	109	48	109	01	117	54	141
37	73	90	90	43	106	96	96	49	104	02	105	55	138
38	86	91	91	44	96	97	111	50	90	03	94	56	107
39	74	92	92	45	82	98	128	51	99	04	125	57	82
40	74	93	93	46	88	99	163	52	95	05	114	58	58
41	76	94	94	47	116	1700	137	53	90	06	98	59	97
42	80	95	95	48	122	01	99	54	80	07	93	60	81
43	96	96	96	49	134	02	85	55	85	08	94	61	107
44	112	97	97	50	119	03	88	56	117	09	94	62	92
45	144	98	98	51	136	04	77	57	112	10	104	63	79

continued

Table 10.1 (*continued*)

Year	Index	Year	Index	Year	Index	Year	Index	Year	Index	Year	Index	Year	Index
46	80	99	99	52	102	05	66	58	95	11	140	64	81
47	54	1600	97	53	72	06	64	59	91	12	121	65	94
48	69	01	80	54	63	07	69	60	88	13	96	66	119
49	100	02	90	55	76	08	125	61	100	14	96	67	118
50	103	03	90	56	75	09	175	62	97	15	130	68	93
51	129	04	80	57	77	10	108	63	88	16	178	69	102
52	100	05	77	58	103	11	103	64	95	17	126		

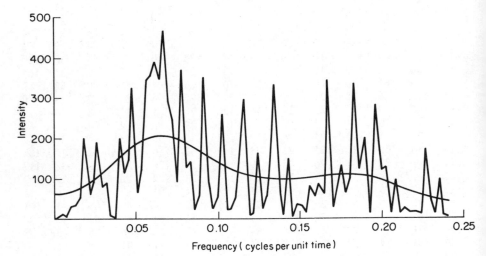

Fig. 10.4 Intensity and smoothed spectrum of the Beveridge wheat – price index series (Table 10.1)

suggesting a maximum at $\alpha = 0.69$ or about 0.11 cycles per unit time. Granger and Hughes (1971) re-analysed the data, pointing out that the detrended series was computed as a ratio of observed values to a 31-point moving average, which could have included cyclical behaviour. They found that the raw data exhibited a strong peak in the spectrum with a period of about 13.3 years or 0.075 cycles per unit time; Fig. 10.4 suggests a somewhat larger cycle.

Smoothing the estimates

10.20 We now consider the process of smoothing and define the smoothed estimator as

$$\hat{w}_A(\alpha) = \Sigma \, h(\alpha - \alpha_j)\hat{w}(\alpha_j), \qquad (10.45)$$

where $\alpha_j = 2\pi j/n$ and h is a weighting function chosen such that $(u = \alpha - \alpha_j)$:

$$h(u) = h(-u) \leqslant h(0)$$
$$\Sigma\, h(u) = 1$$
$$h(u) = 0, \qquad |u| \geqslant m.$$

These choices are made on grounds of statistical and computational efficiency as we shall see later. It follows from (10.45) that even for large n

$$E\{\hat{w}_A(\alpha)\} \neq w(\alpha) \tag{10.46}$$

unless $h(u) = 0$ for all $u \neq 0$ or $w(\alpha_j) = w(\alpha_j) = w(\alpha)$ for all α_j with non-zero weights. Thus, we are introducing biased estimators to achieve consistency. Assuming $w(\alpha)$ to be reasonably smooth, a small bias requires averaging over only a few adjacent frequencies whereas a low variance is achieved by averaging over a considerable number of values. This trade-off is achieved through the weighting function $h(u)$, known as a *kernel* or *spectral window*.

10.21 In large samples, we may choose m to be of order, say $n^{1/2}$, so that the estimators are both asymptotically unbiased and consistent. It then follows that

$$\mathrm{var}\{\hat{w}_A(\alpha)\} \doteq w^2(\alpha)\Sigma\, h^2(u) \tag{10.47}$$

or

$$\mathrm{var}\{\log \hat{w}_A(\alpha)\} \doteq \Sigma\, h^2(u). \tag{10.48}$$

These expressions may be used to construct approximate confidence intervals for $\hat{w}$, (10.48) having the advantage that the bounds are of constant width. When the logarithm is plotted, it is sometimes desirable to omit the lower part of the w-axis to improve the graphical representation, as the estimates may be very small.

10.22 An alternative specification of the smoothing process is to express $\hat{w}$ in terms of the serial correlations so that

$$\hat{w}_A(\alpha) = \sum_j h(\alpha - \alpha_j)\left\{1 + 2\sum_{k=1}^{n-1} r_k \cos k\alpha_j\right\}. \tag{10.49}$$

Changing the order of summation leads to

$$\hat{w}_A(\alpha) = \lambda_0 + 2\sum_{k=1}^{n-1} \lambda_k r_k \cos k\alpha, \tag{10.50}$$

where

$$\lambda_k = h(0) + 2\sum_{u=1}^{q} h(u)\cos uk. \tag{10.51}$$

Similarly, for $-\pi \leqslant u \leqslant \pi$,

$$h(u) = (2\pi)^{-1}[\lambda_0 + 2\,\Sigma\,\lambda_k \cos uk]. \tag{10.52}$$

The set consisting of the weights λ_k is known as the *lag window*. Often the λ_k are chosen directly with the constraint that $\lambda_k = 0$, $k > M$; M is known as the *truncation point*.

10.23 A considerable variety of lag and spectral windows has been constructed over the years and we now review those in common use.

Daniell window (Daniell 1946). If we set

$$h(u) = 1/m, \qquad u = 0, \pm 1, ..., \pm q, \quad m = 2q + 1, \tag{10.53}$$

we find from (10.51) that

$$\lambda_0 = 1$$

and

$$\lambda_k = \frac{\sin(\pi mk/n)}{m \sin(\pi k/n)}, \qquad k = 1, 2, ..., n - 1. \tag{10.54}$$

This window is designed for use in the frequency domain with (10.45), since there is no truncation in the lag window. It always produces non-negative estimates, as is clear from (10.45).

Tukey window (Blackman and Tukey 1958). We start with

$$\lambda_k = (1 - 2a) + 2a \cos(\pi k/M), \qquad k = 0, 1, ..., M. \tag{10.55}$$

When $a = 0.25$, this is called 'hanning' (after Julius Von Hann) and when $a = 0.23$, 'hamming' (after R. W. Hamming). This lag window may be represented in the frequency domain as

$$\hat{w}_A(\alpha) = a\hat{w}_1(\alpha - \pi/M) + (1 - 2a)\hat{w}_1(\alpha) + a\hat{w}_1(\alpha + \pi/M). \tag{10.56}$$

for

$$\alpha = 2\pi j/M, \qquad j = 1, ..., M - 1$$

with

$$\hat{w}_A(0) = (1 - 2a)\hat{w}_1(0) + 2a\hat{w}_1(\pi/M)$$

$$\hat{w}_A(\pi) = 2a\hat{w}_1(\pi - \pi/M) + (1 - 2a)\hat{w}_1(\pi)$$

and

$$\hat{w}_1(\alpha) = 1 + 2 \sum_{k=1}^{M} r_k \cos \alpha k. \tag{10.57}$$

That is, we estimate the spectrum using only the first M autocorrelations as in (10.57), and then smooth further using (10.56). The Tukey windows have the advantage of easy application in both domains; there is little practical difference between 'hanning' and 'hamming'. It is, however, possible for the resulting estimates to be negative for some frequencies.

Parzen window (Parzen 1961; 1963). Setting

$$\lambda_k = \begin{cases} 1 - 6\left(\dfrac{k}{M}\right)^2 + 6\left(\dfrac{k}{M}\right)^3, & 0 \le k \le \tfrac{1}{2}M \\ 2(1 - k/M)^3, & \tfrac{1}{2}M \le k \le M \end{cases} \tag{10.58}$$

corresponds, approximately, to a spectral window of the form

$$h(u) \propto \left\{ \frac{\sin(Mu/4)}{\sin(u/4)} \right\}^4; \tag{10.59}$$

The positive values in (10.59) ensure non-negative estimates.

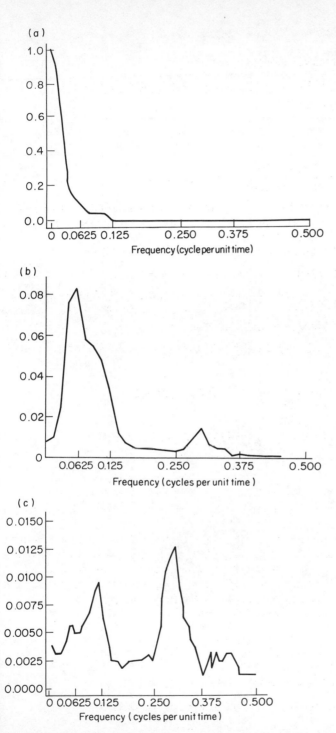

Fig. 10.5 Smoothed spectra for the US immigration data of table 1.4, transformed to logarithms (a) with no trend adjustment; (b) detrended by a 17 term moving average (Table 10.2) (c) detrended by use of first differences

Table 10.2 Residuals of logarithms of immigration data of Table 1.4, after removal of trend by a simple 17-point moving average. Values increased by 2

1828	2.1395	1870	2.1981	1912	2.1482
	1.9973		2.1004		2.3036
			2.1762		2.3413
1830	1.9754		2.1999		1.7889
	1.9039		2.0037		1.7653
	2.1679		1.8484		1.7853
	2.1922		1.7093		1.3817
	2.1800		1.6217		1.5165
	1.9737		1.6126		
	2.1627		1.7150	1920	2.0273
	2.1435				2.3278
	1.7849	1880	2.1118		1.9536
	1.9688		2.2747		2.2463
			2.3465		2.4339
1840	2.0015		2.2151		2.1182
	1.9354		2.1254		2.1919
	2.0087		1.9833		2.2630
	1.7894		1.8925		2.2600
	1.7860		2.0429		2.2739
	1.9087		2.0736		
	1.9958		2.0010	1930	2.2737
	2.1363				1.9115
	2.0929	1890	2.0389		1.5264
	2.1826		2.1527		1.4049
			2.1751		1.5707
1850	2.1702		2.0566		1.7101
	2.2743		1.8565		1.7898
	2.2451		1.7892		1.9839
	2.2375		1.8995		2.1388
	2.3080		1.7109		2.2386
	1.9764		1.6850		
	1.9792		1.7918	1940	2.1550
	2.0760				1.9775
	1.7643	1900	1.9414		1.6609
	1.7562		1.9712		1.5275
			2.0728		1.5560
1860	1.8668		2.1651		1.6421
	1.6451		2.1122		2.0608
	1.6444		2.1812		2.1608
	1.9340		2.1694		2.1900
	1.9564		2.2275		2.1938
	2.0432		2.0134		
	2.1462		2.0064	1950	2.2749
	2.1266				2.1353
	2.0623	1910	2.1862		2.1845
1869	2.1682	1911	2.1510		1.9344
				1954	1.9700

10.24 Several other windows have been proposed such as that of Bartlett (1950) with

$$\lambda_k = 1 - k/M; \tag{10.60}$$

for a detailed review, see Priestley (1981, pp. 437–49). The empirical evidence suggests that is reasonable to use either the Tukey or Parzen windows; the Parzen window was used to produce the smoothed estimate in Fig. 10.4. It is also advisable to employ several different truncation points. Comparison of these several plots will enable the investigator to determine whether a peak at a particular frequency appears to be a real phenomenon or is simply an artifact of the smoothing process. Referring back to the discussion in Section 10.21 choices of M such as 1, 2 and 4 times $\sqrt{n}$ give a reasonable range of possibilities.

10.25 Although we are recommending the spectrum primarily as a data-analytic device, there are several practical issues we must take into account. First of all, our analysis assumes that the series is stationary: non-stationarity, particularly in the mean, may seriously distort the sample spectrum; indeed, a linear trend is treated as a harmonic with $\alpha \to 0$. Major trends must be removed before the spectrum is computed. The particular method used for trend removal will also affect the results, but will not affect the higher frequencies provided we use a *high-pass filter* (see Section 10.31). Figure 10.5 shows the spectrum for the US immigration data of Table 1.4, first without de-trending and then with two trend-removal procedures. The de-trended series using the 17-term moving average is given in Table 10.2. It is clear that the original spectrum is totally distorted by the trend, whereas the de-trended series show peaks corresponding to cycles of sixteen and eight years respectively.

Side-bands

10.26 If the series contains an underlying harmonic component with angular frequency, β, there will be a peak in the unsmoothed spectrum not only at β but at either side of it determined by local maxima of the sine function. In general, smoothing with a variety of truncation points, as suggested above, will avoid wrong interpretations. However, it may happen that a known (e.g. seasonal) harmonic could distort the estimates, and we may prefer to remove the harmonic before developing the sample spectrum. Suitable test procedures are given by Hannan (1961) and Priestley (1962), although this pre-testing procedure needs to be handled with care; see Bhansali (1979).

Echo effects

10.27 If there is a latent harmonic in the series with angular frequency α, giving a spectral peak at α, we may also expect to see peaks at frequencies 2α, 3α and so on. In particular, a seasonal or 12-monthly cycle may produce echoes at 6, 4 and 3 months as illustrated in Fig. 10.6. These effects should not pose a problem provided the analyst is aware that they can occur and interprets them accordingly. The effects of seasonal adjustment on the spectrum are discussed in Section 10.32.

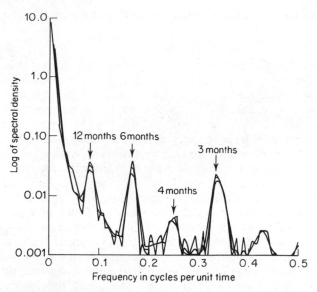

Fig. 10.6 Power spectrum of detrended US bank clearing data (monthly, 1875–1958), showing peaking effects a 2-monthly, 3-monthly etc. intervals (exponential linear trend removed). (source: Granger 1963)

Fast Fourier transform

10.28 For series of length, say $n < 1000$, which are most common to many statisticians, the computational procedures described in Sections 10.20–24 are entirely adequate. However, in experimental settings where automated data-logging devices are increasingly common, series of several thousand observations may be available. Methods based upon (10.50) require of the order of nM or, say, $n^{3/2}$ calculations, which begins to get rather expensive.

To cut computational costs, we may use a procedure developed originally in the 1900s, but rediscovered and exploited effectively by Cooley and Tukey and by Sande; see Cooley *et al.* (1967). This method is known as the Fast Fourier Transform (FFT). We may develop this in the following way. Suppose that n is even and factors into $n = rs$, where r is even. Now consider the direct evaluation of

$$J(\alpha) = a(\alpha) + \mathrm{i}b(\alpha)$$

$$= \frac{1}{\sqrt{(n\pi)}} \sum_{0}^{n-1} u_t e^{\mathrm{i}\alpha t}, \tag{10.61}$$

where $u_t = y_t - \bar{y}$ and now t is indexed over 0 to $n-1$ for mathematical convenience. Let

$$t = rt_1 + t_2,$$

where $t_1 = 0, 1, \ldots, s-1$ and $t_2 = 0, 1, \ldots, (r/2) - 1$; similarly, let

$$\alpha = \frac{2\pi j}{n} = \frac{2\pi}{n}(sj_2 + j_1) = s\alpha_2 + \alpha_1,$$

where $j_1 = 0, 1, \ldots, s - 1$ and $j_2 = 0, 1, \ldots, (r/2) - 1$. Using these expressions, (10.61) becomes

$$\sum_{t_2=0}^{r-1} e^{i\alpha t_2} \sum_{t_1=0}^{s-1} e_t^{i\alpha r t_1}. \tag{10.62}$$

Now

$$\exp(i\alpha r t_1) = \exp\{2\pi i (s j_2 + j_1) r t_1 / n\}$$
$$= \exp\{2\pi i j_1 r t_1 / n\} = \exp(i r \alpha_1 t_1 / n),$$

so that the inner summation in (10.62) does not depend on α_2. On inspection, we see that there are a total of rs distinct terms to be evaluated, each requiring s complex multiplications and additions. Finally, the $rs/2$ different $J(\alpha)$ terms each require r further operations in (10.62). Thus, that total computational effort is of the order of $rs(s + r/2)$ or $n(s + r/2)$ operations.

This process can be repeated as long as n can be factored. In the best case, $n = 2^p$ and we finish up needing of the order of $n \log_2 n$ calculations.

10.29 Since nature only rarely is kind enough to make n a power of 2, we can achieve this same end by adding zeros to each end of the series until the next largest power of 2 is achieved. When this is done, it is recommended that the data at the ends of the series be *tapered*, a concept originally due to Blackman and Tukey (1958, p. 169). The aim is to give a smooth transition between the ends of the actual series and the strings of zeros. For example, one recommended scheme is to set

$$u_t^* = \begin{cases} t u_t / (m + 1) & t = 1, \ldots, m \\ u_t & t = m + 1, \ldots, n - m \\ (n + 1 - t) u_t / (m + 1) & t = n - m + 1, \ldots, n, \end{cases} \tag{10.63}$$

where $m \doteq 0.05n$ in long series.

10.30 Several other estimation procedures for the spectrum have been proposed; for a review, see Robinson (1983). Also, robust estimators have been developed to handle possible outliers in the data; see Martin (1983).

Filtering and prewhitening

10.31 We have mentioned the need to remove trends and also the fact that the smoothing operation relies upon the spectrum being reasonable flat so that the bias contained in the smoothed estimates is not excessive. We define a linear *filter* as

$$v_t = \sum a_j y_{t-j}; \tag{10.64}$$

it is evident that the moving averages discussed in Chapter 2 are linear filters as are the various orders of differencing. A filter is *one-sided* if $a_j = 0$ whenever $j < 0$, a case of natural interest when we are concerned with prediction. It follows from (10.35) that the spectrum of v is

$$w_v(\alpha) = w_y(\alpha) T(\alpha) T(-\alpha), \tag{10.65}$$

where $T(\alpha) = \sum a_j e^{i\alpha j}$. The function $T(\alpha)$ is known as the frequency response function or (*frequency*) *transfer* function.

Example 10.7   Suppose that

$$y_t = \phi y_{t-1} + \varepsilon_t$$

and we apply the filter

$$v_t = y_t - a y_{t-1}. \tag{10.66}$$

It follows from (10.35) that

$$w_v(\alpha) = \text{const.} \frac{(1 - az)(1 - az^{-1})}{(1 - \phi z)(1 - \phi z^{-1})},$$

where $z = e^{i\alpha}$. This reduces to

$$\frac{1 + a^2 - 2a \cos \alpha}{1 + \phi^2 - 2\phi \cos \alpha} \tag{10.67}$$

which gets progressively flatter as a gets close to ϕ and, of course, produces a flat spectrum corresponding to white noise when $a = \phi$.

Examination of the numerator of (10.67) indicates the pattern of values in Table 10.3. Thus, when a is near $+1$, the filter (10.66) is called a *high-pass* filter since it 'passes on' the high-frequency signals but severely dampens the signal at the low frequencies. Conversely, when a is near -1, (10.66) represents a *low-pass* filter. We can see from (10.67) that if a series is non-stationary in the mean, its spectrum will be sharply peaked (theoretically infinite) in the neighbourhood of the origin, stressing the need for trend-removal prior to estimation.

More generally, filters such as (10.66) may be used to remove dominant autocorrelation structures in the series; such a process is known as *pre-whitening* since it produces a series closer to white noise. From the viewpoint of estimating the spectrum, the advantage of pre-whitening is that the bias of the estimates can be reduced. Parzen (1969; 1983) recommended using a high-order AR scheme to remove the dominant structure in the series; standard methods may then be used to estimate the spectrum of the residuals. The spectrum for the original series is then obtained using (10.65) and may be expected to show less bias. We shall make heavy use of pre-whitening in Chapter 11 when it is required for a somewhat different purpose.

Table 10.3

Value of a	Value of α	
	near zero	near π
near $+1$	near zero	near one
near -1	near one	near zero

Complex demodulation

10.32 If interest focuses upon a particular frequency, such as the seasonal frequency, we may 'shift' the series to a new frequency origin by the method of *complex demodulation*. That is, we define the new series

$$Z(t) = y_t \exp(-i\alpha_0 t), \tag{10.68}$$

where α_0 is the frequency of interest. Typically, we would centre the series about its mean before making the transformation; note that $Z(t)$ is complex rather than real-valued. A low-pass filter may then be applied to $Z(t)$ prior to estimating its spectrum. This has the effect of enhancing the signal in the region of frequency α_0, much as we enhance a radio signal at the desired frequency. This procedure generates improved estimates of the spectrum is the neighbourhood of α_0. For further details, see Hasan (1983).

Seasonality and harmonic components

10.33 We can now return to a topic which we left on one side in Chapter 4, namely the effects of trend-removal procedures upon the cyclical components of the series. Consider a symmetric moving average of order $2m + 1$ which we may represent with weights a_{-m} to a_m. Its frequency-response function reduces to

$$T(\alpha) = a_0 + 2 \sum_{j=1}^{m} a_j \cos j\alpha. \tag{10.69}$$

The spectrum for the de-trended series is given by

$$w_v(\alpha) = w_y(\alpha)\{1 - T(\alpha)\}\{1 - T(-\alpha)\} \tag{10.70}$$

since the de-trended values are

$$v_t = y_t - \Sigma \, a_j y_{t-j}.$$

For example, consider a centred moving average of 12 terms and a Spencer 15-point average. Table 10.4 shows their frequency response functions. For example, with an angular frequency of $\pi/4$ or $45°$, the value of (10.69) for the centred moving average is

$$\tfrac{1}{24} \, [2 + 2\{\cos 45° + 2 \cos 90° + 2 \cos 135° + 2 \cos 180°$$
$$+ 2 \cos 225° + \cos 270°\}] = \tfrac{1}{24} \{2 + 2(-2 - \sqrt{2})\} = -0.201.$$

The values of the transfer function of the centred average fall to 10 percent, or less after an angular frequency of $45°$, corresponding to cycles per unit (month) of $\tfrac{1}{8}$ or a wavelength of 8 months. For cycles of shorter wavelength, the spectrum of the series after trend removal is not greatly affected; for larger cycles the effect is substantial and they would be mostly removed with the trend, if any. The Spencer 15-point is of the same kind and on the whole distorts the residual spectrum rather less. The general effect of either is to remove with the trend some or all of the long cycles, but to leave the shorter ones largely untouched.

Table 10.4 Transfer functions of a centred 12-point average and a Spencer 15-point average (Burman, 1965)

Angular frequency (degrees)	TF centred 12-point	TF Spencer 15-point	Angular frequency (degrees)	TF centred 12-point	TF Spencer 15-point
0	1.000	1.000	95	−0.038	−0.002
5	0.955	1.000	100	−0.061	−0.007
10	0.824	1.003	105	−0.064	−0.012
15	0.633	0.984	110	−0.051	−0.015
20	0.409	0.952	115	−0.027	−0.016
25	0.188	0.895	120	0	−0.013
30	0	0.809	125	0.022	−0.008
35	−0.133	0.696	130	0.034	−0.003
40	−0.198	0.564	135	0.034	0
45	−0.201	0.425	140	0.026	0.001
50	−0.155	0.293	145	0.013	0
55	−0.080	0.180	150	0	−0.003
60	0	0.094	155	−0.009	−0.005
65	0.065	0.037	160	−0.013	−0.005
70	0.103	0.006	165	−0.011	−0.004
75	0.109	−0.005	170	−0.006	−0.003
80	0.086	−0.005	175	−0.002	−0.001
85	0.045	−0.002	180	0	0
90	0	0			

10.34　With a suitable trend-removal method, therefore, we can proceed to consider the seasonal component secure in the knowledge that the higher frequencies are only slightly affected. Moreover, since by definition seasonality is strictly periodic, we expect to be able to represent it by a sum of harmonics, namely with regular frequencies $\alpha_k = 2\pi k/12$ and wavelengths 12, 6, 4, 3, 2.4, and 2 months. Thus, if x_t is the deviation from trend of the tth month in a set of $12p$ months, we have

$$a_j = \frac{2}{12p} \sum_{t=1}^{12p} x_t \cos \alpha_j t, \qquad j = 1, 2, ..., 5 \qquad (10.71)$$

$$b_j = \frac{2}{12p} \sum_{t=1}^{12p} x_t \sin \alpha_j t, \qquad j = 1, 2, ..., 5 \qquad (10.72)$$

$$a_6 = \frac{1}{12p} \sum_{t=1}^{12p} x_t \cos \alpha_6 t. \qquad (10.73)$$

The seasonal movement will then be represented by a set of 11 constants.

10.35　Now that the trend and seasonal components have been filtered out, we may estimate the spectrum to look for other cyclical behaviour. If the spectrum is estimated without making any seasonal adjustments, we expect a major peak at the seasonal frequency plus echoes at higher frequencies, as in Fig. 10.6. A slight modification enables us to take account of moving seasonal effects. We may, in fact, analyse the series by year, taking $p = 1$ in (10.71)–(10.73), and hence obtain 11 annual series of the 11 constants. Each of these series can be smoothed and extrapolated and the resulting values used to estimate the seasonal component for any given year. Since the Fourier constants are uncorrelated, the smoothing can proceed independently for each series. Burman (1965; 1966) developed such a method but using a 13-point average instead of the Spencer 15-point. Missing values at the ends of the series, caused by taking the moving average, were fitted using exponential smoothing. More recently, Burman (1980) has developed a signal extraction procedure for seasonal adjustment, based upon a decomposition of the spectrum. This was briefly discussed in Section 9.18. Also, it should be noted that the treatment of seasonal patterns in structural models is based upon the systematic updating of (10.71)–(10.73); see (9.48).

10.36 Spectral methods are also useful in determining the performance of seasonal adjustment procedures such as the US Bureau of the Census X-11 method. Nerlove (1964) showed the then current procedure (X-10) to have certain defects which have since been remedied. In general, any seasonal adjustment process will tend to remove 'too much' from the seasonal frequencies (Grether and Nerlove, 1970), although this does not seem to be a major problem in practice for the procedures now employed. For further discussion of seasonal adjustment procedures, see Cleveland (1983).

Evolutionary spectra

10.37 In Section 10.30 we suggested computing harmonic components a year at a time. This notion has been generalised and extended by Priestley (1965) to what is known as the *evolutionary spectrum*. For details and further references, see Kendall *et al.* (1983, 49.37–49.38).

Inverse autocorrelations

10.38 Starting with the spectrum in (10.35) for an *ARMA* scheme we may define its inverse as

$$wi(\alpha) = [w(\alpha)]^{-1}. \tag{10.74}$$

The inverse autocorrelations, introduced in Section 6.25, are then defined as

$$\rho i(k) = \gamma i(k)/\gamma i(0), \tag{10.75}$$

where

$$wi(\alpha) = \sum_{-\infty}^{\infty} \gamma i(j) z^j. \tag{10.76}$$

It follows that $\rho i(k) = \rho i(-k)$.

Example 10.8 The $AR(1)$ scheme has

$$w(\alpha) = \sigma^2/[\gamma_0(1 - \phi z)(1 - \phi z^{-1})].$$

Hence

$$wi(\alpha) = \sigma^{-2}\gamma_0(1 - \phi z)(1 - \phi z^{-1})$$

and

$$\rho i(1) = -\phi, \quad \rho i(k) = 0, \quad k > 1.$$

In general, the inverse *ACF* behaves like the partial *ACF*, but with the signs reversed.

10.39   There are two approaches to estimating the *IACF*. We may estimate the spectrum, and then determine the $\rho i(k)$ by numerical inversion of $wi(\alpha)$ using the analogue of (10.28). Alternatively, we may fit a high-order auto-regression (following Parzen 1969) and then generate the $\rho i(k)$ from (10.75) and (10.76). For further discussion, see Chatfield (1979) and Bhansali (1983).

Forecasting

10.40   Instead of developing the forecast function using a time-domain model, we may work in the frequency domain and then construct the forecast function for the spectrum; this is known as the Wiener–Kolmogorov filter. For details, see Bhansali and Karavellas (1983).

Further reading

10.41   In addition to the references given earlier, the volume of review papers edited by Brillinger and Krishnaiah (1983) is a valuable source of current information and further references.

Exercises

10.1   Sketch, as functions of t,

$$\cos(\pi t/4) \quad \text{and} \quad \cos(9\pi t/4).$$

Observe that the two curves have the same values at $t = 0, \pm 1, \pm 2, \ldots$. (Refer to Section 10.17 for an interpretation.)

10.2   Show that (10.16) follows from (10.15).

10.3   Verify the inversion formula (10.28) by substituting for $w(\alpha)$ using (10.26) and performing the integration. (*Hint:* It follows from (10.6) that $\int_0^\pi \cos rx \cos sx\, dx = 0$, $r \neq s$; $= \pi$, $r = s$.)

10.4   Verify that the spectrum for an $AR(1)$ scheme is given by (10.36).

10.5   Verify that the spectrum for an $AR(2)$ scheme is given by (10.37) and that the spectrum shows a peak at α given by (10.38).

10.6   Find the spectrum for an $MA(2)$ and show that it has a maximum (minimum) at $\cos \alpha = -\theta_1(1 - \theta_2)/4\theta_2$ if $\theta_2 > 0$ ($\theta_2 < 0$).

10.7   Find the inverse *ACF* for the $MA(1)$ scheme.

10.8   Use an available statistical package to estimate the spectrum for one or more of the following series:
(a) the sheep data (Table 1.2);
(b) the airline data (Table 1.3);
(c) the Financial Times Index (Table 1.6);
(d) any of the series in Appendix A, Tables A1–A8.
Recall that the series should be *stationary*; you should try estimation with and without differencing and also try different levels of smoothing.

11

Transfer functions: models and identification

11.1 Thus far, we have considered only relationships between successive terms of a single series. By contrast, traditional correlation measures and regression models deal only with cross-sectional relationships between variables that ignore the time dimension. Our purpose in this chapter is to develop measures of association and statistical models that incoporate both time-series and cross-sectional components of dependence. For the most part, we restrict attention to only two series, y_{1t} and y_{2t}, although the extension to $m > 2$ series is straightforward. The association measures are then used to develop model identification procedures.

Cross correlations

11.2 Assume that y_{1t} and y_{2t} are stationary series such that

$$E(y_{jt}) = \mu_j, \tag{11.1}$$

and

$$\text{var}(y_{jt}) = \gamma_j = \sigma_j^2, \qquad j = 1, 2. \tag{11.2}$$

We define the *cross-covariance* between y_{1t} and $y_{2,t-k}$ as

$$\text{cov}(y_{1t}, y_{2,t-k}) = E(y_{1t} y_{2,t-k}) - \mu_1 \mu_2 = \gamma_{12}(k). \tag{11.3}$$

The *cross-correlation* is then

$$\rho_{12}(k) = \gamma_{12}(k)/\sigma_1 \sigma_2. \tag{11.4}$$

Note that the order of the subscripts is important in this notation, for

$$\rho_{12}(k) = \rho_{21}(-k) \tag{11.5a}$$

but

$$\rho_{12}(k) \neq \rho_{21}(k). \tag{11.5b}$$

If $\rho_{12}(k)$ were the only non-zero cross-correlation, we could say that y_2 *leads*

y_1 by k periods, since knowing the value of y_2 at $(t-k)$ furnishes information about y_1 at t. Likewise, y_1 *lags* y_2 by k time periods.

The *cross-correlation function* (*CCF*) is defined as

$$\rho_{12}(k), \qquad k = 0 \pm 1, \pm 2, \ldots$$

and will be plotted as a useful summary measure just as with the *ACF*, now denoted by $\rho_{jj}(k)$, $k = 0, 1, 2, \ldots$.

11.3   The corresponding sample quantities are

$$c_{12}(k) = \left\{ \sum_{t=k+1}^{n} y_{1t} y_{2,t-k} - (n-k)\bar{y}_1 \bar{y}_2 \right\} /(n-1) \tag{11.6}$$

$$c_{jj} = s_j^2 = \left\{ \sum_{t=1}^{n} y_{jt}^2 - n\bar{y}_j^2 \right\} /(n-1) \tag{11.7}$$

and

$$r_{12}(k) = c_{12}(k)/s_1 s_2. \tag{11.8}$$

As for the serial correlations, a variety of alternatives exist to (11.6)–(11.8), but these versions are probably the most widely used.

11.4   A large sample expression for the covariance between $r_{12}(k)$ and $r_{12}(k+s)$ was developed by Bartlett (see Kendall *et al.* 1983, Section 51.10). This is rather cumbersome but simplifies considerably, when the two series are uncorrelated with each other, to

$$\mathrm{cov}\{r_{12}(k), r_{12}(k+s)\} = \frac{1}{n-k} \sum_{-\infty}^{\infty} \rho_{11}(i)\rho_{22}(i+s), \tag{11.9}$$

and

$$\mathrm{var}\{r_{12}(k)\} = \frac{1}{n-k} \sum_{-\infty}^{\infty} \rho_{11}(i)\rho_{22}(i). \tag{11.10}$$

Example 11.1   Suppose that y_{1t} and y_{2t} are both $AR(1)$ schemes with parameters ϕ_1 and ϕ_2, respectively. It follows from (11.10) that

$$\mathrm{var}\{r_{12}(k)\} = \frac{1+\phi_1\phi_2}{(n-k)(1-\phi_1\phi_2)}.$$

Further, when $\phi_1 = \phi_2 = \phi$, we find from (11.9) and (11.10) that

$$\mathrm{corr}\{r_{12}(k), r_{12}(k+s)\} = \phi^2\{(s+1) - (s-1)\phi^2\}/(1+\phi^2). \tag{11.11}$$

Table 11.1

Value of ϕ	0	0.5	0.9		-0.5	-0.9
$(n-k)\mathrm{var}(r)$	1	1.67	9.53		1.67	9.53
(11.11), $s=1$	0	0.80	0.994		-0.80	-0.994
(11.11), $s=5$	0	0.13	0.900		-0.13	-0.900

The numerical values in Table 11.1 indicate the behaviour of these functions (when $\phi_1 = \psi_2 = \phi$).

These figures illustrate the very high correlations between successive terms in the *CCF*, a problem we shall address in Section 11.9. Note that $\text{var}(r) \simeq 1/(n - k)$ when both series are white noise.

Cross-spectra

11.5 We may define the *cross-spectrum* between the two series as

$$w_{12}(\alpha) = \Sigma \, \rho_{12}(k)e^{i\alpha k}, \tag{11.12}$$

where $0 < \alpha < \pi$ and $w_{12}(\alpha) = w_{21}(-\alpha)$. This has the inverse transformation

$$\rho_{12}(k) = \frac{1}{\pi} \Sigma \, w_{12}(\alpha)e^{-i\alpha k}, \tag{11.13}$$

by analogy with (10.32). The cross-spectrum is a complex function and is not readily interpretable in its original form. However, it may be decomposed into its real and imaginary parts as follows:

$$\begin{aligned} w_{12}(\alpha) &= \rho_{12}(0) + \Sigma \, \{\rho_{12}(k)\cos \alpha k + \rho_{12}(-k)\cos \alpha k\} \\ &\quad + i \, \Sigma \, \{\rho_{12}(k)\sin \alpha k - \rho_{12}(-k)\sin \alpha k\} \\ &= c(\alpha) + iq(\alpha), \end{aligned} \tag{11.14}$$

where

$$c(\alpha) = 1 + \Sigma \, \{\rho_{12}(k) + \rho_{12}(-k)\}\cos \alpha k \tag{11.15}$$

$$q(\alpha) = \Sigma \, \{\rho_{12}(k) - \rho_{12}(-k)\}\sin \alpha k. \tag{11.16}$$

The quantity $c(\alpha)$ is known as the co-spectrum or co-spectral density, whereas $q(\alpha)$ is known as the quadrature spectrum. The sum of squares $c^2(\alpha) + q^2(\alpha)$ is called the *amplitude* of the spectrum. If we standardise by division by the separate spectral densities of the two series we obtain the *coherence*, namely

$$C(\alpha) = \frac{c^2(\alpha) + q^2(\alpha)}{w_1(\alpha)w_2(\alpha)} = \frac{|w_{12}(\alpha)|^2}{w_1(\alpha)w_2(\alpha)}. \tag{11.17}$$

The phase relationship of these quantities also requires attention. The so-called *phase* diagram is the plot of $\psi(\alpha)$ against α, where

$$\psi(\alpha) = \text{arc tan} \, \frac{q(\alpha)}{c(\alpha)}. \tag{11.18}$$

The *gain* diagrams plots $R_{12}^2(\alpha)$ against α, where

$$R_{12}^2(\alpha) = \frac{w_1(\alpha)}{w_2(\alpha)} C(\alpha). \tag{11.19}$$

11.6 The coherence $C(\alpha)$ may be thought of as the correlation between the α-frequency components of y_1 and y_2, whereas the gain is the slope of the regression equation for the α-frequency component of y_1 on that of y_2.

11.7 In order to estimate the cross-spectrum we may work with (11.12), replacing $\rho_{12}(k)$ by $r_{12}(k)$. As is to be expected, the unsmoothed estimator is inconsistent, but may be made consistent by use of one of the window estimators described in Section 10.23. Also, if y_2 appears to lead y_1 by k time periods, the quality of the estimators is improved if we first align the series; that is, consider y_{1t} along with $y_{2t}^* = y_{2,t-k}$.

11.8   Except in series involving a small number of major harmonics, the interpretation of the coherence and related quantities is rather difficult since a high coherence may be unimportant if the corresponding components of the respective spectra are small. The following example illustrates the problems of interpretation, although for a fuller treatment of the subject the reader should consult Granger and Hatanaka (1964) or Priestley (1981).

Example 11.2   Table 11.2, for comparison with Table 11.7, gives the UK production of motor vehicles for each quarter of 1960–71. There are no seasonal components in these two series. In Fig. 11.1, we show the two series (which happen to have about the same range for the dependent variable) after removal of the irregular component by the Census X-11 method. The smoothing operation makes it easier to appreciate the similarities in the series in this exploratory analysis; it is not used in the final analysis in Section 12.12. Thus, the series as graphed comprise only trend and short-term oscillatory movements. A glance at the diagram suggests that, although the two series pursue far from identical courses, there is some similarity in the time of occurrence of peaks and troughs.

There are various ways in which we might proceed to eliminate the trend, which is not very marked in either case. For simplicity in this example a linear trend was removed from both series. Figure 11.2 shows the correlograms of

Table 11.2 UK production of cars, seasonally adjusted (thousands)

Year and quarter		Car production	Year and quarter		Car production	Year and quarter		Car production
1960	1	374.241	1964	1	460.397	1968	1	435.000
	2	375.764		2	462.279		2	435.000
	3	354.411		3	434.255		3	465.000
	4	249.527		4	475.890		4	483.000
1961	1	206.165	1965	1	439.365	1969	1	423.000
	2	258.410		2	431.666		2	427.990
	3	279.342		3	399.160		3	425.490
	4	264.824		4	449.564		4	438.230
1962	1	312.983	1966	1	435.000	1970	1	407.000
	2	300.932		2	435.000		2	416.000
	3	323.424		3	408.000		3	332.000
	4	312.780		4	330.000		4	469.320
1963	1	363.336	1967	1	369.000	1971	1	403.750
	2	378.275		2	390.000		2	416.046
	3	414.457		3	381.000		3	435.236
	4	459.158		4	414.000		4	455.897[*]

[*] First two months expressed at a quarterly rate.

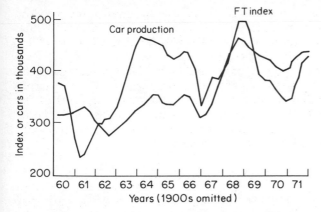

Fig. 11.1 FT index and car production in UK (both smoothed by removal of irregular component)

the de-trended series. In both cases the oscillatory effects are marked. The cross-correlogram is shown in Fig. 11.3. Here, at zero lag, the correlation is relatively small but positive, being about 0.33. At FT index leading the car production by about seven quarters it is negative and about −0.45. Again, at the car production leading FT index by about seven quarters it has a minimum of −0.22.

In Fig. 11.4, we have replotted the original series but have reversed the sign of the FT index and led it seven quarters in front of car production. The coincidence of peaks and troughs is now brought out more sharply. The indication is that a peak in the FT index is followed seven quarters later (or thereabouts) by a trough in car production.

The interesting question, of course, is what is implied by this result. It would be premature to conclude that a peak in the FT index *causes* a slump in car production about two years later. We may have an economy with an

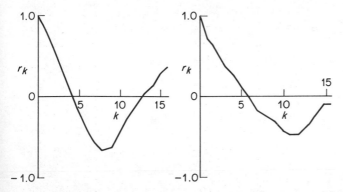

Fig. 11.2 Correlograms of FT index (left) and car production (right): original series with irregular component and linear trend removed

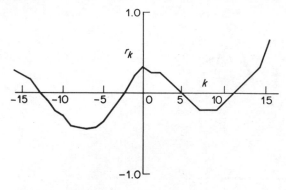

Fig. 11.3   Cross-correlogram, FT index vs, car production: original series irregular component and linear trend removed

oscillation in each series averaging about four years in duration, the oscillations being to some extent in step (as measured by the correlation at zero lag). The analysis is suggestive of a relationship to the extent that a stock market index is widely regarded as a leading indicator of economic activity. Relationships of this kind, if stable, can be used for forecasting, as we shall see in the next chapter, even when the causal linkage is not overt. In any case, we have here a series of only 48 terms extending over 12 years. Nor do ordinary tests of significance of individual correlations help very much, because we have deliberately picked out the biggest for attention.

Let us consider the relationship between the series from the point of view of spectrum analysis. Figure 11.5 shows the spectra of the two series, and Fig. 11.6 their coherence. The individual spectra of Fig. 11.5 are not very informative. There is no evidence of 'cycles' in any strict sense. The higher values on the left indicate mean periods of about 16 quarters for the FT index (a frequency of about 0.06) and rather longer for car production which confirms the picture presented by the correlograms. In the coherence diagram

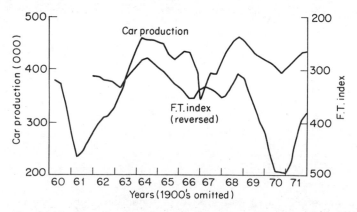

Fig. 11.4   Car production, with FT index reversed and leading 7 quarters

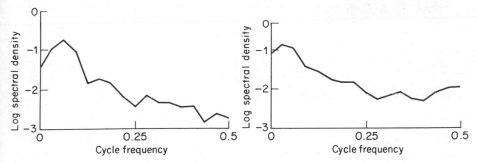

Fig. 11.5 Spectrum of FT index (left) and car production (right): original series, irregular component and linear trend removed; Bartlett smoothing

two peaks are suggestive, one at frequency about 0.09, the other at about 0.27, corresponding to mean periods about 11 and four quarters, respectively. The interpretation of these results is more difficult. It must always be remembered that they are sample values and therefore liable to considerable fluctuation. The one-yearly component seems to imply some common movement within the year, notwithstanding that one series (car production) is supposed to be seasonally adjusted. The other is presumably a reflection of the fact that both series have an oscillatory element of several years, and these elements, within the time period of 12 years, are to some extent in step.

In series generated by physical processes, especially where fundamental generators of a harmonic kind are operating, the interpretation of spectra is usually more straightforward. In economic series there are considerable difficulties, especially where the series are short.

Nevertheless, as we shall now demonstrate, we may hope to gain rather more from such analyses if we take account of the autocorrelation structures within the series before examining their interrelationships.

11.9 In order to understand the need for care in modelling interrelations between time series, we first present two examples, one theoretical and the other empirical. We then proceed to develop a general framework for such models and the means to select a specific model from within the general class.

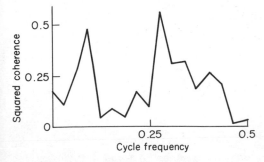

Fig. 11.6 Coherence of FT index and car production: original series, irregular component and linear trend removed

Example 11.3   Suppose that

$$y_t = x_t + \varepsilon_t \tag{11.20}$$

and

$$x_t = \phi x_{t-1} + \delta_t \tag{11.21}$$

from which we can eliminate x_t between the equations to obtain

$$y_t = \phi y_{t-1} + \varepsilon_t - \phi \varepsilon_{t-1} + \delta_7 \tag{11.22}$$

which is expressible as an $ARMA(1, 1)$ scheme. Thus, univariate analysis for each of x and y would indicate a strong dependence structure when the sample ACF was plotted. Following along similar lines to Example 11.1, the CCF is found to have a peak at lag zero, but to exhibit strong cross-correlations at positive and negative lags. Since the true relationship between x and y is purely contemporaneous, these spurious cross-correlations impair our ability to model the processes and some adjustments must be made.

Example 11.4   We generated 100 observations for each of the independent processes

$$y_t = y_{t-1} + \varepsilon_t$$
$$x_t = x_{t-1} + \delta_t.$$

The CCF is plotted in Fig. 11.7(a), that for y on ∇x in Fig. 11.7(b) and that for ∇y on ∇x in Fig. 11.7(c). The CCF for y on x shows a strong but spurious cross-correlation at all lags, whereas Fig. 11.7(b) shows evidence of non-

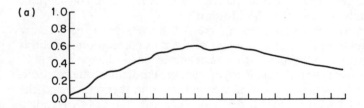

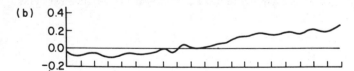

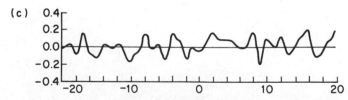

Fig. 11.7   CCF for two independent white noise series (a) Y and X; (b) Y and ∇X; (c) ∇Y and ∇X

stationarity in the very slow decay. As expected, Fig. 11.7(c) indicates that there is no serial correlation.

With these examples to guide us, we now develop a general approach to modelling interrelationships. First we develop an appropriate class of models and then consider procedures for model identification.

Transfer functions

11.10 A natural extension of the simple regression model

$$y_t = \alpha + \beta x_t + \varepsilon_t \tag{11.23}$$

is to include all possible lags for the explanatory variable:

$$y_t = a + \beta_0 x_t + \beta_1 x_{t-1} + \cdots + \beta_k x_{t-k} + \varepsilon_t. \tag{11.24}$$

Ignoring the error term for the moment, the set of coefficients $\{\beta_0, \beta_1, ..., \beta_k\}$ are termed the *impulse response function* since the set of values

$$x_t = 0, \quad t \neq 0, \quad x_0 = 1$$

produces the set of values for y_t (with $\alpha = 0$): $y_0 = \beta_0$, $y_1 = \beta_1$, $y_2 = \beta_2$ and so on. That is, these y-values represent the y-response over time to a unit impulse or change in x at time zero. Again, if we had the sequence of inputs

$$x_t = 0, \quad t < 0, \quad x_t = 1, \quad t \geq 0$$

the y-values would be

$$y_0 = \beta_0, \ y_1 = \beta_0 + \beta_1, ..., y_t = \beta_0 + \beta_1 + \cdots + \beta_k, \qquad t \geq k.$$

The quantity

$$G = \beta_0 + \beta_1 + \cdots + \beta_k \tag{11.25}$$

is termed the *gain*; that is, the total change in y for a unit change in x. The general term for models such as (11.24) is (time-domain) *transfer functions*. Note that the terminology is similar to that used in our discussion of cross-spectra in Section 11.6, but we are now working in the time domain.

11.11 Suppose now that the error terms are negligible, so that we may concentrate upon the form

$$y_t = \alpha + \beta_0 x_t + \beta_1 x_{t-1} + \cdots + \beta_k x_{t-k}. \tag{11.26}$$

If x_t is a white noise process with variance $\gamma_{xx}(0) = \sigma_x^2$ it follows directly that the covariance between y_t and x_{t-j} is

$$\gamma_{xy}(j) = \beta_j \gamma_{xx}(0)$$

or, for the cross-correlation,

$$\rho_{xy}(j) = \beta_j \sigma_x / \sigma_y, \tag{11.27}$$

where $\text{var}(y) = \sigma_y^2$. That is, when x is a white noise process, the cross-correlations have a clear interpretation as being proportional to the terms of the impulse response function.

However, when the x_t have autocovariances $\text{cov}(x_t, x_{t-j}) = \gamma_{xx}(j)$, the set of

Yule–Walker-like equations resulting from (11.26) are of the form:

$$\left.\begin{aligned}
\gamma_{xy}(0) &= \beta_0\gamma_{xx}(0) + \beta_1\gamma_{xx}(1) + \cdots + \beta_k\gamma_{xx}(k) \\
\gamma_{xy}(1) &= \beta_0\gamma_{xx}(1) + \beta_1\gamma_{xx}(0) + \cdots + \beta_k\gamma_{xx}(k-1) \\
&\cdots\cdots\cdots\cdots\cdots\cdots\cdots\cdots\cdots\cdots\cdots\cdots\cdots\cdots\cdots \\
\gamma_{xy}(k) &= \beta_0\gamma_{xx}(k) + \beta_1\gamma_{xx}(k-1) + \cdots + \beta_k\gamma_{xx}(0)
\end{aligned}\right\}. \tag{11.28}$$

Evidently, the relationship between the *CCF* and the impulse response function is now anything but obvious. Simplicity would be restored, however, if the x_t could be transformed to a white noise process, and we now describe this form of *pre-whitening*.

11.12   Suppose that x_t may be described by an *ARMA* scheme

$$\phi(B)x_t = \theta(B)u_t; \tag{11.29}$$

for convenience, we assume that the constant term is zero and assume any differencing has already been performed on X_t such that $x_t = \nabla^d X_t$ is a stationary process. From (11.29), if follows that

$$u_t = \frac{\phi(B)}{\theta(B)} x_t \tag{11.30}$$

is a white noise process. Rewriting (11.24), with $\alpha = 0$ for convenience, as

$$y_t = \beta(B)x_t \tag{11.31}$$

we may multiply both sides of (11.31) by $\phi(B)/\theta(B)$ to obtain

$$v_t = \frac{\phi(B)}{\theta(B)} y_t = \frac{\phi(B)}{\theta(B)} \beta(B)x_t$$

$$= \beta(B)u_t. \tag{11.32}$$

The *CCF* for v and u will now give the terms of the impulse response function just as we obtained in (11.27) for a white noise input process. That is, in order to obtain estimates of the β_j, we compute the sample *CCF* after pre-whitening.

Form of the transfer function

11.13   The original expression (11.26) may involve a large number of parameters and it behoves us to find more parsimonious representations that would provide better estimates. A natural extension, corresponding to the autoregressive moving-average structure for univariate series, is to formulate the transfer function as

$$y_t = \frac{\omega(B)}{\delta(B)} x_{t-b}. \tag{11.33}$$

or

$$\beta(B) = B^b\omega(B)/\delta(B), \tag{11.34}$$

where

$$\omega(B) = \omega_0 - \omega_1 B - \cdots - \omega_l B^l, \tag{11.35}$$

$$\delta(B) = 1 - \delta_1 B - \cdots - \delta_r B^r, \tag{11.36}$$

Transfer function   Impulse response function

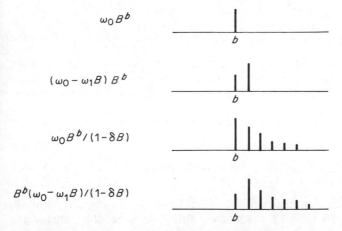

$\omega_0 B^b$

$(\omega_0 - \omega_1 B)\, B^b$

$\omega_0 B^b / (1 - \delta B)$

$B^b(\omega_0 - \omega_1 B)/(1 - \delta B)$

Fig. 11.8 Typical forms of transfer function with corresponding impulse response functions. (Note that for the examples given $\omega_0 > 0$, $\omega_1 < 0$ and $\delta > 0$)

and b denotes the *delay* before x has an impact on y. The alternative version

$$\omega(B) = \omega_0(1 - \omega_1^* B - \cdots - \omega_l^* B^l) \qquad (11.37)$$

is sometimes preferred as it is rather easier to interpret. Some simple examples of different impulse response functions are given in Fig. 11.8

Example 11.5   Suppose that

$$y_t = 2\,\frac{(1 + 0.5B)}{(1 - 0.4B)}\, x_{t-1} \qquad (11.38)$$

and

$$x_t = 1, \qquad t = 0,$$
$$= 0, \qquad t \neq 0.$$

Equation (11.38) may be re-expressed as

$$(1 - 0.4B)y_t = 2(1 + 0.5B)x_{t-1}$$

or

$$y_t = 0.4y_{t-1} + 2x_{t-1} + x_{t-2}.$$

Thus, y_t takes on values as shown in Table 11.3. Similarly if $x_t = 1$, $t \geq 0$, $x_t = 0$, $t < 0$, we have the values in Table 11.4.

Since the value of x is unchanged for $t \geq 1$, the limiting value of y is given by

$$y_\infty = \beta(1)x_\infty$$

Table 11.3

t	$\leqslant 0$	1	2	3	4	5	...	$t \to \infty$
y_t	0	2	1.8	0.72	0.29	0.12	...	0

Table 11.4

t	$\leqslant 0$	1	2	3	4	5	...	$t \to \infty$
y_t	0	2	3.8	4.52	4.81	4.92	...	5

or

$$y_\infty = \{2(1 + 0.5)/(1 - 0.4)\}x_\infty$$

in our example.

11.14 Our model-selection process, therefore, involves two steps. First of all, we develop an *ARIMA* model for x and use this to generate the transformed series u and v. We then compute the sample *CCF* for u and v in order to identify an impulse response function.

A key assumption we have made here is that the error terms can be ignored, at least to start with. From an empirical viewpoint, this is reasonable since the influence of the x-variable should be sufficiently strong to stand out over and above the error process. If the effect of x is completely masked by the errors, it is presumably not very important. Once the initial impulse response function, $\hat{\beta}(B)$, has been estimated, we may estimate the error process as

$$\hat{a}_t = y_t - \hat{\beta}(B)x_t. \tag{11.39}$$

The *ARIMA* structure for the error process

$$\phi(B)a_t = \theta(B)\varepsilon_t \tag{11.40}$$

may then be estimated using the methods developed in Chapter 6. That is, we finish up by identifying a complete model of the form

$$y_t = \frac{\omega(B)}{\delta(B)}\, x_t + \frac{\theta(B)}{\phi(B)}\, \varepsilon_t. \tag{11.41}$$

Linear systems

11.15 Expression (11.41) is an example of a linear system which may be summarised as in Fig. 11.9. This is an *open* system, since x has an impact on y, but y does not feed back to act upon x. In a closed system the chain $x \to y \to x$ would be complete, with suitable time delays. *The identification process we have described depends upon the system being open*; unfortunately, many economic systems involve feedback loops, which precludes the effective use of a transfer-function model. In such circumstances, we must turn to a multivariate system; we defer discussion of this until Chapter 14. However, there are many circumstances when an open-loop model is appropriate and we now continue our investigation of transfer functions with some examples.

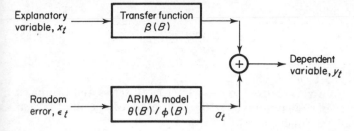

Fig. 11.9 Linear systems representation of the transfer function model

Example 11.6 A sample of $n = 100$ was generated from the model

$$y_t = 10 + 3x_t + 2x_{t-2} + a_t$$
$$x_t = x_{t-1} + \delta_t - 0.6\delta_{t-1}$$
$$a_t = \varepsilon_t - 0.4\varepsilon_{t-1},$$

where ε_t and δ_t are $IIN(0, 1)$.

The sample *CCF* for x_t and y_t without pre-whitening is shown in Fig. 11.10a. It is clear that little can be determined about the form of the model from this diagram; we proceed to follow the steps described in Section 11.12.

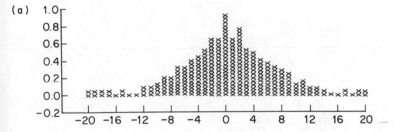

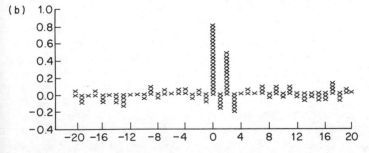

Fig. 11.10 CCfs for simulated series: (a) before prewhitening; (b) after pre-whitening

Step 1:   The fitted model for x_t was

$$\nabla x_t = \delta_t - 0.671\delta_{t-1} \qquad (11.42)$$
$$(8.91)$$

with $Q(11) = 5.1$, $\hat{\sigma} = 0.96$.

Step 2:   Model (11.42) was then used to pre-whiten both series and produced the sample *CCF* shown in Fig. 11.10(b). From this diagram, it is clear that the transfer function is of the form

$$y_t = (\omega_0 - \omega_2 B^2)x_t. \qquad (11.43)$$

Step 3:   The initial residuals from (11.43) were then examined and produced the serial correlations given in Fig. 11.11. From these results, either $AR(1)$ or $MA(1)$ seems appropriate.

The initial fitted model and resulting changes will be presented in Section 12.10 after we have discussed estimation and diagnostic checking.

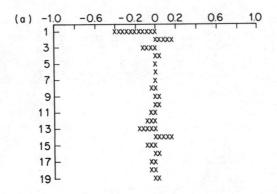

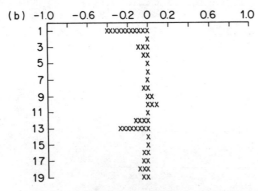

Fig. 11.11   (a) SACF and (b) SPACF for residuals from simulated series fitting preliminary model

Cars and the FT index

11.16 We now turn to the more challenging task of applying these procedures to real data. We shall try to identify a transfer-function model for the cars series given in Table 11.2, using the FT index as an explanatory variable. Recall that the CCF obtained after de-trending, given in Fig. 11.3, did not allow a clear interpretation.

Step 1: From (7.20), the pre-whitening model for the FT index is

$$\nabla x_t = \delta_t + 0.256\delta_{t-1}. \tag{11.44}$$

Model (7.20) was used in preference to (7.33) since the differences are not large and (7.20) preserves eight more observations.

Step 2: The pre-whitened series produced the sample *CCF* shown in Fig. 11.12. From this diagram we tentatively identify the transfer function component as

$$y_t = (\omega_0 - \omega_1 B - \omega_2 B^2)x_t. \tag{11.45}$$

Step 3: The *SACF* and *SPACF* in Fig. 11.13 suggest either an *AR*(1) or an *MA*(1) structure.

This model will be used as the starting point for our model development process in Section 12.12. It is, however, interesting to note that as a result of the pre-whitening process we appear to have come up with a much simpler relationship than looked likely from our initial analysis in Section 11.8. Also, it should be noted that (11.45) does not involve differencing either series, although both series originally displayed non-stationarity; such series are called *co-integrated*; see Granger (1981) and Hendry (1986).

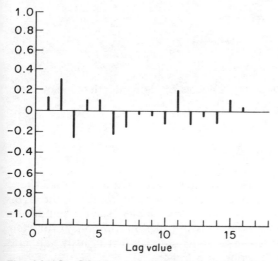

Fig. 11.12 CCF for car production of FT index after prewhitening (positive lags only)

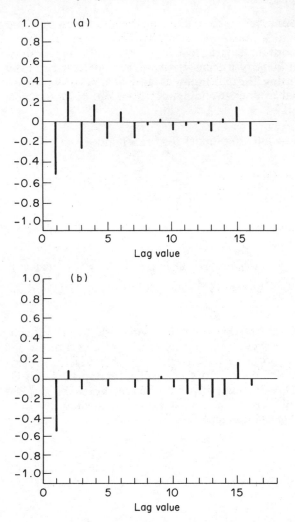

Fig. 11.13   (a) SACF and (b) SPACF for residuals of car production series

Several explanatory variables

11.17   Our identification process was predicated on the assumption that there was a single input variable whose impact was sufficient to dominate the error term. When there are two or more explanatory variables, such assumptions cannot be made. If we try to identify the transfer function component for each variable in turn, we risk getting incorrect results unless it fortuitously happens that these variables are independent one of another.

There are three ways in which we may proceed. One is to develop a common filter for all inputs (Liu and Hanssens, 1982); the second is to find a separate pre-whitening filter for each series and then to apply all the filters to each series (Box and Jenkins, 1976). A third possibility is to filter each series separately (Box and Haugh, 1976).

None of these methods is wholly satisfactory in theoretical terms, but all seem to give reasonable results in practice unless there is a high degree of collinearity between the variables when even straightforward multiple regression runs into difficulties.

Seasonal models

11.18 The extension to seasonal models is straightforward, at least in principle. We may specify the transfer function component as

$$y_t = \frac{\omega(B)\Omega(B^s)}{\delta(B)\Delta(B^s)} x_{t-b} + a_t, \tag{11.46}$$

and the noise component as

$$a_t = \frac{\theta(B)\Theta(B^S)}{\phi(B)\Phi(B^S)} \varepsilon_t, \tag{11.47}$$

where the seasonal polynomials are of the form

$$\Omega(B^S) = 1 - \Omega_1 B^S - \cdots - \Omega_L B^{SL} \tag{11.48}$$

and so on, Δ being of order R, Θ of order Q, and Φ of order P. The pre-whitening and model selection procedures operate in exactly the same manner as before.

Comparison with other approaches

11.19 Model (11.41) clearly reduces to simple linear regression when $\omega(B) = \delta(B) = 1$ and $\theta(B) = \phi(B) = 1$; recall that the constant term was omitted for convenience. However, a variety of other schemes proposed in the literature arise as special cases, as follows:

regression with lagged dependent variables

$$\delta(B)y_t = \omega(B)x_t + \varepsilon_t;$$

distributed lag models (Almon, 1965):

$$y_t = \frac{\omega(B)}{\delta(B)} x_t + \varepsilon_t;$$

regression with autocorrelated residuals:

$$y_t = \omega(B)x_t + \frac{1}{\phi(B)} \varepsilon_t.$$

11.20 Given the general form of the transfer function model and an identification procedure, we are now ready to consider the estimation and diagnostic phases of our model-building paradigm. These are the topics covered in the next chapter.

Exercises

11.1   If

$$y_t = \varepsilon_t + \varepsilon_{t-1} + \varepsilon_{t-2} + \varepsilon_{t-3}$$
$$x_t = \varepsilon_t - 2\varepsilon_{t-1} + 2\varepsilon_{t-2},$$

where the ε are independent and identically distributed with mean zero and variance one, find the form of the *CCF*.

11.2   A company reports the advertising and sales figures shown in Table 11.5 for 12 successive weeks of operations. Find the sample *CCF* for lags $k = 0, \pm 1, ..., \pm 4$.

Table 11.5

Week:	1	2	3	4	5	6	7	8	9	10	11	12
Sales:	100	110	140	180	140	90	90	50	70	70	90	160
Advertising:	100	110	150	150	120	100	80	80	80	100	150	150

11.3 x_t and y_t are uncorrelated $MA(q)$ processes with the same parameters $\theta_1, ..., \theta_q$, find the appropriate correlations between the cross-correlations using (11.9) and (11.10).

11.4   Find the theoretical *CCF* for

$$y_t = x_t + x_{t-1} + \varepsilon_t$$
$$x_t = \delta_t - 0.8\delta_{t-1},$$

where ε and δ are independent with means zero and variances one. Then use the *x*-model to pre-whiten the *y*-series and recompute the *CCF*.

11.5   Generate sample series of length $n = 100$ from the scheme

$$y_t = \phi y_{t-1} + \varepsilon_t$$
$$x_t = \phi x_{t-1} + \delta_t$$

where ε_t and δ_t are independent and identically distributed, $N(0, 1)$. Compare the sample and theoretical *CCFs*. Take $\phi = 0.8$ initially and then try other vlaues.

11.6   Generate sample series of length $n = 100$ from the model described in Exercise 11.4. Compare the sample and theoretical *CCF* before and after pre-whitening. Repeat with $n = 50$ and $n = 25$. Comment on your results.

11.7   Assume that

$$y_t = 0.7y_{t-1} + 2x_t + x_{t-2}$$

and

$$x_t = 1 \quad t \geqslant 0; \quad x_t = 0, \quad t < 0.$$

Find y_t, $t = 1, ..., 5$ and the limiting value of y. Take $y_t = 0$, $t < 0$. Repeat for

$$x_t = 1, t = 0; x_t = 0, t \neq 0.$$

11.8 Estimate (a) the cross-spectrum and coherence and (b) the *CCF* for one or more of the pairs of series listed below:

(a) Lydia Pinkham sales and advertising (Table A13);
(b) Hog supply and prices (Table A14).

11.9 Identify a transfer-function model for one of the pairs of series listed in Exercise 11.8. How well do you think the assumptions are met in this case?

12

Transfer functions: estimation, diagnostics and forecasting

Introduction

12.1 In the previous chapter, we developed transfer-function models and preliminary model-selection procedures. In order to complete the model-building process, we now consider the remaining steps: estimation, diagnostic testing and forecasting. Thus, in large measure, we shall be following the path charted out in Chapters 7 and 8 for univariate processes, and the reader will see that many of the procedures described here are direct extensions of that earlier work.

Estimation procedures are considered in Section 12.2–4 and diagnostic tests in Sections 12.5–12.13. Forecasting is considered in Sections 12.14–19 and a general discussion on time series regression follows in Sections 12.20–26. Automatic modelling procedures are discussed briefly in Sections 12.27–28 and the chapter concludes with an evaluation of the different approaches to modelling and forecasting.

Estimation procedures

12.2 We shall assume that the identification procedures have given rise to the selection of a particular model which we write as

$$y_t = \frac{\omega(B)}{\delta(B)} x_{t-b} + a_t \tag{12.1}$$

$$a_t = \frac{\theta(B)}{\phi(B)} \varepsilon_t \tag{12.2}$$

using the same notation as in (11.41), where the variables have been differenced as often as necessary. As for the univariate *ARIMA* scheme we may develop a least-squares procedure by minimising

$$S(\hat{\omega}, \hat{\delta}, \hat{\theta}, \hat{\phi}) = \sum_{t=1}^{n} e_t^2, \tag{12.3}$$

where the e_t represent the estimated ε_t values, $\hat{\omega}$ is the vector of parameter estimates corresponding to the polynomial function $\omega(B)$, and so on. The sum (12.3) may be minimised by iterative search. That is, given initial estimates of the parameters we may generate $\hat{y}_t$ and then the e_t as linear functions of the parameters, given $\{\hat{y}_t\}$. The estimated form (12.3) is then minimised to produce a new set of estimates, hence new $\hat{y}_t$ and so on.

12.3 We may start the process by considering, from (12.1),

$$\hat{\delta}(B)\hat{y}_t = \hat{\omega}(B)x_{t-b} \tag{12.4}$$

or

$$\hat{y}_t = \hat{\delta}_1\hat{y}_{t-1} + \cdots + \hat{\delta}_r\hat{y}_{t-r} + \hat{\omega}_0 x_{t-b} - \hat{\omega}_1 x_{t-b-1} - \cdots - \hat{\omega}_l x_{t-b-l}. \tag{12.5}$$

If we set $u = \max(r, b + l)$, it is evident that (12.4) can be applied to known values $(y_{u-1}, \ldots, y_{u-r}; x_{u-b}, \ldots, x_{u-b-l})$ to generate $\hat{y}_{u+1}$. Thereafter $\hat{y}_{u+2}$, $\hat{y}_{u+3}, \ldots, \hat{y}_n$ can be obtained successively from (12.4). Then, for $t = u + 1$ to n,

$$\hat{a}_t = y_t - \hat{y}_t. \tag{12.6}$$

We now turn to (12.2) and consider

$$\hat{\theta}(B)e_t = \hat{\phi}(B)a_t$$

or

$$e_t = \hat{\theta}_1 e_{t-1} + \cdots + \hat{\theta}_q e_{t-q} + \hat{a}_t - \hat{\phi}_1 \hat{a}_{t-1} - \cdots - \hat{\phi}_p \hat{a}_{t-p}. \tag{12.7}$$

Since the $\hat{a}_t$ are available only for $t \geqslant u + 1$, the first e_t that may be computed would occur at $t = u + p + 1$, provided the earlier e_t were known. We may avoid this problem by setting these e_t equal to their expected values, namely $E(e_t) = 0$, for $t \leqslant u + p$.

Thus, if we restrict the summation in (12.3) to the range $[u + p + 1, n]$, we can find the estimates by solving iteratively. This is known as the method of conditional least squares (*CLS*) since it proceeds by conditioning on the values $e_1 = \cdots = e_{u+p} = 0$. *CLS* is generally the easiest approach to implement, but may not work too well when either the series is too short or the true parameters are near the boundary of the parameter space.

12.4 More complicated estimation procedures lie beyond the scope of our discussions. However, the principles underlying such techniques are readily explained. One possibility is to use unconditional least squares (ULS) where the distribution of e_t, $t \leqslant u + p$ is specified and used to obtain modified variance expressions for the e_t with $t > u + p$. The method of maximum likelihood may be employed in various approximate and exact versions; see Hillmer and Tiao (1979) for further details. As in the univariate case, the Kalman filter offers a flexible method for developing efficient estimation procedures; see Harvey and Phillips (1979).

Diagnostic procedures

12.5 We now assume that a model has been selected and fitted and turn to the question of diagnostic checks. We must be able to address three main issues:

(a)   Are there any unusual observations?
(b)   Is the transfer function correctly specified?
(c)   Is the error process correctly specified?

Each of these topics in turn raises considerations. For example, unusual observations may indicate the need for a transformation, that an explanatory variable has been omitted, that the model is incorrectly specified or that the observation is an outlier. A full consideration of all these possibilities leads us into the realm of regression diagnostics; see, for example, Cook and Weisberg (1982). To keep the discussion within reasonable bounds, we shall assume that no other variables need be considered and that the Box–Cox procedure discussed in Section 5.4 provides a satisfactory answer to the question of transformations.

Under (b) and (c), we must check whether the order and degree of differencing are appropriate, as well as whether existing terms should be deleted or new terms added to each component.

Many of these diagnostics follow directly from our earlier discussions in Sections 7.12–20 and Chapter 11, as follows:

> unusual observations: plot residuals
> transfer function – underspecified: plot of residual CCF, portmanteau
> test on CCF
> – overspecified: tests on individual coefficients
> error process – underspecified: plots of residual ACF and $PACF$,
> portmanteau test on ACF
> – overspecified: tests on individual coefficients.

We now examine these in turn. To illustrate the procedures, we consider the pair of simulated series introduced in Example 11.6. To add a little excitement to the discussion, we start with the naive model

$$y_t = \alpha + \omega_0 x_t + \varepsilon_t \tag{12.8}$$

which is clearly deficient in terms of both the generating process and the model tentatively identified in Example 11.6. The initial estimates are, with t-ratios shown underneath,

$$\hat{\alpha} = 8.66, \qquad \hat{\omega}_0 = 4.33 \qquad \text{with} \quad s = 2.36 \quad \text{and} \quad R^2 = 0.868. \tag{12.9}$$
$$(19.5) \qquad\qquad (25.4)$$

12.6   A plot of the residuals showed large values (standardised residuals above two standard deviations in absolute value) at $t = 2, 5, 61$ and 93, but none of these exceeded 2.5 standard deviations. Such occasional extremes are in accord with expectations and no action need be taken. The plot is omitted since no features of interest emerge.

The reader may feel that this treatment of outliers is rather cursory, as indeed it is. We shall return to this issue in Section 13.13.

Checking the transfer function

12.7 The check for excess parameters is readily carried out by examining the t-values for each coefficient separately. The overall fit of the model may be examined using an *approximate* F-test of the form

$$F = \frac{R^2/n}{(1 - R^2)/(n - m - 1)} \qquad (12.10)$$

where m = total number of parameters fitted other than the overall mean, and the degrees of freedom are $(m, n - m - 1)$ as usual. When the series is differenced, R_d^2 should be used in place of R^2. Strictly speaking, all we can say is that F in (12.10) is asymptotically distributed as chi-squared with m degrees of freedom, but the more familiar and more conservative F-test is generally preferred.

Because of the tendency to try a variety of different models, nominal significance levels from the F-tables will usually underestimate the probability of a type I error. However, if the F-value is not 'significant' at conventional levels, we may certainly conclude that the model is not useful.

These overall tests have a limited role to play in time-series analysis, but they do help to avoid over-elaborate model building.

In our example, F reduces to t^2 and the single coefficient is highly significant.

12.8 To determine whether we have missed anything, we examine the cross-correlation function of the residuals. Following the discussion in Section 11.12, note that the *CCF* is computed between the residuals (e_t) and x-values (u_t), after pre-whitening both series by the *ARIMA* model for x. Potential additions to the model may be identified as in Fig. 11.8. In addition, a portmanteau statistic, due to Pierce (1972), may be calculated. This is of essentially the same form as the test statistic for the sample *ACF* given in (7.23); that is,

$$Q(K + 1) = M(M + 2) \sum_{k=0}^{K} r_{eu}^2(k)/(M - k), \qquad (12.11)$$

where M denotes the number of residuals used, cf. (12.7). Thus, $M = n - u - p$, where $u = \max(l, r + b)$ and p is the order of the *AR* component of the error process. Further, under the null hypothesis that the model is correct, Q is asymptotically distributed as chi-squared with $(K + 1 - g)$ degrees of freedom, where g represents the number of fitted parameters in the transfer function model. The asymptotic distribution is not affected by fitting an *ARMA* component.

Example 12.1 For the simulated series with model (12.8):

$$n = 100, \quad p = 0, r = 0, l = 0, \quad b = 0 \quad \text{and} \quad g = 1 \qquad (12.12)$$

so that $M = 100$. The first 8 cross-correlations, with their standard errors in brackets, are as shown in Table 12.1.

Hence $Q(7) = 70.3$ with $D.F. = 8$, which is clearly significant. Inspection of the cross-correlations indicates spikes at lags 0 and 2, suggesting the need for a

Table 12.1

k	0	1	2	3	4	5	6	7
r_{xy}	−0.495	−0.189	0.622	0.010	−0.061	0.030	0.005	0.008
	(0.102)	(0.102)	(0.103)	(0.103)	(0.104)	(0.104)	(0.105)	(0.105)

term in x at lag 2. The spike at lag 0 reflects the extent to which the incorrect specification of the initial model has led to a biased estimate of ω_0.

Checking the error structure

12.9 The checks for the error structure are essentially the same as those for univariate *ARIMA* schemes:

(a) check individual coefficients using t-values;
(b) plot the residual *ACF* and *PACF*, and other plots as described in Chapter 7;
(c) consider the portmanteau statistic

$$Q(K) = M(M+2) \sum_{k=1}^{K} r_{ee}^2(k)/(M-k), \qquad (12.13)$$

where D.F. $= K - p - q$, as in (7.23). Fitting a transfer function component does not affect the asymptotic distribution of Q. As before, $M = n - u - p$.

Example 12.2 The first eight terms of the sample *ACF* and *PACF* are as shown in Table 12.2.

Now $M = 100$ and D.F. $= 8$; $Q(8) = 10.0$, suggesting no real need for an *ARMA* noise structure. However, examination of the individual terms in the *ACF* and *PACF* suggests either an $AR(2)$ or an $MA(2)$ scheme. The $SE \approx 0.01$ for each estimate and the second term is significantly different from zero in each case. Such conflicting results are not uncommon and alternative models should be explored.

12.10 Overall, our diagnostics suggest the revised models:

$$\text{M1:} \qquad y_t = \alpha + (\omega_0 - \omega_2 B^2)x_t + \frac{1}{(1 - \phi_1 B - \phi_2 B^2)}\varepsilon_t,$$

$$\text{M2:} \qquad y_t = \alpha + (\omega_0 - \omega_2 B^2)x_t + (1 - \theta_1 B - \theta_2 B^2)\varepsilon_t.$$

Both M1 and M2 are close to the scheme originally identified in Section 11.15.

Table 12.2

k	1	2	3	4	5	6	7	8
r	−0.011	−0.229	0.076	0.130	0.119	0.000	0.035	0.065
$\hat{\phi}$	−0.011	−0.229	0.074	0.084	0.163	0.048	0.087	0.047

In this case, use of a naive starting model cost only one round in the iterative model search process but the penalty may be much greater; for example, a poorly selected starting model may not fit at all because the estimation procedure fails to converge or else it may yield estimates outside the stationary region.

When models M1 and M2 are fitted, it transpires that $\hat{\phi}_2$ and $\hat{\theta}_2$, respectively, are not significant, but all other estimates do differ significantly from zero. Therefore, the revised M1′ and M2′ are reported with $AR(1)$ and $MA(1)$ error structures. It is quite common for an underspecified model like (12.8) to generate conflicting diagnostics if both the transfer function and the error process are incorrect. No harm is done, it just takes a little longer to arrive at an appropriate model.

The revised estimates (with t ratios) and the original process parameters are as shown in Table 12.3. Both models appear equally satisfactory, although the data were generated from M2′.

12.11   There are, inevitably, some gaps in the set of diagnostics just presented. Most importantly, we have not included any checks for stationarity. As for the univariate case, such checks should involve the following:

(a)   plotting the series (to look for trends or marked seasonal patterns);
(b)   computing the *ACF* and *PACF* for several degrees of differencing if non-stationary is suspected;
(c)   performing the test for unit roots described in Section 6.19;
(d)   checking residuals by these methods.

These steps were omitted for the simulated series since they were known to be well-behaved. Such checks should always be carried out for real data.

The other check which should be performed is to see that the polynomials $\phi(B)$ and $\delta(B)$ correspond to stationary processes; that is, their roots must lie outside the unit circle.

Table 12.3

		M1′ ($M = 97$)	M2′ ($M = 98$)	Process
Estimates	$\hat{\alpha}$	9.80	9.83	10.0
	$\hat{\omega}_0$	3.07(35.0)	3.10(34.3)	3.0
	$\hat{\omega}_2$	-1.88(21.4)	-1.86(-20.2)	-2.0
	$\hat{\phi}_1$	-0.426(-4.6)	—	—
	$\hat{\theta}_1$	—	-0.449(4.9)	0.4
	s	1.04	1.05	—
	R^2	0.972	0.973	—
Diagnostics	$Q_c(6)$	8.56	8.23	—
	$Q_A(7)$	2.99	3.17	—

Cars and the FT Index

12.12   We now illustrate the transfer-function modelling approach on the time series for car sales and the FT Index previously considered in

Section 11.17. The pre-whitening model was

$$\nabla x_t = \delta_t + 0.256\delta_{t-1}$$

and the model identification process led to the suggested models:

$$\text{M1:} \qquad y_t = \alpha + (\omega_0 - \omega_1 B)x_t + (1 - \theta_1 B)\varepsilon_t$$

and

$$\text{M2:} \qquad y_t = \alpha + (\omega_0 - \omega_1 B)x_t + (1 - \phi_1 B)^{-1}\varepsilon_t.$$

The estimated models (using the non-automatic form of AUTOBOX) are, with *t*-values in brackets:

$$\text{M1:} \qquad y_t = 144.6 + (0.508 + 0.195B)x_t + (1 + 0.483B)\varepsilon_t,$$
$$\qquad\qquad (1.74) \quad (-0.65) \qquad (-3.65)$$
$$s = 47.2, \; R^2 = 0.54$$

$$\text{M2:} \qquad y_t = 154.1 + (0.419 + 0.257B)x_t + (1 - 0.748B)^{-1}\varepsilon_t$$
$$\qquad\qquad (1.69) \quad (-1.03) \qquad (7.20)$$
$$s = 39.2, \qquad R^2 = 0.69.$$

12.13   The sample *ACF*, *PACF* and *CCF* for each model are summarised in Table 12.4. The portmanteau statistics are

$$\text{M1}(M = 47): \quad Q_c(6) = 6.38, \quad Q_A(7) = 33.59$$
$$\text{M2}(M = 46): \quad Q_c(6) = 3.03, \quad Q_A(7) = 6.37.$$

These results suggest the revised model:

$$\text{M3:} \qquad y_t = \alpha + \omega_0 x_t + (1 - \phi_1 B)^{-1}\varepsilon_t$$

which is estimated as

$$\text{M3:} \qquad y_t = 207.6 + 0.527 x_t + (1 - 0.744B)^{-1}\varepsilon_t,$$
$$\qquad\qquad (2.38) \qquad\quad (7.26)$$
$$s = 38.8, \qquad R^2 = 0.68.$$

The sample *ACF*, *PACF* and *CCF* are given in Table 12.4; they do not reveal any unusual features, a fact confirmed by the portmanteau statistics:

$$\text{M3}(M = 47): \qquad Q_c(7) = 5.25, \; Q_A(7) = 8.38$$

Thus, we may conclude that the FT index provides some guidance on movements of car production, but mainly from the current index. The last entry in Table 12.4, giving the *CCF* for cars leading the FT index, shows no evidence of causality in the reverse direction.

Forecasting

12.14   Once the cycle of diagnosis and model reformulation is complete, the model is available for forecasting. Although formulated as

$$y_t = \alpha + \frac{\omega(B)}{\delta(B)} x_{t-b} + \frac{\theta(B)}{\phi(B)} \varepsilon_t, \qquad\qquad (12.14)$$

Table 12.4

	Lag								
	0	1	2	3	4	5	6	7	8
Model M1									
ACF	—	0.219	0.590	0.210	0.352	0.110	0.162	−0.024	−0.084
(SE)	—	(0.146)	(0.158)	(0.199)	(0.204)	(0.216)	(0.218)	(0.220)	(0.220)
PACF	—	0.291	0.552	−0.044	0.005	−0.056	−0.061	−0.110	−0.208
(SE)	—	(0.146)	(0.146)	(0.146)	(0.146)	(0.146)	(0.146)	(0.146)	(0.146)
CCF	−0.043	−0.068	−0.126	−0.099	−0.139	−0.161	−0.159	−0.122	—
(SE)	(0.149)	(0.151)	(0.152)	(0.154)	(0.156)	(0.158)	(0.160)	(0.162)	—
Model M2									
ACF	—	−0.141	0.213	−0.072	0.149	−0.025	0.124	−0.101	−0.062
(SE)	—	(0.147)	(0.150)	(0.157)	(0.157)	(0.160)	(0.161)	(0.163)	(0.164)
PACF	—	−0.141	0.197	−0.021	0.103	0.021	0.080	−0.075	−0.141
(SE)	—	(0.147)	(0.147)	(0.147)	(0.147)	(0.147)	(0.147)	(0.147)	(0.147)
CCF	−0.050	−0.066	−0.144	−0.069	−0.083	−0.107	−0.086	−0.019	—
(SE)	(0.151)	(0.152)	(0.154)	(0.156)	(0.158)	(0.160)	(0.162)	(0.164)	—
Model M3									
ACF	—	−0.123	0.158	−0.045	0.087	−0.042	0.161	−0.075	−0.118
(SE)	—	(0.152)	(0.155)	(0.158)	(0.159)	(0.160)	(0.160)	(0.164)	(0.165)
PACF	—	−0.123	0.145	−0.011	0.060	−0.020	0.138	−0.036	−0.185
(SE)	—	(0.152)	(0.152)	(0.152)	(0.152)	(0.152)	(0.152)	(0.152)	(0.152)
CCF	−0.078	0.226	−0.197	−0.007	0.171	−0.044	−0.171	−0.062	—
(SE)	(0.156)	(0.158)	(0.160)	(0.162)	(0.164)	(0.167)	(0.169)	(0.171)	—
CCF*	−0.078	0.054	0.060	0.008	−0.062	0.163	−0.074	−0.051	—

*For cars leading FT Index; *SE*s are as for regular *CCF*.

we cross-multiply to arrive at

$$\delta(B)\phi(B)y_t = \alpha^* + \omega(B)\phi(B)x_{t-b} + \theta(B)\delta(B)\varepsilon_t, \qquad (12.15)$$

where

$$\alpha^* = \alpha\delta(1)\phi(1),$$

so that only a finite number of past values is required. If we are forecasting from the origin n, we follow our conventions in Chapter 8 and set

$$
\begin{aligned}
y_n(j) &= y_n(j) &&\text{if } j > 0 \\
&= y_{n+j} &&\text{if } j \leqslant 0 \\
\varepsilon_n(j) &= 0 &&\text{if } j > 0 \\
&= \varepsilon_{n+j} &&\text{if } j \leqslant 0
\end{aligned}
$$

Clearly, when $j \leqslant 0$, we also set

$$x_n(j) = x_{n+j};$$

however, two possibilities arise for x when $j > 0$.

x-known: The set of future x-values may be known. This may arise because the forecasting exercise is based on historical data and the relevant observations were withheld at the fitting stage or because x is a policy variable which can be specified by the investigator. Known x-values may also be deemed to arise in 'what-if' forecasting, where it is desired to determine the effect of certain x-patterns upon y. This last possibility is particularly useful in assessing different policy options even though the forecasts are not testable in that the particular x-sequence may never occur. In general, when the x-values are known, we term this *ex-post* forecasting.

x-unknown: In 'pure' forecasting applications, it is more common for the x-values to be unknown. In accordance with the open-loop model formulated in Chapter 11 we may estimate the values for x using its *ARIMA* model and substitute these estimates, $x_n(j)$. The resulting forecasts for y are known as the *ex-ante* forecasts.

12.15 A point worth making but often overlooked is that if x is truly representable as an *ARIMA* process, at least in principle, an identified univariate model for y will perform as well as the ex-ante forecasts and, indeed, may be equivalent. To see this, consider the scheme

$$y_t = \nu(B)x_t + \psi(B)\varepsilon_t \qquad (12.16)$$

$$x_t = \psi_x(B)\delta_t. \qquad (12.17)$$

It follows directly that

$$y_t = \nu^*(B)a_t + \psi(B)e_t; \qquad (12.18)$$

where $\nu^*(B) = \nu(B)\psi_x(B)$. Since ε_t and δ_t are white-noise processes, (12.18) may be reformulated as

$$y_t = \psi^*(B)\varepsilon^*(t) \qquad (12.19)$$

following the argument in Section 9.5.

Example 12.3 Let

$$y_t = (\omega_0 - \omega_1 B)x_t + (1 - \theta_{11}B)\varepsilon_t$$

$$x_t = \frac{(1 - \theta_{21}B)}{(1 - \phi_1 B)} \delta_t,$$

whence $(1 - \phi_1 B)y_t = (\omega_0 - \omega_1 B)(1 - \theta_{21}B)\delta_t + (1 - \theta_{11}B)(1 - \phi_1 B)\varepsilon_t$ which is representable as an $ARIMA(1, 0, 2)$ scheme. However, there is no guarantee that this implied process is invertible. This can be seen by taking, for example,

$$\omega_0 = 1, \quad \omega_1 = 2, \quad \theta_{11} = \theta_{21} = \phi_1 = 0.$$

Also, we note that, in general, the univariate scheme will be rather more complicated. Even so, it is well to keep this equivalence in mind when comparing univariate and transfer function forecasts; cf. Sections 12.29–32.

Example 12.4 Generate the forecasts for $t = 21, 22, 23$ from the model:

$$y_t = 10 + \frac{(2 + B)}{(1 - 0.6B)} x_{t-1} + \frac{1}{(1 - 0.8B)} \varepsilon_t \qquad (12.20)$$

$$(1 - 0.5B)x_t = 1 + \delta_t \qquad (12.21)$$

given

$$y_{20} = 17, \quad y_{19} = 15, \quad e_{20} = 1,$$
$$x_{20} = 4, \quad x_{19} = 3, \quad x_{18} = 2.$$

First, from (12.15), we have

$$(1 - 0.6B)(1 - 0.8B)y_t = 10(1 - 0.6)(1 - 0.8)$$
$$+ (2 + B)(1 - 0.8B)x_{t-1} + (1 - 0.6B)\varepsilon_t \qquad (12.22)$$

or

$$y_t = 0.8 + 1.4y_{t-1} - 0.48y_{t-2} + 2x_{t-1} - 0.6x_{t-2} - 0.8x_{t-3} + \varepsilon_t - 0.6\varepsilon_{t-1}$$

so that

$$y_{20}(1) = 0.8 + 1.4(17) - 0.48(15) + 2(4) - 0.6(3) - 0.8(2) + 0 - 0.6(1)$$
$$= 21.4.$$

In order to obtain $y_{20}(2)$, we first need

$$x_{20}(1) = 1 + 0.5x_{20} = 3;$$

similarly $x_{20}(2) = 1 + 0.5x_{20}(1) = 2.5$ and so on. Then

$$y_{20}(2) = 0.8 + 1.4(21.4) - 0.48(17) + 2(3) - 0.6(4) - 0.8(3)$$
$$= 23.8$$

and

$$y_{20}(3) = 0.8 + 1.4(23.8) - 0.48(21.4) + 2(2.5) - 0.6(3) - 0.8(4)$$
$$= 23.848.$$

12.16 When the process is stationary, we may find the mean (or long-term forecast) by setting $B = 1$ in (12.14) and replacing each term by its expected value.

Example 12.5   From (12.21),

$$(1 - 0.5)\mu_x = 1 \quad \text{or} \quad \mu_x = 2.$$

In turn, from (12.22),

$$(0.4)(0.2)\mu_y = 10(0.4)(0.2) + (3)(0.2)2,$$

so $0.08\mu_y = 2$ or $\mu_y = 25$.

Prediction intervals

12.17   In order to develop prediction intervals for the point forecasts, we must first evaluate the forecast mean square error (*FMSE*). Referring back to (12.18), we see that y_{n+k} may be expressed as

$$y_{n+k} = (v_0 a_{n+k} + \cdots + v_k a_n + \cdots)$$
$$+ (\psi_0 \varepsilon_{n+k} + \cdots + \psi_k \varepsilon_n + \cdots) \tag{12.23}$$

following the same approach as in Section 8.8; v_j and ψ_j represent the coefficients B^j in the expansions of $v(B)$ and $\psi(B)$, respectively (we have dropped the * from v for notational ease). Likewise, the k-step ahead *ex-ante* forecast will be:

$$y_n(k) = (v_k a_n + v_{k+1} a_{n-1} + \cdots)$$
$$+ (\psi_k \varepsilon_n + \psi_{k+1} \varepsilon_{n-1} + \cdots) \tag{12.24}$$

so that the forecast error is

$$e_n(k) = y_{n+k} - y_n(k)$$
$$= (v_0 a_{n+k} + \cdots + v_{k-1} a_{n+1})$$
$$+ (\psi_0 \varepsilon_{n+k} + \cdots + \psi_{k-1} \varepsilon_{n+1}). \tag{12.25}$$

Since the as and εs are zero-mean white-noise processes with variances σ_a^2 and σ_ε^2, if follows that

$$E[e_n(k)] = 0 \tag{12.26}$$

$$V_k = V[e_n(k)] = \sigma_a^2 \sum_{j=0}^{k-1} v_j^2 + \sigma_\varepsilon^2 \sum_{j=0}^{k-1} \psi_j^2. \tag{12.27}$$

If the forecasts are developed *ex-post*, the as in (12.24) are specified so that the error reduces to

$$e_n(k) = \psi_0 \varepsilon_{n+k} + \cdots + \psi_{k-1} \varepsilon_{n+1} \tag{12.28}$$

as for the univariate case; see Section 8.8. Conditionally upon the x-values, therefore, the errors have zero expectations and the mean square error reduces to

$$V_k = V[e_n(k)] = \sigma_\varepsilon^2 \sum_{j=0}^{k-1} \psi_j^2. \tag{12.29}$$

as in (8.13).

12.18 Using (12.27) or (12.29) as appropriate, or some intermediate form if some of the xs are known, we can specify an approximate $100(1 - \alpha)$ percent prediction interval in ususal way as

$$y_n(k) \pm z_{\alpha/2}(V_k)^{1/2}. \tag{12.30}$$

However, the apparent simplicity of this result must not blind us to the heroic assumptions made to obtain it. In addition to the standard assumptions of stationarity and normality made in order to arrive at the expressions, we are supposing that the estimation errors for the coefficients are negligible. Particularly when the series are short, the failure to include estimation errors will lead to overly narrow prediction intervals. The nature of this difficulty can be seen by considering the simple linear regression model

$$y_t = \beta_0 + \beta_1(x_t - \bar{x}) + \varepsilon_t. \tag{12.31}$$

It is well known that the prediction mean square error for some new value x_{n+k} is

$$V_k = \sigma_e^2 \left[1 + \frac{1}{n} + \frac{(x_{n+k} - \bar{x})^2}{S_{xx}} \right], \tag{12.32}$$

where $S_{xx} = \Sigma (x_t - \bar{x})^2$. In (12.30), we are, in effect, ignoring the second and third terms in (12.32) and failing to make a small sample correction by using the t-distribution in place of the normal. Further, as noted in Section 8.9, the errors may not be normally distributed.

12.19 Returning to the car production and FT index series, we refitted the selected model using only the first 11 years data. The revised estimates were:

$$y_t = 193.3 + 0.474x_t + (1 - 0.771B)^{-1}\varepsilon_t$$
$$\nabla x_t = \delta_t + 0.26\delta_{t-1}.$$

The forecasts for 1971, together with 95 percent prediction intervals, are as shown in Table 12.5
The *ex-post* forecasts were based on the actual values of the FT index in 1971. Since the relationship between the two series is not overly strong, these forecasts show little improvement over the *ex-ante* forecasts. Such findings are not uncommon and we discuss this further in Section 12.31. The actual values fall well inside the prediction intervals, reflecting the rather stable behaviour of the series at this time.

Table 12.5

		Actual	Ex-post	Ex-ante
1971	Q1	404	448 ± 78	452 ± 82
	Q2	416	451 ± 99	439 ± 102
	Q3	435	458 ± 110	428 ± 112
	Q4	456	459 ± 116	420 ± 118

Time series regression

12.20   We now step back a little from the general models we have been discussing and ask how the approach of this chapter relates to regression analysis, a topic we touched on briefly in Section 11.18.

The simple model

$$y_t = \beta_0 + \beta_1 x_t + \varepsilon_t \tag{12.33}$$

is clearly a special case, as noted in Section 12.5. It main virtues are its simplicity (in specification and in estimation) and the ability to carry out wide-ranging diagnostic tests for departures from the assumptions of the least-squares model, such as heteroscedasticity, autocorrelated errors and (with several variables) multicollinearity. In addition, observations that are outliers or have considerable influence on the estimates may be identified (cf. Cook and Weisberg, 1982). The key weaknesses are, of course, the total failure to recognise any temporal dependence in the relationship between x and y or to capture any autocorrelation among the errors. A very brief historical review of the steps taken to remedy this will indicate why transfer functions are a natural framework for modelling.

The issue of correlated errors was first tackled by Cochrane and Orcutt (1949). They proposed adding to (12.33) a noise model of the form

$$\varepsilon_t = \rho \varepsilon_{t-1} + u_t \tag{12.34}$$

or

$$\varepsilon_t = (1 - \rho \beta)^{-1} u_t. \tag{12.35}$$

Combining (12.33) and (12.35) we see that the Cochrane–Orcutt scheme is a special case of the transfer-function model.

A somewhat different scheme is the mixed autoregressive–regressive model

$$y_t = \alpha + \sum_{j=1}^{p} \phi_j y_{t-j} + \sum_{i=1}^{k} \beta_i x_{it} + \varepsilon_t. \tag{12.36}$$

Durbin (1960) extended the Mann–Wald theorem to show that the least-squares estimators for model (12.36) are consistent and asymptotically unbiased even for non-normal errors. Further extensions of this result justify the least-squares estimators considered in Sections 12.2–4.

In econometrics particularly, the need arose to incorporate multiple lags of input variable into the model. This was handled by the introduction of distributed lag models such as

$$y_t = \alpha + \frac{\omega}{(1 + \delta B)} x_t + \varepsilon_t; \tag{12.37}$$

see Griliches (1967) and Almon (1965). The various extensions of this approach clearly lead to the general transfer function form for the x-relationship. The combination of the ideas underlying (12.34) and (12.36) also leads us to the general transfer-function formulation.

12.21   Approaching the issue from the other direction, a major strength of regression analysis is its set of diagnostic procedures. Perhaps the best-known procedure for time series regression is the test for residual autocorrelation, proposed by Durbin and Watson (1950; 1951; 1971).

The Durbin–Watson statistic d is defined in terms of the residuals by

$$d = \frac{\sum_{t=2}^{n} (e_t - e_{t-1})^2}{\sum_{t=1}^{n} e_t^2} \tag{12.38}$$

This, apart from the end effect, is $2(1 - r_1)$, where r_1 is the first serial correlation of the residuals. Alternatively, it can be looked on as the sum of squares of first differences of the residuals divided by their sum of squares, as in the variate-difference method. If the residuals are highly positively correlated, the value of d is near zero; if they are uncorrelated it is near 2.

12.22   The distribution of d, which is necessary for an exact test of the null hypothesis that the errors are uncorrelated, is complicated. However, d lies between two statistics d_L and d_U whose distributions are related to those of R. L. Anderson's distribution of the first serial correlation coefficient (cf. Section 6.11). The tables in Appendix B give the significance points of d_L and d_U for the 1 percent and the 5 percent levels, for $n = 15$ to 100, and k' (the number of regressors in the regression equation apart from the constant term) from 1 to 5. If an observed d falls below the tabulated value of d_L, we reject the hypothesis that the original residuals are uncorrelated, and if it falls above the tabulated value of d_U the hypothesis is accepted. For cases of indecision where d falls between d_L and d_U, various procedures have been suggested. Durbin and Watson (1971) favour approximating the distribution of d by a statistic based on d_U, namely

$$d^* = a + bd_U,$$

where a and b are constants chosen so that d^* and d have the same mean and variance. Ali (1984) presents a more accurate, but more complex, approximation based on Pearson curves.

For quarterly series, Wallis (1972) provides an extension of d to examine fourth-order autocorrelation, replacing the numerator in (12.37) by $\sum (e_t - e_{t-4})^2$.

12.23   A drawback of the test procedure based on the tables in Appendix B is that it does not apply to models which contain autoregressive terms as well as explanatory variables. To test for autocorrelated errors in this case, Durbin (1970) recommends use of the statistic

$$h = (1 - \tfrac{1}{2}d)\left[\frac{n}{1 - nV_1}\right]^{1/2}, \tag{12.39}$$

where V_1 is the estimated variance of $\hat{\phi}_1$ in the model

$$y_t = \alpha + \phi_1 y_{t-1} + \sum \beta_j x_{jt} + \varepsilon_t. \tag{12.40}$$

For large samples, h is approximately $N(0, 1)$. This test clearly requires $nV_1 < 1$; should this condition be violated, an instrumental variables test due to Wickens (1972) may be used.

12.24   The spirit of these procedures is similar to that underlying the model identification procedures of Sections 11.13–16. The difference lies in the fact that the regression model is pre-specified, so that exact tests are possible. By contrast, our procedures have used the data to screen a large number of potential models so that exact tests are seldom possible; although see Anderson (1971) for exact tests of the order of an *AR* scheme.

12.25   Notwithstanding the lack of exact procedures, the conclusion that arises from this discussion is that the transfer-function approach is to be preferred unless

(a)   the series are too short for full identification procedures to be effective; or
(b)   the model is already well-specified and of simple form; or
(c)   there are too many explanatory variables to be handled by existing identification procedures.

Case (a) offers scant hope to the researcher unless accompanied by a strong dose of (b). In case (c), the best approach might appear to be to run a stepwise regression with a reasonable set of lagged values for each explanatory variable. This idea was developed by Coen *et al.* (1969). It is open to the objection that it may be unsound when we seek to identify a model involving a number of highly (auto-)correlated variables, as noted by Box and Newbold (1971). Indeed, it was this problem that led us to develop more structured identification procedures. However, a relatively minor modification opens up a possible approach along these lines that is reasonably straightforward to implement.

12.26   The essential idea behind Coen *et al.*'s (1969) work was to de-trend and de-seasonalise the variables before looking for relationships between them. We can modify this to the extent of *pre-whitening* each series beforehand; in essence, this is the approach of Box and Haugh (1976) described in Section 11.16. A stepwise search can then be performed using the pre-whitened series.

This procedure tends to over-emphasize the autoregressive structure in y at the expense of structural relations between y and x; nevertheless, it often provides an effective screening for situations like (c) above and gives rise to a reasonable starting model of the form

$$y_t = \alpha + \Sigma \, v_{ji} x_{j,t-i} + \psi(B)\varepsilon_t, \tag{12.41}$$

where ψ is the pre-whitening filter for y and the sum is taken over all inputs (j) at selected lags (i).

Example 12.6   We shall now consider the data on UK imports and related marcoeconomic variables given in Table 12.6.

Two alternative models will be developed, one a 'purely predictive' relationship based upon

$$(I_{t-i}, S_{t-i}, D_{t-i}, F_{t-i}, i = 1, 2, 3, 4) \tag{12.42}$$

and a scheme allowing contemporaneous variation which adds (S_t, D_t, F_t) to the set (12.42).

Straightforward stepwise regressions, using backward and forward elimination with threshold values of $F = 2.0$ both to enter and to leave, were run using MINITAB. Other decision rules may, of course, produce different results as with any identification procedure. The pre-whitening models are (with Q

Table 12.6 UK imports for each quarter of the years 1960–1970

Year	Quarter	I	S	D	F
1960	1	1 382	149	370	1 088
	2	1 417	168	342	1 081
	3	1 432	161	332	1 103
	4	1 438	150	307	1 146
1961	1	1 457	153	327	1 184
	2	1 403	102	331	1 205
	3	1 389	35	329	1 241
	4	1 379	37	313	1 217
1962	1	1 408	7	316	1 197
	2	1 426	12	361	1 221
	3	1 460	53	336	1 222
	4	1 442	−6	350	1 189
1963	1	1 414	−3	353	1 070
	2	1 472	36	398	1 247
	3	1 520	−1	414	1 278
	4	1 540	159	416	1 317
1964	1	1 611	132	417	1 373
	2	1 612	170	429	1 417
	3	1 632	141	439	1 459
	4	1 659	185	440	1 476
1965	1	1 581	92	453	1 486
	2	1 643	83	426	1 474
	3	1 672	112	428	1 471
	4	1 686	89	417	1 529
1966	1	1 722	86	447	1 500
	2	1 681	76	468	1 509
	3	1 726	86	409	1 551
	4	1 642	6	375	1 552
1967	1	1 777	60	392	1 587
	2	1 787	64	428	1 671
	3	1 779	−1	467	1 645
	4	1 850	61	492	1 621
1968	1	1 948	−104	550	1 721
	2	1 903	76	408	1 693
	3	1 945	85	434	1 714
	4	1 937	101	460	1 722
1969	1	1 992	122	400	1 710
	2	1 980	76	425	1 672
	3	1 966	57	444	1 705
	4	2 024	91	432	1 690
1970	1	2 026	−16	433	1 646
	2	2 130	116	457	1 759
	3	2 078	117	476	1 726
	4	2 197	112	480	1 755

I = Imports of goods and services

S = Value of physical increases in stocks and work in progress

D = Consumer expenditure on durables

F = Gross domestic fixed capital formation

Data from *Monthly Digest of Statistics*; see also Gudmundsson (1971).
All data are seasonally adjusted.
Figures in £ million.

based on 12 lags):

$$\nabla I_t = 26.6 - 0.472\nabla I_{t-1} + \varepsilon_{1t}, \quad Q_A(11) = 3.2, \, s = 44.1$$
$$(3.20)$$

$$S_t = 28.2 + 0.341 S_{t-1} + 0.331 S_{t-2} + \varepsilon_{2t}, \quad Q_A(10) = 9.5$$
$$(2.31) \qquad (2.25)$$

$$\nabla D_t = 2.8 + \varepsilon_{3t} - 0.358\varepsilon_{3,t-1}, \quad Q_A(11) = 8.0$$
$$(2.45)$$

$$\nabla F_t = 15.6 + \varepsilon_{4t} - 0.302 e_{4,t-1}, \quad Q_A(11) = 14.7.$$
$$(2.02)$$

The tentatively identified models are summarised in Table 12.7; stepwise regression was employed both with and without pre-whitening. The pre-whitening involves differencing all but S_t and this should be borne in mind in comparing the results. Since the pure autoregressive scheme for I_t yields $s = 45.3$, it is evident that stepwise procedures based on only the lagged variables produce only marginal gains. When current values are added, some further improvement is possible although the choice of terms is not clear-cut. Note that the stepwise schemes without pre-whitening have coefficients that leave the model outside the stationary region (lagged only) or close to the boundary (lagged plus current). However, Tsay and Tiao (1984) have shown that such estimates remain consistent so this does not present a problem at the estimation stage. In this example, the models selected without pre-whitening are quite similar to those chosen after pre-whitening. This is largely due to the dominance of the autoregressive elements, and pre-whitening should be employed in general.

Automatic transfer function modelling

12.27 At this stage, it would be possible to press ahead with the Box–Jenkins and Liu–Hanssens procedures for model identification described in Section 11.16. In fact, the relatively short series and high correlations mean that no dramatic changes are achieved by such attempts. Instead, we shall apply these procedures in an 'automatic' mode using the transfer-function component of AUTOBOX. The logic of this program for the univariate case was shown in Fig. 7.7; the process is essentially similar for transfer functions; see Reilly and Dooley (1987).

12.28 The results from the AUTOBOX analyses are presented in Table 12.8. Because of its multiple pre-whitening operations, the Box–Jenkins model is based on only 37 observations so the smaller value for s is open to question. The common filter method of Liu–Hanssens failed to produce an estimable model, probably because S alone was a stationary process. With F only, Liu–Hanssens produced rather similar results to the stepwise scheme after pre-whitening. Further analysis is certainly possible and the reader is encouraged to try to improve on these results. However, the principal conclusion is that the level of imports is strongly related to its level in the two previous quarters and somewhat related to current levels of other economic indicators.

12.29 An economist's approach to modelling imports would start from a

Table 12.7 Model identification for UK imports using stepwise regression (computed using MINITAB)

	Model form	s
ARIMA	$\nabla I_t = \quad 26.6 - 0.47\nabla I_{t-1} + \varepsilon_t$ $\qquad\qquad (3.20)$	44.1
Stepwise—no pre-whitening autoregressive only	$I_t = -44.5 + 0.53I_{t-1} + 0.51I_{t-2}$ $\qquad\qquad (3.47) \qquad\ (3.31)$	45.3
lagged only	$I_t = -24.9 + 0.46I_{t-1} + 0.58I_{t-2} - 0.21S_{t-2}$ $\qquad\qquad (3.01) \qquad\ (3.73)$	44.1
lagged plus current S, D, F	$I_t = -103.2 + 0.40I_{t-1} + 0.57I_{t-2}$ $\qquad\qquad (2.78) \qquad\ (4.01)$ $\qquad\qquad + 0.50D_t \quad - 0.21S_{t-1}$ $\qquad\qquad (2.85) \quad\ (-1.95)$	40.8
Stepwise—using pre-whitened variables denoted by P superscript lagged only	$I_t^{P} = \quad 0.784 - 0.30F_{t-4}^{P} + 0.39D_{t-4}^{P}$ $\qquad\qquad\ (-1.89) \qquad\ (1.80)$	43.7
lagged plus current S, D, F	$I_t^{P} = -0.436 + 0.42F_t^{P} \quad - 0.22S_{t-4}^{P}$ $\qquad\qquad (3.03) \qquad (-1.63)$	40.7

Notes:
(1) s for the *ARIMA* model is based upon $n' = 42$ observations; those for the stepwise schemes are based on $n' = 40$; the *ARIMA* scheme with $n' = 40$ has $s = 44.7$.
(2) The *ARIMA* model may be re-expressed as

$$I_t = 26.6 + 0.53I_{t-1} + 0.47I_{t-2}.$$

Table 12.8 Model identification for UK imports using AUTOBOX

Method	Final model
Box–Jenkins ($n' = 37$)	$\nabla I_t = 0.577\nabla F_t - 0.195\nabla_2 S_t + (1 + 0.478B)^{-1}\varepsilon_t$ (4.40) (−2.11) (−2.85) $R^2 = 0.967,\ s = 42.5$
Liu–Hanssens[*] ($n' = 43$)	$\nabla I_t = 26.4 - 0.472\nabla I_{t-1} + \varepsilon_t$ (−3.16) $R^2 = 0.965,\ s = 44.6$
Liu–Hanssens: F only ($n' = 42$)	$\nabla I_t = 0.499\nabla F_t + (1 + 0.374B)^{-1}\varepsilon_t$ (3.75) (−2.36) $R^2 = 0.966,\ s = 44.1$

[*]Common filter could not be estimated.

causal model and then move to estimation. Unfortunately, the classical econometric approaches pay insufficient attention to either the transfer function or the error structure and tend to yield poor forecasts as a result (e.g. Nelson 1972; Cooper and Nelson 1975). Conversely, a purely empirical approach may also be criticised (cf. Box and Newbold 1971). The answer lies in a synthesis, as in the work of Hendry and Richard (1983). Theory, economic or otherwise, should take us as far as possible in suggesting possible relationships, and identification procedures can then be used to determine lag structures.

12.30   At this stage, we must recognise that we are dealing with an active research area where many questions remain to be answered. However, some tentative conclusions about model building can be drawn.

(1)   Standard regression procedures are often inappropriate for time series data. The modelling process must include identification of the lag structures for the regressor variables and the error terms.

(2)   Identification procedures must allow for trends and autocorrelations in the regression variables. Some form of pre-whitening is strongly recommended.

(3)   The fact that we are, at least implicitly, choosing one of many competing models means that the error rates for the various tests of hypothesis that we perform will be much higher than the nominal levels. This should be borne in mind during model-building and, whenever possible, models checked by making forecasts outside the time frame used for estimation. Further, the test procedures that are available are typically justified by asymptotic results rather than exact distribution theory and so may be rather inaccurate for short series.

(4)   We have assumed that the regressor variables are measured without error. Errors-in-variables models, if often more appropriate, are very difficult to handle; an account is given in Kendall and Stuart (1979, Chapter 29). Some progress, however, is possible and is considered briefly in Section 13.13.

(5) Spurious effects may be introduced by aggregation as noted in Example 5.2 in Section 5.12. Many series in the economic and social sciences are aggregated by necessity and we must always proceed with caution in variable selection.

12.31 If we negotiate the rapids of model-building successfully, what then of forecasting performance? Two obvious hypotheses are that

(a) transfer-function models should outperform univariate models;
(b) *ex-post* forecasts should be closer on average than *ex-ante* forecasts.

These claims have been examined by several authors; a detailed review and evaluation of this research is given in Armstrong (1985, Chapter 8). Also in her Ph.D. thesis, Pamela Texter (see Texter and Ord, 1991) compared the forecasting performance of a variety of univariate schemes with models ranging from simple regression to transfer functions, including both those identified 'automatically' and 'manually'. The tentative conclusions from these studies may be summarised as follows:

(1) Properly identified transfer function models perform only marginally better than univariate *ARIMA* schemes. In part, this is because much of the cross-series activity may be captured by the univariate process, as noted in Section 12.15.
(2) Models selected without proper regard for lag structures and correlated residuals (e.g. simple regression) fare much worse than *ARIMA* and other autoprojective methods such as exponential smoothing. Proper specification of the relationships is critical.
(3) There is little difference between *ex-ante* and *ex-post* performance in many cases. This may reflect either accurate forecasting of the input variables or relatively poor identification of the structure of the process. Indeed, these two effects may well occur together, as follows from the discussion in Section 12.15. In essence, we have a design problem; although the series may not be experimental in nature, the information they provide may not admit a clear delineation between $x \rightarrow y$ effects and feedback from past y-values.

12.32 These comments may appear to support the use of univariate models rather than those with a more causal basis. However, it should be recognized that the studies on which these comparisons are based are looking at short- to medium-term effects, rather than the medium to longer term. Also, there is an implicit assumption that the input variable is generated by an *ARIMA* scheme rather than subject to manipulation or intervention. Typically, planning decisions require analyses based upon 'what if' forecasts obtained by manipulating the input variables(s). Such questions can be addressed only by regression-based models. Further, such models often perform better in the longer term when autoprojective models have either reverted to a long-run mean or developed prodigiously wide prediction intervals. Long-run trends may be relatively unimportant in making short-term predictions, but can dominate as the time frame is expanded; models of the stock market form an obvious example. In conclusion, therefore, we note that both the time horizon and the purpose to which the forecasts will be put are important when selecting a modelling approach.

Exercises

12.1 Generate simulated series with n = 110 from the model

$$x_t = 10 + 0.7x_{t-1} + \delta_t$$
$$y_t = 20 + 2x_{t-1} + x_{t-2} + \varepsilon_t,$$

where $\delta_t \sim N(0, 1)$ and $\varepsilon_t \sim N(0, 1)$ are independent white noise error terms. Using the first 100 observations:

(a) develop a univariate *ARIMA* model for y_t;
(b) try to identify a transfer function model for y_t without pre-whitening;
(c) identify a transfer function model for y_t with pre-whitening.
(d) for each model in (1)–(3), develop forecasts for $t = 101$, 102, ..., 110 using $t = 100$ as forecast origin.

Compare your results. Repeat the exercise with different parameter values or starting models to develop a feel for transfer-function modelling.

12.2 Generate the forecasts for $t = 51, 52, 53$ from the model

$$y_t = 50 + \frac{(2 - 3B)}{(1 + 0.5B)} x_t + \frac{(1 + 0.4B)}{(1 - 0.8B)} \varepsilon_t$$
$$\nabla x_t = 10 + (1 - 0.6B)\delta_t$$

given

$$x_{49} = 20, \quad x_{50} = 30, \quad \delta_{50} = 5$$
$$y_{49} = 70, \quad y_{50} = 75, \quad \varepsilon_{49} = -5, \quad \varepsilon_{50} = 10.$$

12.3 Note that, in Exercise 12.2,

$$x_t(k) = x_t(1) + 10(k - 1), \qquad k \geqslant 1.$$

Hence develop the forecast function $y_t(k)$ as a function of k.

12.4 Generate 'what if' forecasts for $t = 51, 52, 53$ for the model in Exercise 12.2 using

$$x_{51} = 30, \qquad x_{52} = 100, \qquad x_{53} = 100.$$

12.5 Using the pair of series selected from Exercise 11.8, complete the development of a transfer-function model. Also, generate forecasts for the last four time periods after fitting the model without these observations. Compute both *ex-post* and *ex-ante* forecasts and compare the results.

13

Intervention analysis and structural modelling

Introduction

13.1 In the two previous chapters, we developed transfer-function modelling as an extension of time-series regression under two restrictive assumptions. The first was that there was no feedback from Y to X, an assumption we shall relax in the next chapter where multivariate models are considered. The second restriction was that X should follow an $ARIMA$ process so that the model can be identified using the cross-correlation function (CCF) after pre-whitening. The $ARIMA$ model for X is also used in the generation of ex-ante forecasts for Y but this is not a requirement at the model-building stage.

13.2 It is evident that X may well not follow an $ARIMA$ process. In particular, X may be a policy variable, such as a prime interest rate set by the government or an advertising budget set by a company's board of directors. In some circumstances, the X-series may nevertheless behave rather like an $ARIMA$ process and pre-whitening will provide a satisfactory basis for model selection. Whether the model for x would then provide an adequate basis for forecasting is open to question and must depend upon the context of the problem. In other situations, the system may be subjected to an external shock or *intervention*, as when a strike is called or a competitor launches a rival brand. The impact of such interventions may be permanent or temporary and either gradual or abrupt, but their key feature is that they are essentially 'one-off' experiences. *Intervention analysis*, developed originally by Box and Tiao (1975), is concerned with model building in these circumstances as described in Sections 13.4–13.

13.3 An alternative approach which encompasses both transfer functions and intervention analysis is structural modelling, developed by Harvey and Durbin (1986). The univariate case described in Sections 9.14–18; the inclusion of regressor variables is discussed in Section 13.14.

Intervention analysis

13.4 We shall consider intervention models in the context of univariate *ARIMA* models; the inclusion of transfer-function elements is discussed in Section 13.8. Further, we suppose that only a single intervention is to be considered. These restrictions serve to simplify the discussion without losing any essential principles of the approach.

Therefore, we assume that the external shock to the system, be it strike, policy change, or whatever, occurs at time t_0, which we take to be known. The two simplest cases are, following Fig. 11.8, that the shock is purely temporary, producing a pulse of the form

$$\begin{aligned} P_t &= 1, \qquad t = t_0 \\ &= 0, \qquad t \neq t_0 \end{aligned} \tag{13.1}$$

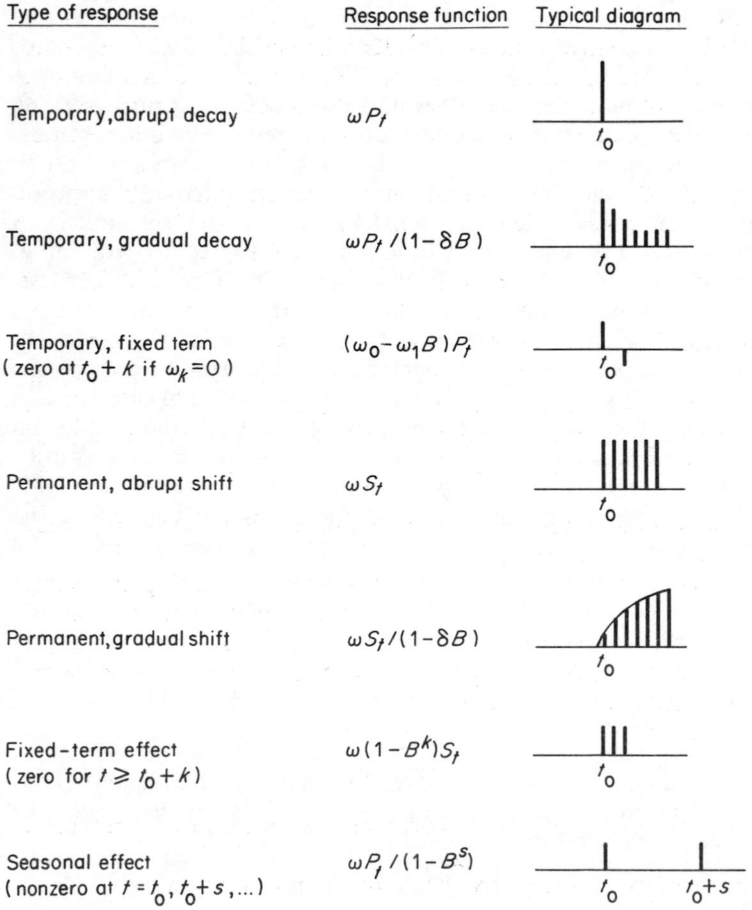
Type of response
Response function
Typical diagram
Temporary, abrupt decay
ωP_t
t_0
Temporary, gradual decay
$\omega P_t/(1-\delta B)$
t_0
Temporary, fixed term
(zero at t_0+k if $\omega_k=0$)
$(\omega_0-\omega_1 B)P_t$
t_0
Permanent, abrupt shift
ωS_t
t_0
Permanent, gradual shift
$\omega S_t/(1-\delta B)$
t_0
Fixed-term effect
(zero for $t \geqslant t_0+k$)
$\omega(1-B^k)S_t$
t_0
Seasonal effect
(nonzero at $t=t_0, t_0+s, \ldots$)
$\omega P_t/(1-B^s)$
t_0
t_0+s

Fig. 13.1 Forms of intervention response

or that it is permanent, producing the step change

$$S_t = 1, \quad t \geqslant t_0$$
$$= 0, \quad t < t_0. \tag{13.2}$$

In accordance with Fig. 11.8, we note that these simple structures are capable of considerable extension by incorporating different impulse response functions, as indicated in Fig. 13.1.

An infinite variety of more complex patterns may be constructed, but these examples cover most of the cases used in practice. It is assumed that t_0 is known; a delay of b periods may be incorporated before the intervention has an effect.

Model identification

13.5 Model building for a process containing an intervention cannot proceed by pre-whitening since the input variable is predetermined and the shock occurs but once. The investigator must consider whether the effect is likely to be permanent or temporary, whether the onset will be subject to a delay and whether the impact will be sudden or gradual. As in other cases, the choice is not irrevocable, but a good initial specification will help to ensure convergence of the estimation procedure.

Using x_t to denote P_t or S_t as appropriate and $v(B)$ to denote its impulse function, the model becomes

$$y_t = v(B)x_t + \frac{\theta(B)}{\phi(B)} \varepsilon_t. \tag{13.3}$$

13.6 The remaining step in model identification is the usual one of selecting the *ARIMA* scheme, and the available tools are those discussed in Chapter 6. However, a problem arises in that a major jump caused by x_t would distort estimates of the autocorrelations. One of four possible approaches may be used, depending on the particular series:

(a) use only observations prior to the intervention or, less commonly, only some time after the intervention;
(b) use observations both before and after but exclude all pairs using observations in the interval $[t_0, t_0 + k]$, which represents a 'cordon sanitaire' in which the effects of the intervention are deemed to have worked themselves out (this option is not available in existing software packages but could easily be implemented);
(c) estimate the residuals

$$e_t = y_t - \hat{v}(B)x_t, \tag{13.4}$$

ignoring any error structure, and then plot the sample autocorrelation functions;
(d) use the whole series, ignoring the effects of the intervention.

Each method has its attractions, depending upon the behaviour of the phenomenon under consideration; use of (c) is advised whenever the effects of the intervention are large and fairly sudden, whereas (b) is probably best for rather short series.

Example 13.1   Suppose that the intervention is determined to be a pure step function and the error structure is $ARIMA(0, 1, 1)$. Then (13.3) becomes

$$y_t = \omega S_t + \frac{(1 - \theta B)}{(1 - B)} \varepsilon_t \tag{13.5}$$

or

$$(1 - B)y_t = \omega(1 - B)S_t + (1 - \theta B)\varepsilon_t$$
$$= \omega P_t + (1 - \theta B)\varepsilon_t, \tag{13.6}$$

since $P_t = (1 - B)S_t$.

Either form (13.5) or (13.6) may be used, but it is recommended that the initial model be formulated in terms of the original variable y, as in (13.5), so that the intervention is correctly specified.

13.7   Once a tentative model has been identified, estimation proceeds in essentially the same manner as for transfer functions. However, it should be noted that the intervention parameters are typically estimated from only a few, or even only one, observation(s). Even when the series is very long, therefore, the standard errors may remain sizeable. Formally speaking, consistent estimators are not available, although this is not a cause for concern in practical terms since the question is usually whether the intervention had a sizeable impact, which can be decided legitimately using the t-ratio for the estimate.

Multiple effects

13.8   Applications of the intervention approach often involve several interventions and one or more input variables with associated transfer functions. The extended model may be written as

$$y_t = \sum_{j=1}^{k} v_j(B)x_{jt}^I + \sum_{j=k+1}^{k+m} v_j(B)x_{jt}^T + \frac{\theta(B)}{\phi(B)} \varepsilon_t, \tag{13.7}$$

where the superscripts I and T refer to the k intervention and m transfer variables, respectively. Identification for each intervention variable proceeds as in Section 13.5. Each stochastic input, x^T, should be pre-whitened with its own filter as discussed in Section 11.16. Depending on the length of series and the timing of the interventions, one of the options (a)–(d) in Section 13.6 should be selected and the resulting (sub)series pre-whitened with its own filter. The *CCF* between the pre-whitened y and $x_j(j = k + 1, ..., k + m)$ may then be used to suggest initial transfer-function specifications. In these circumstances, model identification is not easy and the results may be rather disappointing, particularly when the series are short. There is room for further technical improvement, but we must not expect miracles from a limited data base.

Estimation

13.9 The general form of the estimation procedure is the same as for transfer function modelling, as discussed in Section 12.2–4. For further details, consult Box and Tiao (1975), Tiao and Box (1981).

The airline example

13.10 To illustrate the approach, we consider the airlines data presented in Table 1.3. Models for this series were developed in Sections 7.24–25, when it was noted that outliers occurred in June/July 1968 and April 1969 because of strikes by pilots.

It seems reasonable to argue that customers lost during the strikes would either travel by other carriers or cancel their trips. It is possible that there may be a long-term loss of goodwill leading to loss of customers, but this is unlikely to affect flight schedules in the short run. Therefore, we define two intervention variables:

$$x_{1t} = 1, \quad \text{if } t = 66 \text{ or } 67 \text{ (June, July 1968)}$$
$$= 0, \quad \text{otherwise};$$
$$x_{2t} = 1, \quad \text{if } t = 76 \text{ (April 1969)}$$
$$= 0, \quad \text{otherwise}.$$

13.11 Since the model has already been developed for these data, we could follow option (d) in Section 13.6 and use the same scheme. Instead, however, we follow option (a) and use the first 60 observations to identify the model. The *SACF* and *SPACF* for $\nabla_{12} y_t$ are given in Table 13.1. Identification of the model is not clear-cut, but we begin with the rather conservative choice

$$(1 - \phi_1 B - \phi_3 B^3)\nabla_{12} y_t = \delta + A_t, \tag{13.8}$$

Table 13.1 *SACF* and *SPACF* airlines data, seasonally differenced, using first 60 observations

Lags	Sample autocorrelations							
1–8	0.307	0.249	0.554	0.003	−0.013	0.175	−0.234	−0.205
9–16	−0.220	−0.386	−0.285	−0.289	−0.242	−0.161	−0.119	−0.070
17–14	0.002	−0.057	0.054	0.055	0.001	0.082	0.116	−0.035

Lags	Sample partial autocorrelations							
1–8	0.307	0.171	0.498	−0.384	−0.126	−0.025	−0.118	−0.100
9–16	−0.352	0.001	−0.028	−0.008	0.057	−0.097	0.134	−0.149
17–24	0.029	−0.294	0.026	−0.093	−0.051	−0.018	0.026	−0.063

where A_t will now include the intervention terms, so that

$$A_t = v_1 x_{1t} + v_2 x_{2t} + \varepsilon_t. \tag{13.9}$$

The estimated model showed all parameters significant with significant terms in the $SACF$ and $SPACF$ at lags 7 and 12. A seasonal MA term at lag 12 was added, but no action was taken on lag 7. Thus, the final model was estimated as

$$(1 - 0.262B - 0.420B^3)\nabla_{12} y_t = 219 + A_t \tag{13.10}$$
$$\quad (2.52) \quad\;\; (4.05) \qquad\qquad (2.42)$$

$$A_t = -2344 x_1 t - 2965 x_{2t} + (1 - 0.385B^{12})\varepsilon_t \tag{13.11}$$
$$\quad (-8.62) \quad (-8.36) \qquad\quad (2.93)$$

$$s = 450, \quad R^2 = 0.957, \quad R_D^2 = 0.758.$$

Here, R_D^2 is computed using $\nabla_{12} y$ as the dependent variable. We now find that the residual diagnostics are satisfactory and all terms in the model are significant. It is evident that recognition of the aberrant observations has improved the fit considerably and reduced the error variance by almost two-thirds. Further, the two model-selection options lead to the same results in this case.

13.12   Although the final model is likely to provide better forecasts, as aberrations in the seasonal pattern have been removed, such forecasts are based upon the assumption that the service will not be disrupted; that is, no further interventions will occur. The estimates and 95 percent confidence intervals for aircraft miles lost are, from (13.11),

June/July 1968: 4688 ± 1066 (i.e. 2×2344)

April 1969: 2965 ± 695

and these values might be used to estimate the impact of future strikes, given estimates of duration.

Automatic intervention detection

13.13   For the airlines data, the extreme values were identified by our analysis and subsequently found to correspond to pilots' strikes. This suggests that it may be possible to screen the data by fitting an initial model and then checking all the observations for pulse or step changes, comparing observed values with predicted. This procedure has been developed by Bell (1983) and is implemented in the AUTOBOX system. Such a method must be applied with care as the resulting estimate of the standard deviation may be downwards biased, leading to overly narrow prediction intervals. Nevertheless, it is an effective screening device that may often uncover data errors or omissions in the modelling process.

Example 13.2   The airlines series (all 8 years) was run using the automatic AUTOBOX procedure. Five outliers were detected:

(1)   a step at $t = 76$ (April 1969), $\hat{v}_1 = 4230$
(2)   a step at $t = 77$ (May 1969), $\hat{v}_2 = -1826$

(3) a seasonal step at $t = 69$ (September 1968 onwards), $\hat{v}_3 = 1082$
(4) a seasonal step at $t = 8$ (August 1963 onwards), $\hat{v}_4 = 1167$
(5) a seasonal step at $t = 9$ (September 1963 onwards), $\hat{v}_5 = 985$

The last two are relatively unimportant and will not be considered further. Outlier (1) identifies the second strike but interprets this, together with the second step, as evidence of changes in the mean level of the process. The low value for May 1968 (pre-strike unrest?) followed by the first strike explain how such an identification may have arisen. The pattern could equally be represented by a pulse for April 1969 and a step for May. The third outlier points to a shifting seasonal pattern, notably the emergence of September as the month with most activity. This should certainly be incorporated into the model. Thus, although we may not use the automatically selected model as our final version, it has drawn attention to some interesting features of the data that we had previously missed.

Structural models

13.14 The structural models developed in Section 9.14 readily extend to include regression and intervention terms if we redefined the *observation* equation (9.44) to be

$$y_t = \mu_t + \gamma_t + \sum_{j=1}^{k} v_j x_{jt}^I + \sum_{j=k+1}^{k+m} v_j x_{jt}^T + \varepsilon_t; \qquad (13.12)$$

which may be compared with (13.7). Note that we do not need the transfer function variables to be stationary, before or after differencing; lagged x variables may be included in the scheme. The state equations are unchanged.

Example 13.3 The full structural model was fitted to the airline data with the same two intervention variables as before. The results are summarised in Table 13.2; some results from Table 9.1 are included for ease of comparison. All the hyperparameters are reduced in absolute value, the irregular component by a factor of 10. Nevertheless, we retained the same model for comparative purposes. The level and slope estimates are slightly higher at 12/69, as the depressing effects of the strikes have been removed. The harmonics remain substantially unchanged, indicating that the three-monthly effect ($j = 4$) is real and suggesting perhaps the need for a lagged y-value; recall that an $AR(3)$ term appeared in the $ARIMA$ scheme. The intervention values are very similar to those given by the $ARIMA$ analysis in (13.11).

Calendar adjustments

13.15 Intervention analysis also provides a means of coping with the vagaries of the calendar, such as the timing of Easter, day of the week effects, and so on. For Easter, a 0/1 variable specifying the month containing the Easter weekend will generally suffice, plus an adjustment to cover those occasions where the weekend spans two different months. For days of the week, only six

Table 13.2 Structural model with interventions for airlines data

(a) Variance (or hyperparameter) estimates

Component	No interventions		With interventions	
	Estimate	*t*-ratio	Estimate	*t*-ratio
level	2.24×10^4	1.25	1.09×10^4	1.39
trend	50.1	0.72	47.6	0.76
seasonal	1.89×10^3	2.32	1.41×10^3	1.92
irregular	1.80×10^5	2.74	1.87×10^4	0.89

(b) Component estimates at 12/69

Component	No interventions		With interventions	
	Estimate	*t*-ratio	Estimate	*t*-ratio
level	12 565	44.1	12 747	72.5
slope	75.3	2.14	81.4	2.79
harmonics (j, j^*) significant at $\alpha = 0.05$				
(1)	$-1\ 431$	-7.97	$-1\ 739$	-14.02
(1^*)	$-1\ 239$	-6.80	$-1\ 177$	-8.43
(3^*)	361	2.11	274	2.49
(4)	1 060	6.41	1 088	10.16
interventions				
x_1	—	—	$-2\ 444$	-9.89
x_2	—	—	$-2\ 610$	-6.22

(c) Summary statistics

		No interventions	With interventions
	R^2	0.877	0.956
	R_s^2	0.247	0.733
	s	754	455

dummy variables should be used since the total effects should add to zero. Bell and Hillmer (1983b) give a full account of calender adjustment procedures; Cleveland (1983) discusses these methods in the context of the frequency domain. Indeed, following the discussion in 10.17, we could check whether the quarterly effect in the airlines model is genuine, or an artifact caused by the weekend effect.

13.16 We conclude with a note of caution. It is always possible to think of numerous potential interventions, thereby generating an over-fitted model. It is important to screen such variables and to ensure that only key factors are included in the final model. Also, it should be remembered that the net effect of using intervention variables, especially pulses to adjust additive outliers, will be to reduce the estimated error variance and, in turn, the width of the prediction interval. This implies an optimistic view of the future as being trouble free, or at least intervention free.

Exercises

13.1 The effect of a sales campaign at time t_0 is measured by an intervention variable with response function

$$100P_t/(1 - 0.6B).$$

Find the expected effect of the campaign on sales for $t = 1, 2$ and 5.

13.2 Consider the US Immigration data of Table 1.4. Starting from 1900, develop the form of the response function for

(a) World War I (1914–19)
(b) Great Depression (1930–38)
(c) World War II (1939–46)

and any other events you feel would have a major effect upon immmigration.

13.3 Using your results from Exercise 13.2 develop a model for the immigration series from 1900 on.

13.4 Develop an intervention model for the airlines data (Table 1.3) treating June and July 1968 separately but still including April 1969. Also include a seasonal step for September from 1968 onwards (see Sections 13.10, 13.13).

13.5 Develop intervention models for the following series in Appendix A:

(a) UK unemployment and elections (Table A9);
(b) store sales (Table A10).

The intervention variables are given in those tables.

14

Multivariate series

Introduction

14.1 Our discussion thus far have dealt with only a single dependent variable, yet many realistic applications involve several dependent variables, such as sales for several product lines or quite disparate quantities like gross national product, unemployment, and industrial production. We saw in Chapters 11–13 that it is sometimes possible to deal with such cases on a 'one variable at a time' basis but this requires the assumption of an open-loop system where x may affect y but y cannot then feed back to affect x.

Open-loop systems often exist when the input variable is under the experimenter's control or when x represents a 'policy' variable to be set by management. Even here, caution is advised. For example, management decides the advertising budget and the level of advertising is a determinant of sales. However, the act of setting the budget may well be influenced by recent sales figures. For an example of such effects, see Bhattacharya (1982); the data used in that study are given in Appendix A, Table 13.

14.2 A truly multivariate system involves flows both from x to y and y to x, albeit with time delays before the effects are felt. Multivariate models incorporating such effects are described in Sections 14.3–7 and their basic properties are discussed in Sections 14.8–14. Following the pattern of earlier chapters, we then explore model identification, estimation and forecasting. The chapter concludes with two examples, one using data on hog prices and sales, drawn from the classic study by Quenouille (1968), and the other involving re-analysis of the car production data considered in Chapter 12.

Autoregressive models

14.3 To fix ideas, consider a two-variable model of the form:

$$y_{1t} = \phi_{11}y_{1,t-1} + \phi_{12}y_{2,t-1} + \varepsilon_{1t} \tag{14.1a}$$

$$y_{2t} = \phi_{21}y_{1,t-1} + \phi_{22}y_{2,t-1} + \varepsilon_{2t}; \tag{14.1b}$$

intercept terms would be included in practice, but it simplifies the discussion if we assume that both variables are measured about their means; also we shall assume that the processes are stationary. Inspection of (14.1) reveals that each variable feeds back to affect the other after a one-period delay. The error terms, ε_1 and ε_2, are assumed to have zero means and to be uncorrelated over time, although they may be correlated for the same time period. That is, for $j = 1, 2$,

$$E(\varepsilon_{jt}) = 0 \tag{14.2}$$

$$E(\varepsilon_{jt}\varepsilon_{js}) = \sigma_{jj}, \qquad t = s \tag{14.3}$$
$$= 0, \qquad t \neq s$$

and

$$E(\varepsilon_{1t}\varepsilon_{2s}) = \sigma_{12}, \qquad t = s$$
$$= 0, \qquad t \neq s \tag{14.4}$$

14.4 It is natural to represent (14.1) in matrix terms as

$$\mathbf{y}_t = \boldsymbol{\phi}_1 \mathbf{y}_{t-1} + \boldsymbol{\varepsilon}_t, \tag{14.5}$$

where

$$\mathbf{y}_t = \begin{pmatrix} y_{1t} \\ y_{2t} \end{pmatrix}, \qquad \boldsymbol{\varepsilon}_t = \begin{pmatrix} e_{1t} \\ \varepsilon_{2t} \end{pmatrix} \quad \text{and} \quad \boldsymbol{\phi}_1 = \begin{pmatrix} \phi_{11} & \phi_{12} \\ \phi_{21} & \phi_{22} \end{pmatrix}. \tag{14.6}$$

In turn, (14.2)–(14.4) become

$$E(\boldsymbol{\varepsilon}_t) = \mathbf{0}, \ E(\boldsymbol{\varepsilon}_t\boldsymbol{\varepsilon}_s') = \boldsymbol{\Sigma}, \qquad t = s$$
$$= \mathbf{0}, \qquad \text{otherwise}, \tag{14.7}$$

where

$$\boldsymbol{\Sigma} = \begin{pmatrix} \sigma_{11} & \sigma_{12} \\ \sigma_{21} & \sigma_{22} \end{pmatrix}.$$

Finally, using the backshift operator, $B\mathbf{y}_t = \mathbf{y}_{t-1}$, we may re-express (14.5) as

$$\boldsymbol{\phi}(B)\mathbf{y}_t = \boldsymbol{\varepsilon}_t, \tag{14.8}$$

where

$$\boldsymbol{\phi}(B) = \mathbf{I} - \boldsymbol{\phi}_1 B, \tag{14.9}$$

I being the 2×2 identity matrix.

14.5 Given (14.8) and (14.9), the extension to autoregressive schemes of order $p \ (\geqslant 1)$ for $k \ (\geqslant 2)$ variables may be written in form (14.8) with

$$\boldsymbol{\phi}(B) = \mathbf{I} - \boldsymbol{\phi}_1 B - \boldsymbol{\phi}_2 B^2 - \cdots - \boldsymbol{\phi}_p B^p, \tag{14.10}$$

where $\boldsymbol{\phi}_r$ is a $(k \times k)$ matrix with elements $\phi_{ij}^{(r)}$. Model (14.8) is known as a vector autoregressive scheme of order p, or $VAR(p)$.

Although the notational extension is straightforward enough, such models raise major inferential issues since (14.10) contains pk^2 parameters to which must be added the $\frac{1}{2}k(k + 1)$ parameters of $\boldsymbol{\Sigma}$, defined by obvious extension from (14.7). As a consequence, a considerable amount of effort has gone into identification schemes to reduce the number of parameters. We return to this topic in Section 14.17.

VMA and *VARMA* models

14.6 Corresponding to (14.8) and (14.10) the vector moving average scheme of order q, $VMA(q)$, may be written as

$$\mathbf{y}_t = \boldsymbol{\theta}(B)\boldsymbol{\varepsilon}_t, \tag{14.11}$$

where

$$\boldsymbol{\theta}(B) = \mathbf{I} - \boldsymbol{\theta}_1 B - \boldsymbol{\theta}_2 B^2 - \cdots - \boldsymbol{\theta}_q B^q, \tag{14.12}$$

where $\boldsymbol{\theta}_r$ is a $(k \times k)$ matrix with elements $\theta_{ij}^{(r)}$. The mixed vector or $VARMA$ (p, q) scheme is then

$$\boldsymbol{\phi}(B)\mathbf{y}_t = \boldsymbol{\theta}(B)\boldsymbol{\varepsilon}_t \tag{14.13}$$

with $\boldsymbol{\phi}(B)$ and $\boldsymbol{\theta}(B)$ given by (14.10) and (14.12), respectively. Generally, variables will have different orders of AR and MA terms so p and q should be interpreted as the highest orders for which $\boldsymbol{\phi}_p$ and $\boldsymbol{\theta}_q$ have at least one non-zero element.

14.7 Our specification of $\boldsymbol{\phi}(B)$ and $\boldsymbol{\theta}(B)$ assumes that the leading elements are set equal to the identity matrix, or

$$\boldsymbol{\phi}_0 = \mathbf{I} \qquad \text{and} \qquad \boldsymbol{\theta}_0 = \mathbf{I}.$$

If we attempt to keep general forms for $\boldsymbol{\phi}_0$, $\boldsymbol{\theta}_0$ and $\boldsymbol{\Sigma}$, we have too many parameters; the model is said to be *unidentified*. There are two principal alternatives:

(a) $\boldsymbol{\phi}_0 = \mathbf{I}$, $\boldsymbol{\theta}_0 = \mathbf{I}$, $\boldsymbol{\Sigma}$ unrestricted;
(b) $\boldsymbol{\phi}_0$ has off-diagonal elements unrestricted, $\boldsymbol{\theta}_0 = \mathbf{I}$ and $\boldsymbol{\Sigma} = \mathbf{D}$ is diagonal.

Option (b) has been favoured in the econometric literature; see, for example, Johnston (1984, Chapters 12 and 13) or Theil (1971, Chapter 9). In time-series work, option (a) has been preferred as the least squares estimators then remain consistent. The principal reason for choosing (a) or (b) ultimately reduces to the issue of model specification. Econometricians feel that the simultaneous dependence built into option (b) is an essential ingredient for their model building, whereas time series analysts prefer to have a time delay built into the 'signal', allowing concurrent dependence to be reflected in the error structure. In principle, we may transform either system into the other with $\boldsymbol{\Sigma}$ in (a) and $(\boldsymbol{\phi}_0, \mathbf{D})$ in (b) satisfying the relationship

$$\boldsymbol{\phi}_0 \boldsymbol{\Sigma} \boldsymbol{\phi}_0^{\mathrm{T}} = \mathbf{D}, \tag{14.14}$$

where ^{T} denotes *transpose*, here and elsewhere. That is, multiplying model (b) throughout by $\boldsymbol{\phi}_0^{-1}$ produces a *reduced form* of type (a). The differences arise, of course, in the details of model specification for particular variables.

Stationarity and invertibility

14.8   As for the univariate case, the system is termed *stationary* if it can be represented as a pure *MA* process:

$$\mathbf{y}_t = [\boldsymbol{\phi}(B)]^{-1}\boldsymbol{\theta}(B)\boldsymbol{\varepsilon}_t$$
$$= (\mathbf{I} + \boldsymbol{\psi}_1 B + \boldsymbol{\psi}_2 B^2 + \ldots)\boldsymbol{\varepsilon}_t$$
$$= \boldsymbol{\psi}(B)\boldsymbol{\varepsilon}_t, \tag{14.15}$$

where the $\boldsymbol{\psi}_j$ are $(k \times k)$ matrices. A necessary and sufficient condition for this is that all the roots of the determinantal equation

$$|\boldsymbol{\phi}(x)| = 0 \tag{14.16}$$

should lie outside the unit circle.

Example 14.1   Consider a *VAR*(1) two-variable scheme, as in (14.1). Equation (14.16) becomes

$$\begin{vmatrix} 1 - \phi_{11}x & -\phi_{12}x \\ -\phi_{21}x & 1 - \phi_{22}x \end{vmatrix} = 0$$

or

$$(1 - \phi_{11}x)(1 - \phi_{22}x) - \phi_{12}\phi_{21}x^2 = 0. \tag{14.17}$$

If either $\phi_{12} = 0$ or $\phi_{21} = 0$, (14.17) produces the familiar conditions $|\phi_{11}| < 1$ and $|\phi_{22}| < 1$. However, such conditions are neither sufficient nor necessary in general. For example,

$$\boldsymbol{\phi}_1 = \begin{pmatrix} 0.9 & 0.7 \\ 0.7 & 0.9 \end{pmatrix}$$

leads to $x = 5$ and $x = 0.625$ so that the process is non-stationary. As a second example,

$$\boldsymbol{\phi}_1 = \begin{pmatrix} 0.8 & 0.2 \\ 0.2 & 0.8 \end{pmatrix}$$

produces roots $\frac{5}{3}$ and 1, so is also non-stationary. Note that unit roots arise without there being any differences in the model.

When a *VAR*(1) scheme is stationary, (14.15) reduces to

$$\mathbf{y}_t = (\mathbf{I} - \boldsymbol{\phi}_1 B)^{-1}\boldsymbol{\varepsilon}_t$$
$$= (\mathbf{I} + \boldsymbol{\phi}_1 B + \boldsymbol{\phi}_1^2 B^2 + \cdots)\boldsymbol{\varepsilon}_t$$
$$= \boldsymbol{\varepsilon}_t + \boldsymbol{\phi}_1\boldsymbol{\varepsilon}_{t-1} + \boldsymbol{\phi}_1^2\boldsymbol{\varepsilon}_{t-2} + \cdots \tag{14.18}$$

so that $\boldsymbol{\psi}_j = \boldsymbol{\phi}_1^j$, the matrix analogue of the univariate case.

14.9   In like manner, system (14.13) is *invertible* if and only if the roots of

$$|\boldsymbol{\theta}(B)| = 0,$$

lie outside the unit circle. If the system is invertible, it may be represented as a pure *VAR* scheme of infinite extent:

$$\begin{aligned}
\varepsilon_t &= [\theta(B)]^{-1}\phi(B)\mathbf{y}_t \\
&= (\mathbf{I} - \boldsymbol{\pi}_1 B - \boldsymbol{\pi}_2 B^2 - \cdots)\mathbf{y}_t \\
&= \boldsymbol{\pi}(B)\mathbf{y}_t.
\end{aligned} \tag{14.19}$$

Example 14.2   The *VMA*(1) scheme

$$\mathbf{y}_t = (\mathbf{I} - \boldsymbol{\theta}_1 B)\varepsilon_t$$

may be represented in *VAR* form as

$$\begin{aligned}
\varepsilon_t &= (\mathbf{I} - \boldsymbol{\theta}_1 B)^{-1}\mathbf{y}_t \\
&= (\mathbf{I} + \boldsymbol{\theta}_1 B + \boldsymbol{\theta}_1^2 B^2 + \cdots)\mathbf{y}_t \\
&= \mathbf{y}_t + \boldsymbol{\theta}_1 \mathbf{y}_{t-1} + \boldsymbol{\theta}_1^2 \mathbf{y}_{t-2} + \cdots
\end{aligned} \tag{14.20}$$

so that $\boldsymbol{\pi}_j = -\boldsymbol{\theta}_1^j$.

14.10   When we considered univariate series, the requirement of invertibility was sufficient to ensure a unique model; see Section 5.27. This is no longer true for the multivariate case.

Example 14.3   (cf. Tiao and Tsay, 1989)   Consider the two-variable *AR*(1) scheme with

$$\boldsymbol{\phi}_1 = \begin{pmatrix} 0 & a \\ 0 & a \end{pmatrix}.$$

It follows that

$$(\mathbf{I} - \boldsymbol{\phi}_1 B)^{-1} = \begin{pmatrix} 1 & -aB \\ 0 & 1 \end{pmatrix}^{-1} = \begin{pmatrix} 1 & aB \\ 0 & 1 \end{pmatrix}$$

so that the *VAR*(1) scheme is also representable as an *MA*(1) scheme with

$$\boldsymbol{\theta}_1 = \begin{pmatrix} 0 & -a \\ 0 & 0 \end{pmatrix}.$$

In general, such non-uniqueness arises if $\phi(B)$ may be represented as the product of two matrix polynomials such that

$$\phi(B) = \phi_1(B)\phi_2(B), \tag{14.21}$$

where $|\phi_1(B)| = \text{constant}$. A similar condition applies for $\theta(B)$. The exact conditions for unique model representation are rather awkward. Hannan (1969) gives necessary and sufficient conditions for uniqueness. For a k-series *VARMA*(p, q) scheme, a simpler sufficient condition is that

$$\text{rank}(\boldsymbol{\phi}_p, \boldsymbol{\theta}_q) = k. \tag{14.22}$$

Example 14.4   Consider the *VAR*(1) scheme in Example 14.3 with $k = 2$. The condition reduces to

$$\text{rank}(\boldsymbol{\phi}_1) = 2,$$

which clearly is false. If we had a $VARMA(1, 1)$ scheme with

$$\boldsymbol{\theta}_1 = \begin{pmatrix} b & c \\ d & f \end{pmatrix},$$

the matrix

$$\begin{pmatrix} 0 & a & b & c \\ 0 & 0 & d & f \end{pmatrix}$$

has rank 2 provided at least one of d or f is non-zero.

Autocorrelations

14.11 Provided the system is stationary, we can determine the autocorrelation structure using the same approach as in Sections 5.8–19. Let the covariance matrix of $\mathbf{y}_t$ be $V(\mathbf{y}_t) = \boldsymbol{\gamma}_0$ and the jth autocovariance matrix be

$$E(\mathbf{y}_t \mathbf{y}_{t-j}^T) = \boldsymbol{\gamma}_j. \tag{14.23}$$

For example, given the $VAR(1)$ scheme

$$\mathbf{y}_t = \boldsymbol{\phi}_1 \mathbf{y}_{t-1} + \boldsymbol{\varepsilon}_t \tag{14.24}$$

and assuming that $E(\mathbf{y}_t \, \boldsymbol{\varepsilon}_{t-j}^T) = \mathbf{0}, j \geqslant 1$, we may multiply through by $\mathbf{y}_{t-j}^T$ and take expected values to obtain

$$E(\mathbf{y}_t \mathbf{y}_{t-j}^T) = \boldsymbol{\gamma}_j = \boldsymbol{\phi}_1 \boldsymbol{\gamma}_{j-1}$$

so that

$$\boldsymbol{\gamma}_j = \boldsymbol{\phi}_1^j \boldsymbol{\gamma}_0, \qquad j = 1, 2, \dots. \tag{14.25}$$

Finally, if we multiply each side of (14.24) by its own transpose and take expectations, we arrive at

$$\boldsymbol{\gamma}_0 = \boldsymbol{\phi}_1 \boldsymbol{\gamma}_0 \boldsymbol{\phi}_1^T + \boldsymbol{\Sigma}. \tag{14.26}$$

Thus, (14.26) produces a set of $k(k + 1)/2$ equations that are linear in the elements of $\boldsymbol{\gamma}_0$, so that $\boldsymbol{\gamma}_0$ may be computed, given $\boldsymbol{\phi}_1$ and $\boldsymbol{\Sigma}$.

Example 14.5 Given

$$\boldsymbol{\Sigma} = \begin{pmatrix} 0.92 & 0.94 \\ 0.94 & 2.58 \end{pmatrix} \quad \text{and} \quad \boldsymbol{\phi}_1 = \begin{pmatrix} 0.6 & 0.2 \\ -0.3 & 0.4 \end{pmatrix},$$

find

$$\boldsymbol{\gamma}_0 = \begin{pmatrix} a & b \\ b & c \end{pmatrix}.$$

From (14.26) we obtain three equations such as

$$a = \phi_{11}^2 a + 2\phi_{11}\phi_{12}b + \phi_{12}^2 c + \sigma_{11}. \tag{14.27}$$

For our example, these equations become

$$a = \quad 0.36a + 0.24b + 0.04c + 0.92$$
$$b = -0.18a + 0.18b + 0.08c + 0.94$$
$$c = \quad 0.09a - 0.24b + 0.16c + 2.58$$

whence

$$\gamma_0 = \begin{pmatrix} 2 & 1 \\ 1 & 3 \end{pmatrix}.$$

14.12 Explicit solutions become awkward for higher-order *VAR* schemes, but the general result for *VMA*(q) schemes is expressible as

$$\gamma_j = \sum_{s=j}^{q} (\theta_s \, \Sigma \, \theta_{s-j}^T), \qquad 0 \leqslant j \leqslant q$$

$$= 0, \qquad\qquad\qquad j > q;$$

taking $\theta_0 = -\mathbf{I}$. Thus, (14.27) shows that a *VMA*(q) system has non-zero autocovariances only up to order q.

14.13 Once the autocovariances have been determined, the autocorrelation matrices follow as

$$\rho_j = \mathbf{D}\gamma_j\mathbf{D}, \tag{14.29}$$

where $\mathbf{D}^{-1}$ represents the diagonal matrix of standard deviations, or

$$\mathbf{D}^{-2} = \begin{bmatrix} \gamma_{11}(0) & & & \mathbf{0} \\ & \gamma_{22}(0) & & \\ & & \ddots & \\ \mathbf{0} & & & \gamma_{kk}(0) \end{bmatrix} \tag{14.30}$$

$\gamma_{rr}(0)$ being the diagonal elements of γ_0. The off-diagonal elements $\gamma_{rs}(j)$, $j = 0, \pm 1, \pm 2, \ldots$ are precisely the cross-correlations between series r and s, as defined in Sections 11.2–4. It follows that the parameters may be estimated using the corresponding sample quantities.

14.14 If we consider a *VAR* system of order p and evaluate the autocovariances, we arrive at the set of multivariate Yule–Walker equations

$$\phi_1^{(p)}\gamma_0 + \phi_2^{(p)}\gamma_1 + \phi_3^{(p)}\gamma_2 + \cdots + \phi_p^{(p)}\gamma_{p-1} = \gamma_1 \tag{14.31a}$$

$$\phi_1^{(p)}\gamma_{-1} + \phi_2^{(p)}\gamma_0 + \phi_3^{(p)}\gamma_1 + \cdots + \phi_p^{(p)}\gamma_{p-2} = \gamma_2 \tag{14.31b}$$

and eventually

$$\phi_1^{(p)}\gamma_{1-p} + \phi_2^{(p)}\gamma_{2-p} + \phi_3^{(p)}\gamma_{3-p} + \cdots + \phi_p^{(p)}\gamma_0 = \gamma_p, \tag{14.31c}$$

where $\gamma_{-j} = \gamma_j^T$; see (11.5). When the system is *VAR*(p), it follows that

$$\phi_r^{(j)} = \mathbf{0}, \qquad j > \mathrm{p}, \quad r > \mathrm{p}. \tag{14.32}$$

Equations (14.31) are analogous to the partial autocorrelations in the univariate case and can be reduced to correlation form if we set

$$\rho_j = \mathbf{D}^{-1}\gamma_j\mathbf{D}^{-1}, \phi_r^* = \mathbf{D}^{-1}\phi_r\mathbf{D}. \tag{14.33}$$

Whittle (1963) has given a recursive method of solution for equations (14.30); De Jong (1976) developed the multivariate version of Durbin's algorithm; see Section 5.19. If we replace the γ_j by their sample estimates, we can, at least in principle, use the resulting partial correlation estimates to identify an appropriate model.

Model identification

14.15 The sample autocorrelation and partial autocorrelation coefficients are asymptotically normally distributed and their standard errors may be determined by an extension of the results in Sections 6.10 and 6.12. As a first rough approximation, we may take $SE = n^{-1/2}$ for all the sample coefficients. Model identification procedures based on the sample autocorrelation matrices have been developed by Tiao and Box (1981) among others.

The sheer volume of information makes the arrays of coefficients difficult to assess. To simplify the presentation, we may code the values as follows:

$$> 2SE, \text{ write as } +$$
$$< -2SE, \text{ write as } -$$
$$\text{within } \pm 2SE, \text{ write as } \cdot$$

Example 14.6 Quenouille (1968) gave data for five variables connected with the hog market in the USA:

hog sales (y_1), hog prices (y_2), corn supply (y_3), corn prices (y_4) and farm wages (y_5).

The data refer to the period 1867–1948, $n = 82$ observations in all. The series are listed in Appendix A, Table A.14. Several analyses of these series have appeared

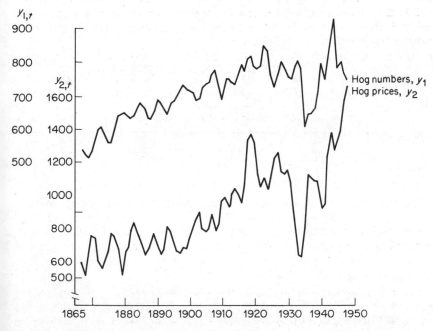

Fig. 14.1 US hog number (y_1) and hog prices (y_2) on logarithmic scale (from Quenouille, 1968)

since Quenouille's original study, notably Box and Tiao (1977) and Tiao and Tsay (1983, 1989). As our aim is expository, we shall consider only the submodel based upon y_1 and y_2. Following Quenouille and later authors, both series are transformed to logarithms before undertaking any analysis.

The two series are plotted in Fig. 14.1. The effects of the Great Depression are clearly seen as is the general upward trend in prices. The series on farm wages might be used as a price deflator in a more complete analysis. Beyond that, it can be seen that the members of the series tend to move together and there is a natural interaction between them. Our analysis is based on $y_1(t)$ and $y_2(t)$ concurrently, although it should be noted that there are some arguments in favour of using $y_1(t)$ and $y_2(t+1)$; see Box and Tiao (1977).

The correlations, for lags 0–3, are summarised in Table 14.1 together with their arrays of coded indicators. There is a slow decay in the cross-correlations but a much more rapid decay in the partials and we conclude that a reasonable initial model is $VAR(2)$.

Table 14.1 Cross-correlation and partial correlation matrices for US hog supply (y_1) and hog prices (y_2).

(a) Cross-correlations

Lag	0		1		2		3	
y_1	1.00	0.63	0.87	0.68	0.72	0.66	0.62	0.60
y_2	0.63	1.00	0.60	0.85	0.64	0.66	0.69	0.54
Coding	+	+	+	+	+	+	+	+
	+	+	+	+	+	+	+	+

(b) Partial correlations

Lag	0		1		2		3	
y_1	1.00	0.63	0.87	0.68	−0.19	−0.06	−0.03	0.01
y_2	0.63	1.00	0.60	0.85	0.34	−0.32	0.09	0.06
Coding	+	+	+	+	.	.	.	.
	+	+	+	+	+	−	.	.

Estimation

14.16 Parameter estimation by least squares or maximum likelihood follows the general principles developed in Sections 7.7–9 and 12.2–4. Hillmer and Tiao (1979) developed an exact likelihood method for $VARMA$ schemes; Spliid (1983) gives another procedure that appears to produce considerable savings in computer time. Ansley and Kohn (1985) give an efficient procedure for checking that estimates satisfy invertibility conditions.

Akaike (1976) developed an estimation procedure using the state space approach and the AIC criterion; see Sections 7.28–29. Harvey and Peters (1984) provide an algorithm based upon the Kalman filter; they note that it is easier to ensure that the parameter estimates are admissible for structural models than it is to check for invertibility in $VARMA$ schemes.

Example 14.7   (Example 14.6 continued)   The model was fitted using the MTS program; see Appendix D. The estimates for the $VAR(2)$ scheme, with t-values in parentheses, are as shown in Table 14.2. Coefficients not significant at the 90 percent level were dropped and are denoted by (·); the model was re-estimated after removing these redundant parameters. The covariance matrix was estimated as

$$\begin{pmatrix} 2.18 & -0.84 \\ -0.84 & 9.81 \end{pmatrix} \times 10^{-3}$$

and the residual cross-correlation matrices are as shown in Table 14.3.

These results suggest that it may be desirable to include a first-order moving average component. However, none of the additional parameters proves to be significant and we conclude that the $VAR(2)$ representation is adequate.

14.17   In the example we used t-tests on individual coefficients to decide whether or not to retain each term in the model. Others (e.g., Akaike, 1976) have recommended the use of AIC or some other information criterion. Tiao and Tsay (1989) develop the notion of scalar canonical models (SCMs) which can be used, in conjunction with a canonical correlation analysis, to arrive at a parsimonious representation. As noted earlier, the number of parameters increases with the square of the number of series, and it is clear that the estimation of even moderate-sized systems ($k \geqslant 10$) depends upon the analyst's skill in discarding large numbers of parameters. When some form of prior information is available, this may be used to impose restrictions on the parameter values; this approach has been adopted by Litterman (1986) for his Bayesian VAR, or $BVAR$, system.

14.18   An alternative approach is to develop the model in block-recursive form so that y_1 depends on x, y_2 on (x, y_1) and so on; x may well include lagged values of both y_1 and y_2. Such an approach reduces the estimation problem

Table 14.2

	$\hat{\phi}_1$		$\hat{\phi}_2$		
y_1	0.885	0.099	-0.178		
	(8.11)	(3.32)	(-1.70)		
y_2	·	1.009	0.621	-0.379	
		(10.1)	(4.13)	(-3.66)	

Table 14.3

lag	0		1		2		3	
	1.00	-0.15	-0.04	0.10	-0.06	-0.00	-0.03	-0.00
	-0.15	1.00	-0.27	0.14	-0.00	0.15	-0.06	-0.05
	+	·	·	·	·	·	·	·
	·	+	-	·	·	·	·	·

from a large one to several of more manageable size, but does require expert knowledge of the system in order to make the appropriate block specifications.

A further possibility is to develop models for each y_j conditionally upon the other *y*-values. That is, we develop univariate schemes

$$y_j \quad \text{on} \quad (y_i(i \neq j), \mathbf{x});$$

and then recombine these univariate schemes for final efficient estimation. This approach is discussed in Ord (1983).

Forecasting

14.19   Forecasts for vector schemes may be generated in the same way as for single series following the conventions given in Sections 8.10 and 12.14. Prediction intervals or regions may then be generated by the appropriate multivariate extensions of the arguments given in Sections 8.9 and 12.17–18.

Example 14.8   (Examples 14.6, 14.7 continued)   After deleting the last four observations, the model was re-estimated, yielding

$$\hat{\phi}_1 = \begin{pmatrix} 0.677 & 0.132 \\ -0.390 & 1.142 \end{pmatrix}, \qquad \hat{\phi}_2 = \begin{pmatrix} \cdot & \cdot \\ 0.800 & -0.462 \end{pmatrix}.$$

The *h*-step ahead forecasts are given by

$$\mathbf{y}_t(h) = \hat{\phi}_1 \mathbf{y}_t(h-1) + \hat{\phi}_2 \mathbf{y}_t(h-2),$$

where $y_t(h) = y_{t-h}$, $h \leqslant 0$. The results are given in Table 14.4. The forecasts

Table 14.4 One to four period ahead forecasts for hog data

Hog supply (y_1)

Year	Actual	Forecast	90 percent prediction interval	
			Lower	Upper
1945	774	885.8	823.3	953.0
1946	787	850.7	778.2	930.0
1947	754	827.1	748.8	913.6
1948	737	813.7	729.9	907.0

Hog price (y_2)

Year	Actual	Forecast	90 percent prediction interval	
			Lower	Upper
1945	1 314	1 130.0	967.7	1 319.4
1946	1 380	1 123.4	883.6	1 428.2
1947	1 556	1 146.4	875.8	1 500.6
1948	1 632	1 151.5	871.5	1 521.5

are not particularly impressive as supply drops sharply at the beginning of the four year period with an accompanying sharp increase in price, a shift reflected in the revised parameter estimates. Given the dramatic changes in market conditions occasioned by the end of World War 2, these results are not surprising.

14.20 As a second example, we re-analysed the data on car production and the FT share index, previously considered in Section 12.12. The correlation matrices are given in Table 14.5, giving a clear indication in favour of a $VAR(1)$ scheme.

The estimates for the initial model appear in Table 14.5(c); they suggest only a contemporaneous relationship between the series. It should be noted that the

Table 14.5 Cross-correlation and partial correlation matrices for UK car production (y_1) and the FT share index (y_2)

(a) Cross-correlations

Lag	0		1		2		3	
y_1	1.00	0.56	0.80	0.49	0.70	0.36	0.57	0.30
y_2	0.56	1.00	0.53	0.87	0.51	0.71	0.45	0.56
Coding	+	+	+	+	+	+	+	+
	+	+	+	+	+	+	+	+

(b) Partial correlations

Lag	0		1		2		3	
y_1	1.00	0.56	0.80	0.49	0.17	-0.21	-0.09	0.14
y_2	0.56	1.00	0.53	0.87	0.07	-0.23	-0.07	-0.00
Coding	+	+	+	+	.	.	.	.
	+	+	+	+	.	.	.	.

(c) Initial estimates

$$\hat{\phi}_1 = \begin{pmatrix} 0.796 & \cdot \\ \cdot & 0.873 \end{pmatrix} \hat{\Sigma} = \begin{pmatrix} 1\,603 & 307 \\ 307 & 609 \end{pmatrix}$$

(d) Residual cross-correlations

Lag	0		1		2		3	
y_1	1.00	0.31	-0.14	0.26	0.18	-0.10	0.14	0.09
y_2	0.31	1.00	0.03	0.29	0.07	0.11	-0.03	-0.10
Coding	+	+	.	+	.	.	.	.
	+	+	.	+	.	.	.	.

(e) Final estimates

$$\hat{\phi}_1 = \begin{pmatrix} 0.750 & \cdot \\ \cdot & 0.873 \end{pmatrix}, \quad \hat{\theta}_1 = \begin{pmatrix} \cdot & -0.422 \\ \cdot & \cdot \end{pmatrix}, \quad \Sigma = \begin{pmatrix} 1\,487 & 234 \\ 234 & 609 \end{pmatrix}$$

$$t\text{-values:} \begin{pmatrix} 8.18 & \cdot \\ \cdot & 12.2 \end{pmatrix} \quad \text{and} \quad \begin{pmatrix} \cdot & -1.70 \\ \cdot & \cdot \end{pmatrix}.$$

series were not differenced as it has been argued by several authors (e.g. Tiao and Box, 1981) that one should not difference individual components prior to vector modelling.

The residual cross-correlations, in Table 14.5(d), suggest the possible addition of a first-order moving average component. The final estimates, in Table 14.5(e), show the lag 1 *MA* component for cars on the FT index to be marginally significant. We may conclude that the FT index has some weak explanatory power for the car production figures, but there is no flow the other way. This concurs with the earlier analysis in Section 12.12.

14.21   *VARMA* schemes provide structural models, a benefit that may be further enhanced by adding explanatory variables. However, whether substantial forecasting gains materialise from using these models is more open to question. McNees (1986) compared the forecasting performance of Litterman's *BVAR* method with that of several major econometric models; the results are generally encouraging. As more empirical comparisons become available, the strengths and weaknesses of vector methods will be better understood. As we have noted previously, the method of analysis ultimately selected will depend on both the purpose of the study and the subject matter.

Exercises

14.1   A *VAR*(1) process has

$$\phi_1 = \begin{pmatrix} 0.8 & 0.7 \\ 0.7 & 0.8 \end{pmatrix}.$$

Is it stationary?

14.2   Invert the *VMA*(1) scheme with

$$\theta_1 = \begin{pmatrix} 0.4 & -0.4 \\ 0.2 & 0.6 \end{pmatrix}$$

as far as $p = 3$. (Check that it is invertible first.)

14.3   Find the cross-correlation function for the *VMA*(1) scheme with $\Sigma = I$ and θ_1 as in Exercise 14.2.

14.4   Carry out a VARMA analysis for one or more of the following data sets:
   (a) UK imports (Table 12.5)
   (b) US flour prices (Appendix A, Table A12)
   (c) Lydia Pinkham sales and advertising (Appendix A, Table A13)
   (d) US hogs (Appendix A, Table A14)

15

Other recent developments

Introduction

15.1 It is inevitable that an introductory book such as this must treat many topics only briefly or not at all. Yet it is often these omitted areas that represent the most exciting parts of the subject in terms of new developments. In order to overcome this difficulty, at least partially, the present chapter gives brief details of recent developments in a variety of areas, and provides references for further reading.

15.2 Our discussion of seasonal adjustment procedures in Chapter 4 was presented before we had considered *ARIMA* models. The fusion of classical adjustment procedures with formal modelling techniques is one of the major recent advances in time series. These developments are discussed in Sections 15.5–7.

15.3 The *ARIMA* models considered so far have assumed complete sets of observations taken at regular intervals. Further, it has been assumed that regular or seasonal differencing will be sufficient to induce stationarity and, finally, that least-squares procedures will generate satisfactory estimates. All of these assumptions may break down; Sections 15.8–11 discuss missing values and unequal spacing, fractional differencing, and robust estimation.

15.4 A further major area that is attracting increasing attention is that of non-linear time-series, which we considered only indirectly when using simple transformations. In the frequency domain, the bispectrum may be used to check for nonlinearity and a variety of time-domain methods have been developed to handle non-linearities. These are described in Sections 15.12–21. The chapter concludes with a brief discussion of multidimensional processes in Sections 15.22–25.

Seasonal adjustment

15.5 Our discussion of seasonal adjustment procedures in Chapter 4 revealed both a strength and a weakness in moving average procedures. The strength lay in their local nature, ensuring that observations in the distant past would be adjusted slightly or not at all when a new observation came to hand. Their weakness was the high variability associated with the adjustments to the most recent observations. This led to the development of the *X11−ARIMA* method (Dagum, 1975) whereby the series is first forecast using a seasonal *ARIMA* model; the *X11* smoothing operations are then applied to the extended series. This procedure has been found to produce smaller revisions to adjusted values than the pure *X11* method (e.g. Dagum 1982). Further, Cleveland and Tiao (1976) showed that the additive *X11* procedure is closely approximated by an *ARIMA* scheme, thereby bringing the debate on seasonal adjustment procedures into the mainstream of modern time series analysis.

15.6   A second issue of major importance to official statisticians is whether adjustments should be *concurrent*; that is, whether the adjustment factors should be recalculated every time a new observation comes to hand. In general, most official procedures have been based upon *periodic* revisions, whereby seasonal adjustment factors are computed once a year and then applied to new observations as they become available. Kenney and Durbin (1982) showed that concurrent adjustment provides major benefits in terms of reducing the magnitude of revisions to the seasonally adjusted series; they also demonstrated that *X11-ARIMA* is to be preferred over pure *X11*. McKenzie (1984) has also demonstrated the benefits to be derived from concurrent adjustment. Pierce (1983) provides a summary of a report produced for the US Federal Reserve, whose recommendations include moving to concurrent adjustment. Many government agencies are now moving to adopt these recommendations: an encouraging sign that sound statistical studies can lead to changes of policy.

15.7   A comprehensive review of current issues in seasonal adjustment is provided by Bell and Hillmer (1983a,b); this paper is followed by comments from several discussants and a reply by the authors.

Missing values and unequal spacing

15.8   Our discussions have always assumed complete data records and equal time intervals between observations. Modest departures from these assumptions such as the different lengths of the months can be ignored if the recorded variable is a *stock* (e.g. total ownership of cars) or adjusted to per-day if it is a *flow* (e.g. production of cars). Such modifications have been found to work well in practice (Granger, 1963).

When unequal spacing of the observations is more marked, a different approach must be employed. For example, if the underlying process is a continuous $AR(1)$ scheme and observations occur at times t_1, t_2, ..., we must consider

$$y(t_k) = a(u_k)y(t_{k-1}) + \varepsilon(u_k), \tag{15.1}$$

where

$$u_k = t_k - t_{k-1}, \qquad a(u_k) = \exp(-\alpha u_k),$$

and

$$E\{\varepsilon(u_k)\} = 0, \qquad \text{var}\{\varepsilon(u_k)\} = \sigma^2 u_k.$$

Model (15.1) reduces to the regular $AR(1)$ scheme when $u_k = 1$ and $\phi = \exp(-\alpha)$. Quenouille (1958) was a pioneer in developing such AR schemes, which received relatively little further attention until 1980 or thereabouts. Since then the Kalman filter has been shown to provide an elegant framework within which such models can be developed and fitted iteratively. In essence the state space component can be updated repeatedly between observations and the observation component is updated as and when another observation becomes available. This same process can be applied to handle regularly spaced series with missing values. Jones (1985) provides an excellent introduction to this procedure.

Wright (1986) provides an extension of Holt's method for forecasting based upon irregularly spaced data.

Fractional differencing

15.9 When we defined the differencing operation in Section 3.14, we always took the power d to be an integer. However, in principle, we could consider fractional values of d, $0 < d < 1$. To see the effects of this operation, consider the model

$$(1 - B)^d y_t = \nabla^d y_t = \varepsilon_t. \tag{15.2}$$

When $d = 0$, (15.2) is a white noise process and when $d = 1$ it corresponds to the random walk. For fractional d, we may express (15.2) in random shock form as

$$y_t = (1 - B)^{-d} \varepsilon_t$$

$$= \left(1 + dB + \frac{d(d+1)}{2} B^2 + \cdots\right) \varepsilon_t$$

$$= \varepsilon_t + d\varepsilon_{t-1} + \frac{d(d+1)}{2} \varepsilon_{t-2} + \cdots. \tag{15.3}$$

The ratio of successive coefficients in (15.3) approaches one, rather than strictly less than one in AR schemes; cf. (5.27). Thus (15.2) allows for *strong persistence* in a time series. Yet, the model is stationary for $d < \frac{1}{2}$, leading to

$$\mathrm{var}(y_t) = \frac{\sigma^2 \Gamma(1 - 2d)}{[\Gamma(1 - d)]^{1/2}}. \tag{15.4}$$

Thus, model (15.2) may describe a stationary process with a very slow rate of decay in the coefficients, termed a long memory process. Conversely, fractional differencing with $d < 1$ may be more appropriate as a way of removing apparent non-stationarity. For further details, see Granger and Joyeux (1980) and Hosking (1981).

Robust estimation

15.10 Throughout this book we have recognised that extreme observations may have a considerable impact upon estimation procedures and subsequent forecasting performance. Two types of extreme observation may be considered, as noted in Section 13.4. In general, innovation outliers (IO) may even be beneficial in estimation but additive outliers (AO) can seriously distort the estimation process. Unfortunately, as indicated in the review by Martin and Yohai (1985), the usual M-type robust estimators (cf. Huber, 1981) are ineffective for the AO-case. Various modifications are possible, but these tend to work only for AR schemes. Instead, Martin and Yohai (1985) recommend the use of *robust smoother–cleaners* whereby extrema are identified and adjusted by means of a modified Kalman filter operation. This notion is quite close to the idea of testing for outliers and adjusting by intervention variables employed in Section 13.13, except that the robust procedure operates in a smooth fashion. In both cases, the resulting estimators may lack desirable large-sample properties such as Fisher consistency, but this is likely to be outweighed in practice by the ability to avoid the bias caused by large outliers of the AO-type.

15.11   Estimates of the spectrum are equally affected by additive outliers. An iterative scheme based on the smoother–cleaner approach may be utilized here also; see, for example, Martin (1983).

Nonlinear models

15.12   With the exception of multiplicative seasonal models of the Holt–Winters type in Section 8.23, our development has relied heavily on the assumption that processes are linear. Unfortunately, nature is not always so understanding. We now explore several ways in which non-linear models may be developed, although it has to be admitted that our understanding of such processes is far from complete.

There are four principal ways in which we can examine non-linear processes:

(1)   consider nonlinear functions of past y_t and ε_t;
(2)   develop intrinsically non-linear schemes that change as certain boundaries or thresholds are crossed;
(3)   introduce random coefficients which enable us to make successive linear approximations to non-linear schemes;
(4)   apply transformations which induce linearity (as with the Box–Cox transform introduced in Section 5.4).

In the following sections we review each of these approaches in turn, but first we consider the bispectrum which provides a frequency description for non-linear processes.

Bispectra

15.13   The regular spectrum uses only the second order, or covariance, structure of a time series. Frequency-domain analysis may be extended to higher orders using *polyspectra*, developed by Brillinger (1965). The case of greatest interest is the *bispectrum* which may be defined in terms of the third-order moments in the following way.

Consider a stationary time series $y(t)$ with

$$E\{y(t)\} = \mu, \qquad \text{cov}\{y(t), y(t-s)\} = \gamma_s,$$

and define the third-order moments

$$\gamma(s_1, s_2) = E\{u(t)u(t - s_1)u(t - s_2)\}, \tag{15.5}$$

where $u(t) = y(t) - \mu$. The concept of stationarity used now extends the assumptions that $E\{y^6(t)\} < \infty$, and that all expectations depend only on relative positions in the series, as in (15.5). Then the bispectrum is given by

$$w(\alpha_1, \alpha_2) = \frac{1}{4\pi^2} \Sigma\Sigma \, \gamma(s_1, s_2)\exp(-i\alpha_1 s_1 - i\alpha_2 s_2), \tag{15.6}$$

where the sums range over $-\infty < s_1, s_2 < \infty$.

The bispectrum may be estimated from the sample version of (15.6), provided an appropriate two-dimensional window is used; see, for example, Brillinger and Rosenblatt (1967a,b). A test for non-linearity using the bispectrum is described in Subba Rao (1983).

Bilinear models

15.14 Several attempts have been made to develop time series models as non-linear functions of past y_t and ε_t values, but many of these foundered on the problem of being able to specify reasonable conditions for stationarity. However, moving average models for proportions, such as

$$\frac{y_t - y_{t-1}}{y_{t-1}} = \varepsilon_t - \theta_1\varepsilon_{t-1} \tag{15.7}$$

or

$$y_t = y_{t-1} + \varepsilon_t y_{t-1} - \theta_1\varepsilon_{t-1}y_{t-1} \tag{15.8}$$

suggest the class of *bilinear* models:

$$y_t - \phi_1 y_{t-1} - \cdots - \phi_p y_{t-p} = \varepsilon_t - \theta_1\varepsilon_{t-1} - \cdots - \theta_q\varepsilon_{t-q}$$
$$+ \sum_{j=1}^{m} \sum_{i=1}^{k} \beta_{ij} y_{t-j}\varepsilon_{t-i}, \tag{15.9}$$

which may be denoted by $BL(p, q, m, k)$. This class has been studied in detail by Granger and Andersen (1978) and Subba Rao (1981, 1983).

Example 15.1 Consider the $BL(1, 0, 1, 1)$ scheme

$$y_t = \phi y_{t-1} + \beta y_{t-1}\varepsilon_{t-1} + \varepsilon_t, \tag{15.10}$$

where the error terms are independent and identically distributed $N(0, 1)$ variables and ε_t is independent of y_{t-j} for all $j > 0$. Then it follows that

$$E(y_t) = \beta/(1 - \phi) = \mu \tag{15.11}$$

$$V(y_t) = \frac{(1 - \phi + 2\beta^2 + 2\phi\beta^2)}{(1 - \phi)(1 - \phi^2 - \beta^2)} - \mu^2 = \sigma^2 \tag{15.12}$$

and the autocorrelations are

$$\rho_1 = \phi + \frac{\beta^2}{\sigma^2(1 - \phi)}, \quad \rho_j = \phi^{j-1}\rho_1, \quad j \geqslant 2. \tag{15.13}$$

Second-order (asymptotic) stationarity follows from the expressions for the mean and variance; we require that

$$|\phi| < 1 \quad \text{and} \quad \phi^2 + \beta^2 < 1.$$

15.15 General conditions for asymptotic stationarity are given by Subba Rao (1981); also, he demonstrates that the $BL(p, 0, p, 1)$ scheme has the same autocorrelation structure as the $ARMA(p, 1)$ process. This can be seen from Example 5.9 and (15.13) when $p = 1$. In consequence, it is not possible to distinguish between BL and $ARMA$ schemes solely on the basis of their second-order properties. Granger and Anderson (1978) recommend consideration of the ACF of e_t^2, the squared residuals. For linear models, all these autocorrelations have near-zero expectations, but features of interest will show up for non-linear schemes. Higher-order frequency-domain properties may be examined using the *bispectrum* defined in Section 15.13; see Subba Rao (1983). Estimation and prediction procedures are considered by Gabr and Subba Rao (1981).

Threshold autoregression

15.16 Many systems may be subject to structural change of a more or less predictable type. For example, an economy may be operating under conditions of either labour shortage (full employment) or labour surplus (underemployment). Again, the population dynamics of a wildlife population are different when resources are plentiful than when resource are scarce or under pressure from that population. In such cases, it is natural to consider using different models for the two regimes, switching from one to the other when some threshold is crossed; hence the use of *threshold autoregressive* (TAR) models.

Example 15.2 A first-order *TAR* with two submodels is

$$y_t = \phi_{11}y_{t-1} + \varepsilon_t, \quad \text{if } y_{t-1} \leqslant c, \tag{15.14a}$$
$$y_t = \phi_{12}y_{t-1} + \varepsilon_t, \quad \text{if } y_{t-1} > c; \tag{15.14b}$$

c is the threshold parameter and the usual assumptions are made concerning the error process. The process is stationary provided $|\phi_{1j}| < 1$, $j = 1, 2$. Extensions to more submodels and higher-order schemes are straightforward. Details and examples are given in Tong and Lim (1980), including a discussion of the lynx data (Appendix Table A4). The frequency domain properties of TAR schemes are considered by Pemberton and Tong (1983).

Random coefficients

15.17 The state-space models of Chapter 9 allow coefficients that are time-dependent; in that discussion, the coefficients were usually taken to be

non-stationary although that is clearly not necessary. In a similar vein, but w
a rather different emphasis, we may consider random coefficients models su
as the following $AR(1)$ scheme:

$$y_t = \phi_{1t} y_{t-1} + \varepsilon_t \qquad (15.15a)$$

$$\phi_{1t} = \phi_1 + \eta_t, \qquad (15.15b)$$

where $\varepsilon_t \sim IIN(0, \sigma^2)$ and $\eta_t \sim IIN(0, \omega^2)$ and (ε_t, η_s) are independent for all t
and s. Model (15.15) may be extended in the usual way to include both
higher-order lags and explanatory variables. A comprehensive discussion is
given in Nicholls and Pagan (1985); our discussion follows their general
development.

Equations (15.15) reduce to

$$y_t = \phi_1 y_{t-1} + \varepsilon_t + \eta_t y_{t-1}, \qquad (15.16)$$

from which it follows that the process is stationary if and only if

$$\phi_1^2 + \omega^2 < 1. \qquad (15.17)$$

Least-squares estimators may be obtained by the following two-step process:

(1)   estimate ϕ_1 by the usual LS estimator

$$\hat{\phi}_1 = \Sigma \, y_t y_{t-1} / \Sigma \, y_{t-1}^2; \qquad (15.18)$$

(2)   compute the residuals

$$u_t = y_t - \hat{\phi}_1 y_{t-1} \qquad (15.19)$$

and fit the regression model

$$v_t = u_t^2 = \alpha_0 + \alpha_1 y_{t-1}^2 + \xi_t, \qquad (15.20)$$

where $\alpha_0 = \sigma^2$ and $\alpha_1 = \omega^2$ since

$$E(u_t^2 \mid y_{t-1}) = \sigma^2 + \omega^2 y_{t-1}^2. \qquad (15.21)$$

A negative value for $\hat{\alpha}_1$ implies that we set $\hat{\omega}^2 = 0$. These estimators are
consistent and easy to use, but inefficient. Maximum likelihood estimators are
given in Nicholls and Pagan (1985).

15.18   Under the assumptions of model (15.15) the one-step ahead prediction
interval for y_{t+1} given y_t is, approximately,

$$\hat{\phi}_t y_t \pm z(\sigma^2 + \omega^2 y_t^2)^{1/2}; \qquad (15.22)$$

these intervals clearly widen as y_t departs increasingly from the mean value of
zero, reflecting the relative lack of knowledge of the process.

Transformations and growth curves

15.19   The Box–Cox transform, introduced in Section 5.4, is a useful
data-analytic device for stabilising variances which also tends to produce more
nearly linear processes. However, the power transform is clearly restricted in
the types of non-linearity it can approximate. In particular, there is no

convenient way of allowing for upper or lower bounds on the range of the random variable, yet such bounds often occur in practice.

A good example is the logistic growth curve

$$y_t = L_0 + \frac{(L - L_0)}{1 + \exp\{\beta(t - t_0)\}}, \tag{15.23}$$

where L_0, L and β are all positive. This curve has the limits

$$y_t \to L_0 \qquad \text{as } t \to -\infty$$

and

$$y_t \to L \qquad \text{as } t \to \infty;$$

also there is an inflection at $t = t_0$ when $y_t = (L + L_0)/2$. The logistic has been used used widely as a model for growth processes since it has finite upper and lower bounds and a natural interpretation for β as a rate-of-change parameter. Suitably scaled, the logistic is very similar to the normal distribution function except in the extreme tails.

Differentiating (15.23) with respect to t we obtain, after some manipulation,

$$\frac{dy_t}{dt} = \beta(y_t - L_0)(L - y_t). \tag{15.24}$$

In this form the logistic is often used to describe epidemiological processes, usually with $L_0 = 0$. In this context, the derivative represents the rate of increase in infectives, y represents the number of infectives, $L - y$ the number of susceptibles and β the rate of new infections per contact between infectives and susceptibles. The rate peaks at $y = L/2$, corresponding to the inflection mentioned above. The logistic has also been used to describe the rate of technological change (Martino, 1983) and market penetration for new products (Bass, 1969).

15.20   The problem that the logistic poses for the time-series analyst is how to incorporate the random error component. For example, Bass (1969) used (15.24), approximating the derivative by a difference and reparameterising to the form

$$\nabla y_t = \beta_0 + \beta_1 y_{t-1} + \beta_2 y_{t-1}^2 + \varepsilon_t. \tag{15.25}$$

Unfortunately, this model often fails to perform satisfactorily in this form as a peak is reached and then y_t is projected to decline, contrary to the known nature of the process (Heeler and Hustad, 1980).

15.21   In growth curve studies, the four-parameter form (15.23) is often used with an additive error term satisfying the usual assumptions; that is, $\varepsilon \sim IN(0, \sigma^2)$. This appears to work quite well in those circumstances where overall fit is more critical than extrapolation.

For time series studies, it is more appropriate to use local trends than global ones, as for linear processes. When processes are cumulative, it is better to consider the increments over time, thereby using the previous level as the new starting point. A variety of such schemes have been suggested, see Ord and Young (1989), but one particular approach that seems promising is to consider the transformation

$$h_t = h(y_t) = \ln\{(y_t^\gamma - L_0^\gamma)/(L^\gamma - y_t^\gamma)\} \tag{15.26}$$

and to set

$$h_t = \beta_0 + \beta_1 t + \varepsilon_t', \qquad (15.27)$$

or

$$\nabla h_t = \beta_1 + \varepsilon_t''. \qquad (15.28)$$

When $\gamma = 1$, (15.26) corresponds to the logistic curve and when $\gamma = 0$, to the Gompertz curve:

$$y_t = L \exp(-\exp(-\beta_0 - \beta_{1t})), \qquad (15.29)$$

another popular model for such processes. Thus, (15.26) offers a family of transformations similar to the Box–Cox transform for the unbounded case; see Section 5.4. Further, (15.27) can easily handle unequally spaced data if the error variance is made proportional to the time between observations; (15.28) can be adapted in similar fashion. An attraction of (15.28) is that it allows the investigator to consider the whole range of *ARMA* schemes for modelling the error process. For further details, see Ord and Young (1989).

Multidimensional processes

15.22 Processes may be defined spatially as well as, or in place of, the time dimension. Assumptions of stationarity may then be applicable in some or all dimensions. For example, consider a two-dimensional spatial process observed on the regular grid of locations

$$(i, j), \qquad i = 1, \ldots, N_1, \quad j = 1, \ldots, N_2.$$

An array of autocorrelations may be defined by

$$\rho(s_1, s_2) = \gamma(s_1, s_2)/\gamma(0, 0), \qquad (15.30)$$

where γ represents the covariances. The sample analogue is

$$r(s_1, s_2) = c(s_1, s_2)/c(0, 0) \qquad (15.31)$$

where the covariance terms are given by

$$(N - s_1)(N - s_2)c(s_1, s_2) = \Sigma\Sigma \, u(i, j)u(i - s_1, j - s_2), \qquad (15.32)$$

where $u(i, j) = y(i, j) - \bar{y}$ and the sums are taken over $i = s_1 + 1$ to N_1, $j = s_2 + 1$ to N_2. Guyon (1982) showed that the corrections $(N_i - s_i)$ are necessary for two- and higher-dimensional schemes. The spectrum is

$$w(\alpha_1, \alpha_2) = \frac{1}{4\pi^2} \Sigma\Sigma \, \gamma(s_1, s_2)\exp(-i\alpha_1 s_1 - i\alpha_2 s_2), \qquad (15.33)$$

where the sums ranges over $-\infty < s_1, s_2 < \infty$. As in other cases, the sample analogue must be smoothed in order to obtain consistent estimators. Ripley (1981, pp. 81–7) gives several examples of spatial spectra.

15.23 Considerable attention has been devoted to *isotropic* or direction invariant spatial processes which depend only on the distance between two points, $y(\mathbf{x}_1)$ and $y(\mathbf{x}_2)$, defined as

$$d_{12} = \{(x_{11} - x_{21})^2 + (x_{12} - x_{22})^2\}^{1/2}, \qquad (15.34)$$

where $\mathbf{x}_i^T = (x_{i1}, x_{i2})$.

The autocovariance term then becomes

$$c(d) = \Sigma \ u(\mathbf{x}_i)u(\mathbf{x}_j)/N(d),\qquad\qquad(15.35)$$

where the sum is taken over all pairs (i, j) such that $d_{ij} = d$ or, because of data limitations, all pairs that fall in some range $d_L \leqslant d \leqslant d_U$, there being $N(d)$ such pairs. The detailed analysis of spatial dependence, for both regular lattice and irregularly located data, is discussed in Cliff and Ord (1981).

15.24   Spatio-temporal processes that assume both spatial and temporal dependence are considered by Aroian and his co-workers in a series of papers (see Aroian, 1980) and by Pfeiffer and Deutsch (1980, and references cited therein); see also the special issue of *Communications in Statistics*, series B (1980) devoted to this topic.

15.25   It is often much more difficult either to justify the assumption of spatial stationarity or to induce stationarity by suitable filtering operations. It is then more appropriate to consider the process as multivariate time series schemes (Bennett, 1979) or a spatial econometric model (Anselin, 1988).

Appendix A

Data sets and references

Table A.1 Values of series $u_t = 1.1u_{t-1} - 0.5u_{t-2} + \varepsilon_t$ where ε_t is a rectangular random variable with range -9.5 to 9.5, rounded off to nearest unit

Number of term	Value of series	Number of term	Value of series	Number of term	Value of series
1	7	23	−4	45	−13
2	6	24	−5	46	1
3	−6	25	−9	47	6
4	−4	26	−4	48	4
5	3	27	−4	49	11
6	−4	28	3	50	15
7	−5	29	9	51	9
8	−1	30	4	52	8
9	10	31	−8	53	4
10	10	32	−6	54	−1
11	6	33	−3	55	4
12	−4	34	−2	56	7
13	−4	35	0	57	11
14	−7	36	−1	58	0
15	−2	37	−3	59	1
16	6	38	3	60	0
17	17	39	−1	61	−5
18	24	40	−8	62	−11
19	17	41	−3	63	−8
20	4	42	−8	64	−3
21	1	43	−10	65	5
22	−5	44	−16		

Source: Kendall, 1946.

Table A.2 Gross domestic product, at constant factor prices, for United Kingdom (1980 = 100)

Year	Value	Year	Value
1948	48.0	1968	85.8
1949	50.2	1969	87.5
1950	52.6	1970	88.9
1951	53.8	1971	90.7
1952	53.3	1972	93.2
1953	56.3	1973	98.1
1954	58.8	1974	92.8
1955	60.9	1975	92.6
1956	61.4	1976	96.1
1957	62.5	1977	97.2
1958	62.4	1978	100.0
1959	65.2	1979	102.1
1960	68.6	1980	100.0
1961	69.9	1981	98.9
1962	70.7	1982	99.9
1963	73.1	1983	103.7
1964	77.3	1984	105.5
1965	79.6	1985	109.5
1966	81.0	1986	112.8
1967	82.2	1987	117.6

Source: Central Statistical Office.

Table A.3 Wölfer sunspot numbers: yearly, based on 100 observations

Year	Value	Year	Value	Year	Value	Year	Value
1770	101	1795	21	1820	16	1845	40
1771	82	1796	16	1821	7	1846	62
1772	66	1797	6	1822	4	1847	98
1773	35	1798	4	1823	2	1948	124
1774	31	1799	7	1824	8	1949	96
1775	7	1800	14	1825	17	1850	66
1776	20	1801	34	1826	36	1851	64
1777	92	1802	45	1827	50	1852	54
1778	154	1803	43	1828	62	1953	39
1779	125	1804	48	1829	67	1954	21
1780	85	1805	42	1830	71	1855	7
1781	68	1806	28	1831	48	1856	4
1782	38	1807	10	1832	28	1857	23
1783	23	1808	8	1833	8	1858	55
1784	10	1809	2	1834	13	1859	94
1785	24	1810	0	1835	57	1860	96
1786	83	1811	1	1836	122	1861	77
1787	132	1812	5	1837	138	1862	59
1788	131	1813	12	1838	103	1863	44
1789	118	1814	14	1839	86	1864	47
1790	90	1815	35	1840	63	1865	30
1791	67	1816	46	1841	37	1866	16
1792	60	1817	41	1842	24	1867	7
1793	47	1818	30	1843	11	1868	37
1794	41	1819	24	1844	15	1869	74

Source: Schuster (1906).
Discussions: Box and Jenkins (1976), who also cite earlier studies.

Table A.4 Numbers of lynx trapped in Mackenzie River district of NW Canada from 1821 to 1934

1821−40	1841−60	1861−80	1881−1900	1901−20	1921−34
269	151	236	469	758	229
321	45	245	736	1 307	399
585	68	552	2 042	3 465	1 132
871	213	1 623	2 811	6 991	2 432
1 475	546	3 311	4 431	6 313	3 574
2 821	1 033	6 721	2 511	3 794	2 935
3 928	2 129	4 254	389	1 836	1 537
5 943	2 536	687	73	345	529
4 950	957	255	39	382	485
2 577	361	473	49	808	662
523	377	358	59	1 388	1 000
98	225	784	188	2 713	1 590
184	360	1 594	377	3 800	2 657
279	731	1 676	1 292	3 091	3 396
409	1 638	2 251	4 031	2 985	
2 285	2 725	1 426	3 495	3 790	
2 685	2 871	756	587	674	
3 409	2 119	299	105	81	
1 824	684	201	153	80	
409	299	229	387	108	

Source: Elton and Nicholson (1942).
Discussions include: Tong (1977), Pemberton and Tong (1983).

Table A.5 US Government Treasury Bill Rate 1974−80: monthly average figure (per cent)

Month	Year						
	1974	1975	1976	1977	1978	1979	1980
Jan.	7.76	6.49	4.96	4.60	6.45	9.35	12.04
Feb.	7.06	5.58	4.85	4.66	6.46	9.27	12.81
Mar.	7.99	5.54	5.05	4.61	6.32	9.46	15.53
Apr.	8.23	5.69	4.88	4.54	6.31	9.49	14.00
May	8.43	5.32	5.18	4.94	6.43	9.58	9.15
June	8.14	5.18	5.44	5.00	6.71	9.05	7.00
July	7.75	6.16	5.28	5.15	7.07	9.26	8.13
Aug.	8.74	6.46	5.15	5.50	7.04	9.45	9.26
Sept.	8.36	6.38	5.08	5.77	7.84	10.18	10.32
Oct.	7.24	6.08	4.93	6.19	8.13	11.47	11.58
Nov.	7.58	5.47	4.81	6.16	8.79	11.87	13.89
Dec.	7.18	5.50	4.35	6.06	9.12	12.07	15.66

Source: US Dept of Commerce
Discussion: Kendall, Stuart and Ord (1983).
Note:
There was a policy change in late 1979; it is suggested that initial analysis be confined to 1974−79.

Table A.6 Sales Data for company X

Year	Jan.	Feb.	Mar.	Apr.	May	June	July	Aug.	Sept.	Oct.	Nov.	Dec.
1965	154	96	73	49	36	59	95	169	210	278	298	245
1966	200	118	90	79	78	91	167	169	289	347	375	203
1967	223	104	107	85	75	99	135	211	335	460	488	326
1968	346	261	224	141	148	145	223	272	445	560	612	467
1969	518	404	300	210	196	186	247	343	464	680	711	610
1970	613	392	273	322	189	257	343	404	677	858	895	664
1971	628	308	324	248	272							

Source: Chatfield and Prothero (1973).
Discussion: In and following Chatfield and Prothero (1973); Raveh (1985).

Table A.7 UK unemployment, quarterly, 1949–80

Year	Quarter 1	2	3	4
1949	1.45	1.20	1.02	1.12
1950	1.38	1.21	1.22	1.37
1951	1.92	1.56	1.27	1.55
1952	2.07	2.21	2.16	2.49
1953	2.78	2.29	2.01	2.04
1954	2.12	1.68	1.46	1.67
1955	1.86	1.46	1.46	1.79
1956	2.14	2.02	2.11	2.46
1957	3.67	2.51	2.21	2.17
1958	2.12	1.71	1.60	1.59
1959	1.71	1.44	1.39	1.49
1960	1.52	1.30	1.33	2.30
1961	2.65	2.33	2.26	2.57
1962	2.69	2.49	2.43	2.51
1963	2.68	2.40	2.48	2.55
1964	2.80	2.62	2.70	2.68
1965	3.29	3.45	3.82	4.13
1966	4.20	3.83	3.87	3.50
1967	3.33	2.80	2.56	2.26
1968	2.73	2.43	2.70	2.75
1969	3.37	3.66	4.67	5.00
1970	5.73	5.57	6.30	5.80
1971	6.00	5.93	6.87	6.30
1972	6.30	6.00	6.57	5.83
1973	6.03	5.53	5.97	5.63
1974	6.13	6.46	8.17	8.90
1975	10.17	10.70	12.13	12.27
1976	12.73	12.60	13.73	13.44
1977	13.60	12.87	12.87	12.89
1978	13.33	12.83	13.13	13.40
1979	13.67	13.33	12.96	11.87
1980	12.13	11.87	11.97	11.67

Source: Central Statistical Office.
Discussion: Koot, Young and Ord (1992).

Table A.8 Thousand hectolitres of home-produced whisky, UK

	Jan.	Feb.	Mar.	Apr.	May	June	July	Aug.	Sept.	Oct.	Nov.	Dec.
1980	34.6	59.1	82.5	9.3	12.2	19.5	28.5	29.3	35.1	58.9	79.8	52.8
1981	53.5	67.9	50.5	6.6	12.3	18.5	24.0	29.6	34.6	53.8	73.3	52.6
1982	31.5	57.4	61.9	7.1	12.4	21.9	19.7	26.8	30.9	49.7	72.4	55.9
1983	10.4	50.6	73.6	8.7	16.2	23.4	20.9	27.2	31.4	50.2	82.8	49.3
1984	10.7	54.7	78.3	7.5	12.9	17.6	23.5	26.9	26.9	56.2	73.8	44.8
1985	17.1	33.4	90.7	7.6	16.8	21.7	25.7	29.6	32.2	56.3	78.2	51.6
1986	12.3	35.7	91.1	7.6	14.4	22.3	24.3	28.1	34.5	53.8	77.9	54.2
1987	11.7	32.5	52.1	14.0	21.5	30.6	32.3	29.0	35.2	53.6	80.6	53.0

Source: HM Customs and Excise.
Note:
Figures prior to April 1983 included other home-produced spirits.

Table A.9 US unemployment, quarterly, 1949–80, plus intervention variables

Quarter	Unemployment	WAR	PELE	CELE	Quarter	Unemployment	WAR	PELE	CELE
1949.1	5.49	0	0	0	1965.1	5.42	1	0	0
1949.2	5.86	0	0	0	1965.2	6.67	1	0	0
1949.3	6.41	0	0	0	1965.3	4.16	1	0	0
1949.4	5.96	0	0	0	1965.4	3.74	1	0	0
1950.1	7.52	0	0	0	1966.1	4.22	1	0	0
1950.2	5.60	0	0	0	1966.2	4.00	1	0	0
1950.3	4.44	1	0	0	1966.3	3.63	1	0	0
1950.4	3.64	1	0	1	1966.4	3.85	1	0	1
1951.1	4.10	1	0	0	1967.1	3.78	1	0	0
1951.2	3.15	1	0	0	1967.2	3.66	1	0	0
1951.3	3.06	1	0	0	1967.3	3.85	1	0	0
1951.4	2.96	1	0	0	1967.4	4.10	1	0	0
1952.1	3.60	1	0	0	1968.1	3.66	1	0	0
1952.2	3.04	1	0	0	1968.2	4.02	1	0	0
1952.3	3.04	1	0	0	1968.3	3.55	1	0	0
1952.4	2.46	1	1	1	1968.4	3.59	1	1	1
1953.1	3.16	1	0	0	1969.1	3.16	1	0	0
1953.2	2.66	1	0	0	1969.2	3.61	1	0	0
1953.3	2.54	1	0	0	1969.3	3.41	1	0	0
1953.4	3.29	0	0	0	1969.4	3.66	1	0	0
1954.1	6.15	0	0	0	1970.1	3.35	1	0	0
1954.2	5.85	0	0	0	1970.2	4.49	1	0	0
1954.3	5.49	0	0	0	1970.3	4.68	1	0	0
1954.4	4.74	0	0	1	1970.4	5.18	1	0	1
1955.1	5.57	0	0	0	1971.1	5.41	1	0	0
1955.2	4.51	0	0	0	1971.2	5.82	1	0	0
1955.3	3.79	0	0	0	1971.3	5.95	1	0	0
1955.4	3.74	0	0	0	1971.4	6.47	1	0	0
1956.1	4.74	0	0	0	1972.1	5.54	1	0	0
1956.2	4.33	0	0	0	1972.2	6.32	1	0	0
1956.3	3.81	0	0	0	1972.3	5.60	1	0	0
1956.4	3.67	0	1	1	1972.4	5.59	1	1	1
1957.1	_(cut off)_				1973.1	4.95			

Period		WAR	CELE	PELE
1957.3	3.87	0	0	0
1957.4	4.42	0	0	0
1958.1	7.40	0	0	0
1958.2	7.41	0	0	0
1958.3	6.70	0	0	0
1958.4	5.71	0	1	0
1959.1	6.81	0	0	0
1959.2	5.18	0	0	0
1959.3	4.89	0	0	0
1959.4	5.04	0	0	0
1960.1	5.99	0	0	0
1960.2	5.31	0	0	0
1960.3	5.15	0	0	0
1960.4	5.70	0	1	1
1961.1	7.86	0	0	0
1961.2	6.99	0	0	0
1961.3	6.30	0	0	0
1961.4	5.63	0	0	0
1962.1	6.46	0	0	0
1962.2	5.49	0	0	0
1962.3	5.18	0	0	0
1962.4	5.05	0	1	0
1963.1	6.61	0	0	0
1963.2	5.77	0	0	0
1963.3	5.21	0	0	0
1963.4	5.10	0	0	0
1964.1	6.15	0	0	0
1964.2	5.34	0	0	0
1964.3	4.73	0	0	0
1964.4	4.52	1	1	1
1973.3	4.82	0	0	0
1973.4	4.45	0	0	0
1974.1	5.59	0	0	0
1974.2	5.12	0	0	0
1974.3	5.57	0	0	0
1974.4	6.15	0	1	0
1975.1	9.08	0	0	0
1975.2	8.66	0	0	0
1975.3	8.31	0	0	0
1975.4	7.79	0	0	0
1976.1	8.51	0	0	0
1976.2	7.38	0	0	0
1976.3	7.60	0	0	0
1976.4	7.33	0	1	1
1977.1	8.23	0	0	0
1977.2	6.95	0	0	0
1977.3	6.83	0	0	0
1977.4	6.28	0	0	0
1978.1	6.86	0	0	0
1978.2	5.86	0	0	0
1978.3	5.99	0	0	0
1978.4	5.57	0	1	0
1979.1	6.32	0	0	0
1979.2	5.61	0	0	0
1979.3	5.82	0	0	0
1979.4	5.65	0	0	0
1980.1	6.76	0	0	0
1980.2	7.19	0	0	0
1980.3	7.53	0	0	0
1980.4	7.07	0	1	1

Notes:

Intervention variables.

WAR = 1, if US at war
    = 0, otherwise

PELE = 1, if presidential election
    = 0, otherwise

CELE = 1, if congressional election
    = 0 otherwise

Source: US Dept. of Labor.

Discussion: Koot, Young and Ord (1992).

Table A.10 Retail sales of variety stores in US, 1967–79

	Jan.	Feb.	Mar.	Apr.	May	June	July	Aug.	Sept.
1967	296	303	365	363	417	421	404	436	421
1968	331	361	402	426	460	457	451	476	436
1969	345	364	427	445	478	492	469	501	459
1970	370	378	453	470	534	510	485	527	536
1971	394	411	482	484	550	525	494	537	513
1972	393	425	503	529	581	558	547	588	549
1973	463	459	554	576	615	619	589	637	601
1974	490	490	598	615	681	654	637	694	645
1975	489	511	612	623	726	692	623	734	662
1976	503	537	636	560	607	585	559	608	556
1977	427	450	573	579	615	601	608	617	550
1978	438	458	548	584	639	616	614	647	588
1979	483	483	593	620	672	650	643	702	654

Oct.	Nov.	Dec.
429	499	915
464	525	939
494	548	1 022
553	621	1 122
521	596	1 069
593	649	1 191
642	737	1 279
684	749	1 245
684	781	1 386
596	665	1 229
616	673	1 199
648	713	1 261

Source: US Bureau of the Census.
Discussion: Bell (1983).
Notes:
The intervention variable is defined with respect to April 1976 when a large chain (W. T. Grant) went out of business. The data have been adjusted for trading day and holiday effects as discussed by Hillmer *et al.* (1983).

t	$x_{1,t}$	$x_{2,t}$	$x_{3,t}$	t	$x_{2,t}$	$x_{1,t}$	$x_{3,t}$	t	$x_{1,t}$	$x_{2,t}$	$x_{3,t}$	t	$x_{1,t}$	$x_{2,t}$	$x_{3,t}$
1	118	155	150	26	-18	11	-18	51	199	162	84	76	-82	-16	-11
2	102	128	114	27	7	-15	-3	52	137	139	142	77	-74	-34	-1
3	131	148	122	28	20	28	-58	53	126	100	142	78	-117	-22	13
4	129	99	132	29	-29	6	27	54	73	130	70	79	-97	-10	-13
5	72	97	114	30	-20	39	53	55	76	124	121	80	-51	-64	4
6	78	63	123	31	11	-12	68	56	46	152	124	81	-42	-101	-34
7	117	33	91	32	57	7	-20	57	31	173	135	82	-29	-90	-131
8	134	44	67	33	65	-3	-37	58	-20	165	160	83	-1	-88	-105
9	174	73	12	34	98	9	43	59	-54	109	162	84	-28	-29	-71
10	146	63	104	35	102	19	35	60	-29	60	52	85	-1	-46	-48
11	94	54	38	36	139	61	-15	61	-32	68	64	86	26	17	-39
12	62	27	42	37	128	58	27	62	-82	72	95	87	62	70	-29
13	56	2	-17	38	101	64	85	63	-110	3	55	88	81	107	33
14	39	37	2	39	132	91	85	64	-76	-34	-23	89	77	149	103
15	36	0	70	40	95	89	95	65	-69	-83	-15	90	57	172	126
16	8	-42	45	41	43	94	73	66	-92	-51	-35	91	52	156	169
17	-25	-79	-54	42	43	41	91	67	-90	-100	-46	92	54	72	172
18	17	-104	-103	43	62	-6	0	68	-71	-122	-116	93	82	11	19
19	38	-65	-109	44	108	-11	16	69	-45	-137	-88	94	45	45	2
20	42	2	-18	45	88	-19	2	70	-74	-125	-132	95	40	53	54
21	49	6	-10	46	134	36	-2	71	-30	-92	-73	96	33	11	21
22	52	21	38	47	139	51	-8	72	-46	-70	-43	97	46	46	-28
23	51	-20	29	48	153	69	52	73	-48	-93	-72	98	-7	108	48
24	11	-16	-19	49	178	108	82	74	-88	-66	-78	99	4	94	94
25	2	-29	-26	50	181	127	129	75	-70	-48	-15	100	37	61	48

Source: Quenouille (1968).
Notes:
These series were generated from the model

$$x_{1,t} = x_{1,t-1} - 0.1 x_{2,t-2} + \varepsilon_{1,t}$$
$$x_{2,t} = 0.2 x_{1,t-1} + x_{2,t-1} - 0.3 x_{3,t-1} + \varepsilon_{2,t}$$
$$x_{3,t} = 0.9 x_{2,t-1} + \varepsilon_{3,t}$$

Table A.12 Monthly flour price indices in 3 US cities (August 1972–November 1980)

Buffalo	Minneapolis	Kansas City	Buffalo	Minneapolis	Kansas City
107.1	106.5	110.9	143.5	137.3	134.2
113.5	112.4	114.6	135.6	129.7	126.1
112.7	111.8	115.5	135.4	128.4	124.2
114.7	113.3	117.0	134.5	126.9	122.7
123.4	124.5	135.0	136.1	128.8	123.5
123.6	124.3	132.8	135.6	126.5	118.3
116.3	116.5	122.6	122.8	116.6	112.3
118.5	118.6	123.8	119.0	113.4	105.7
119.8	119.6	128.9	108.5	102.8	97.7
120.3	119.4	126.7	113.3	107.7	105.8
127.4	128.6	139.3	114.8	109.4	106.9
125.1	126.3	135.7	120.9	114.9	110.0
127.6	126.8	135.6	123.7	117.5	114.3
129.0	125.7	146.0	127.8	120.0	118.8
124.6	120.8	140.7	125.4	117.6	117.2
134.1	127.9	147.0	131.5	124.0	126.1
146.5	147.6	163.9	127.7	119.7	120.5
171.2	169.8	194.3	131.2	125.0	125.6
178.6	177.6	200.8	145.2	141.1	132.0
172.2	172.5	193.4	141.9	137.0	134.6
171.5	170.1	190.3	139.3	132.3	130.3
163.6	171.3	188.0	141.1	134.8	137.0
185.6	189.9	196.1	135.9	129.7	136.6
198.8	206.9	215.0	136.5	128.7	137.0
195.7	197.4	201.6	137.2	129.9	138.4
190.3	195.0	203.4	143.8	137.2	142.9
207.9	214.2	222.1	138.7	132.8	140.4
212.8	219.2	228.7	133.9	127.5	136.0
199.9	205.6	216.1	137.7	131.2	140.1
185.3	193.4	200.2	143.8	137.1	148.2
183.0	185.1	189.6	140.8	135.5	146.4
173.5	174.0	173.3	153.4	147.1	158.5
172.2	173.2	169.7	157.5	151.6	163.5
165.3	164.5	161.0	179.5	173.7	187.1
159.9	158.9	151.7	177.5	171.6	181.7
170.3	169.7	167.1	178.0	170.8	181.5
172.2	174.4	174.4	176.8	172.4	181.9
184.5	186.2	189.7	179.8	174.9	190.9
185.0	184.7	187.4	174.2	168.1	186.9
177.7	176.4	178.4	171.1	164.7	180.1
169.1	167.6	165.8	175.9	170.0	184.8
174.7	170.9	164.9	172.2	164.9	174.8
169.4	168.3	171.8	164.7	157.9	169.0
177.8	176.4	175.4	175.7	169.2	178.4
170.1	168.6	165.9	177.4	168.6	175.3
167.1	164.6	157.3	187.5	179.8	178.2
171.4	170.1	161.4	190.7	179.0	182.0
172.3	169.4	159.2	190.4	179.2	188.6
152.6	149.6	142.8	192.4	181.4	190.8
141.1	139.5	138.5	192.9	181.8	192.2

Source: Tiao and Tsay (1989), (and discussion).

Table A.13 Lydia Pinkham sales and advertising figures (January 1954–June 1960)

Sales (in $00)

	Jan.	Feb.	Mar.	Apr.	May	June	July	Aug.	Sept.	Oct.	Nov.	Dec.
1954	1 295	1 318	1 728	1 539	1 324	1 264	1 169	1 479	1 631	1 546	1 459	1 087
1955	1 171	1 406	1 619	1 508	1 521	1 341	1 247	1 262	1 419	1 558	1 222	1 053
1956	1 242	1 361	1 660	1 717	1 371	1 293	1 285	1 210	1 142	1 586	1 441	1 262
1957	1 267	1 278	1 544	1 534	1 332	1 200	1 314	1 180	1 264	1 318	1 018	1 438
1958	772	902	1 265	1 229	1 318	1 195	1 105	1 095	1 298	1 482	1 163	1 072
1959	1 052	1 102	1 355	1 323	1 296	1 127	1 170	1 059	1 116	1 214	966	1 089
1960	814	1 087	1 180	1 167	1 210	1 092						

Advertising (in $00)

	Jan.	Feb.	Mar.	Apr.	May	June	July	Aug.	Sept.	Oct.	Nov.	Dec.
1954	1 280	1 350	982	919	87	39	72	467	1 170	917	701	128
1955	1 014	1 274	1 388	1 071	537	123	60	351	1 061	791	138	77
1956	1 000	1 182	1 225	936	625	60	61	169	946	1 306	426	88
1957	1 104	1 093	1 080	1 012	745	78	66	94	774	971	536	150
1958	580	1 121	974	1 002	138	72	59	270	986	673	304	209
1959	838	994	1 020	865	819	83	56	224	881	436	160	68
1960	749	857	898	705	489	59						

Source: Palda (1964).
Discussion: Bhattacharya (1982).

Table A.14 US hog, corn and wage series, 1867–1948

t	$x_{1,t}$	$x_{2,t}$	$x_{3,t}$	$x_{4,t}$	$x_{5,t}$	t	$x_{1,t}$	$x_{2,t}$	$x_{3,t}$	$x_{4,t}$	$x_{5,t}$
1867	538	597	944	900	722*	1908	766	777	813	1 409	971*
1868	522	509	841	964	719*	1909	720	810	790	1 417	982
1869	513	663	911	893	716	1910	682	957	712	1 455	987
1870	529	751	768	1 051	724*	1911	743	970	831	1 394	991
1871	565	739	718	1 057	732*	1912	743	903	742	1 469	1 004
1872	594	598	634	1 107	740*	1913	730	995	847	1 357	1 013
1873	600	556	735	1 003	748*	1914	723	1 022	850	1 402	1 004
1874	584	594	858	1 025	756	1915	753	998	830	1 452	1 013
1875	554	667	673	1 161	748*	1916	782	928	1 056	1 385	1 053
1876	553	776	609	1 170	740*	1917	760	1 073	1 163	1 464	1 149
1877	595	754	604	1 181	732	1918	799	1 294	1 182	1 388	1 248
1878	637	689	457	1 194	744*	1919	808	1 346	1 180	1 428	1 316
1879	641	498	612	1 244	756	1920	779	1 301	805	1 487	1 384
1880	647	643	642	1 232	778	1921	770	1 134	714	1 467	1 190
1881	634	681	849	1 095	799	1922	777	1 024	865	1 432	1 179
1882	629	778	733	1 244	799*	1923	841	1 090	911	1 459	1 228
1883	638	829	672	1 218	799*	1924	823	1 013	1 027	1 347	1 238
1884	662	751	594	1 289	799	1925	746	1 119	846	1 447	1 246
1885	675	704	559	1 313	801*	1926	717	1 195	869	1 406	1 253
1886	658	633	604	1 251	803*	1927	744	1 235	928	1 418	1 253
1887	629	663	678	1 205	806	1928	791	1 120	924	1 426	1 253
1888	625	709	571	1 352	806*	1929	771	1 112	903	1 401	1 255
1889	648	763	490	1 361	806	1930	746	1 129	777	1 318	1 223
1890	682	681	747	1 218	810*	1931	739	1 055	507	1 411	1 114
1891	676	627	651	1 368	813	1932	773	787	500	1 467	982
1892	655	667	645	1 278	810	1933	793	624	716	1 380	929
1893	640	804	609	1 279	806	1934	768	612	911	1 161	978
1894	668	782	705	1 208	771	1935	592	800	816	1 362	1 013
1895	678	707	453	1 404	771	1936	633	1 104	1 019	1 178	1 045
1896	692	653	382	1 427	780*	1937	634	1 075	714	1 422	1 100
1897	710	639	466	1 359	789*	1938	649	1 052	687	1 406	1 097
1898	727	672	506	1 371	799	1939	699	1 048	754	1 412	1 090
1899	712	669	525	1 423	820	1940	786	891	791	1 390	1 100
1900	708	729	595	1 425	834*	1941	735	921	876	1 424	1 188
1901	705	784	829	1 234	848*	1942	782	1 193	962	1 487	1 303
1902	680	842	654	1 443	863	1943	869	1 352	1 050	1 472	1 422
1903	682	886	673	1 401	884*	1944	923	1 243	1 037	1 490	1 498
1904	713	784	691	1 429	906*	1945	774	1 314	1 104	1 458	1 544
1905	726	770	660	1 470	928*	1946	787	1 380	1 193	1 507	1 582
1906	729	783	643	1 482	949	1947	754	1556	1 334	1 372	1 607
1907	752	877	754	1 417	960*	1948	737	1 632	1 114	1 557	1 629

*Obtained by interpolation.
Source: Quenouille (1968).
Discussions: Quenouille (1968), Box and Tiao (1977), Tiao and Tsay (1983, 1989), Velu, Reinsel and Wichen (1986).

Table A.15 Radial cell diameters (in μ) for cells along an entire file of tracheids

Cell numbers

1–20	21–40	41–60	61–80	81–100	101–20	121–40	141–59
39.00	37.00	42.75	43.50	46.75	35.50	27.75	15.00
41.50	28.75	40.75	40.50	52.75	36.50	26.50	11.88
34.75	32.50	40.50	45.50	38.25	36.50	24.50	13.75
31.75	30.25	45.50	41.50	37.50	31.25	22.50	15.75
35.00	32.00	40.75	49.50	46.25	30.00	28.50	14.75
37.00	39.25	39.50	47.75	41.25	36.00	29.50	12.75
41.00	40.75	34.00	46.75	46.00	34.00	31.50	6.75
32.50	43.25	30.00	46.75	40.75	31.00	32.50	6.75
31.00	49.25	32.00	39.75	44.75	38.00	32.50	5.75
28.00	47.25	32.50	34.25	40.00	36.00	36.50	8.75
34.25	33.25	44.25	29.00	35.50	34.00	28.50	12.75
35.75	36.00	43.75	35.00	28.00	39.00	28.50	13.75
40.00	32.00	44.50	30.00	30.00	32.75	28.50	6.75
34.50	30.00	39.00	37.00	34.00	22.25	29.50	6.75
30.50	29.75	36.00	34.00	35.75	28.00	16.50	8.75
40.00	34.75	37.75	39.25	31.75	24.00	16.50	7.75
37.00	38.75	34.25	41.75	31.00	25.00	16.50	10.75
44.00	36.50	35.25	45.00	33.00	27.00	15.25	10.75
35.75	36.50	41.50	50.75	30.75	26.00	17.00	10.75
33.75	38.50	35.50	43.50	31.50	23.00	15.00	9.13

Source: E. D. Ford (unpublished)
Discussion: Ord (1979).

Table A.16 Percent of US households with Cable Antenna Television (CATV)

Year	Households (%)	Year	Households (%)
1955	0.5	1969	6.1
1956	0.9	1970	7.7
1957	0.9	1971	8.8
1958	1.1	1972	9.7
1959	1.3	1973	11.3
1960	1.4	1974	12.7
1961	1.5	1975	13.8
1962	1.7	1976	14.8
1963	1.9	1977	16.1
1964	2.1	1978	17.1
1965	2.4	1979	18.2
1966	2.9	1980	19.2
1967	3.8	1981	22.2
1968	4.4		

Source: Martino (1983, p. 364).
Discussion: Young and Ord (1985), Ord and Young (1989).

Appendix B

Tables of the Durbin–Watson Statistic

Table B.1 Significance points of the Durbin–Watson statistics d_L and d_U: 1 percent

n	$k'=1$		$k'=2$		$k'=3$		$k'=4$		$k'=5$	
	d_L	d_U	d_L	d_U	d_L	d_U	d_L	d_U	d_L	d_U
15	0.81	1.07	0.70	1.25	0.59	1.46	0.49	1.70	0.39	1.96
16	0.84	1.09	0.74	1.25	0.63	1.44	0.53	1.66	0.44	1.90
17	0.87	1.10	0.77	1.25	0.67	1.43	0.57	1.63	0.48	1.85
18	0.90	1.12	0.80	1.26	0.71	1.42	0.61	1.60	0.52	1.80
19	0.93	1.13	0.83	1.26	0.74	1.41	0.65	1.58	0.56	1.77
20	0.95	1.15	0.86	1.27	0.77	1.41	0.68	1.57	0.60	1.74
21	0.97	1.16	0.89	1.27	0.80	1.41	0.72	1.55	0.63	1.71
22	1.00	1.17	0.91	1.28	0.83	1.40	0.75	1.54	0.66	1.69
23	1.02	1.19	0.94	1.29	0.86	1.40	0.77	1.53	0.70	1.67
24	1.04	1.20	0.96	1.30	0.88	1.41	0.80	1.53	0.72	1.66
25	1.05	1.21	0.98	1.30	0.90	1.41	0.83	1.52	0.75	1.65
26	1.07	1.22	1.00	1.31	0.93	1.41	0.85	1.52	0.78	1.64
27	1.09	1.23	1.02	1.32	0.95	1.41	0.88	1.51	0.81	1.63
28	1.10	1.24	1.04	1.32	0.97	1.41	0.90	1.51	0.83	1.62
29	1.12	1.25	1.05	1.33	0.99	1.42	0.92	1.51	0.85	1.61
30	1.13	1.26	1.07	1.34	1.01	1.42	0.94	1.51	0.88	1.61
31	1.15	1.27	1.08	1.34	1.02	1.42	0.96	1.51	0.90	1.60
32	1.16	1.28	1.10	1.35	1.04	1.43	0.98	1.51	0.92	1.60
33	1.17	1.29	1.11	1.36	1.05	1.43	1.00	1.51	0.94	1.59
34	1.18	1.30	1.13	1.36	1.07	1.43	1.01	1.51	0.95	1.59
35	1.19	1.31	1.14	1.37	1.08	1.44	1.03	1.51	0.97	1.59
36	1.21	1.32	1.15	1.38	1.10	1.44	1.04	1.51	0.99	1.59
37	1.22	1.32	1.16	1.38	1.11	1.45	1.06	1.51	1.00	1.59
38	1.23	1.33	1.18	1.39	1.12	1.45	1.07	1.52	1.02	1.58
39	1.24	1.34	1.19	1.39	1.14	1.45	1.09	1.52	1.03	1.58
40	1.25	1.34	1.20	1.40	1.15	1.46	1.10	1.52	1.05	1.58
45	1.29	1.38	1.24	1.42	1.20	1.48	1.16	1.53	1.11	1.58
50	1.32	1.40	1.28	1.45	1.24	1.49	1.20	1.54	1.16	1.59
55	1.36	1.43	1.32	1.47	1.28	1.51	1.25	1.55	1.21	1.59
60	1.38	1.45	1.35	1.48	1.32	1.52	1.28	1.56	1.25	1.60
65	1.41	1.47	1.38	1.50	1.35	1.53	1.31	1.57	1.28	1.61
70	1.43	1.49	1.40	1.52	1.37	1.55	1.34	1.58	1.31	1.61
75	1.45	1.50	1.42	1.53	1.39	1.56	1.37	1.59	1.34	1.62
80	1.47	1.52	1.44	1.54	1.42	1.57	1.39	1.60	1.36	1.62
85	1.48	1.53	1.46	1.55	1.43	1.58	1.41	1.60	1.39	1.63
90	1.50	1.54	1.47	1.56	1.45	1.59	1.43	1.61	1.41	1.64
95	1.51	1.55	1.49	1.57	1.47	1.60	1.45	1.62	1.42	1.64
100	1.52	1.56	1.50	1.58	1.48	1.60	1.46	1.63	1.44	1.65

Table B.2 Significance points of the Durbin–Watson statistics d_L and d_U: 5 percent

n	$k'=1$		$k'=2$		$k'=3$		$k'=4$		$k'=5$	
	d_L	d_U	d_L	d_U	d_L	d_U	d_L	d_U	d_L	d_U
15	1.08	1.36	0.95	1.54	0.82	1.75	0.69	1.97	0.56	2.21
16	1.10	1.37	0.98	1.54	0.86	1.73	0.74	1.93	0.62	2.15
17	1.13	1.38	1.02	1.54	0.90	1.71	0.78	1.90	0.67	2.10
18	1.16	1.39	1.05	1.53	0.93	1.69	0.82	1.87	0.71	2.06
19	1.18	1.40	1.08	1.53	0.97	1.68	0.86	1.85	0.75	2.02
20	1.20	1.41	1.10	1.54	1.00	1.68	0.90	1.83	0.79	1.99
21	1.22	1.42	1.13	1.54	1.03	1.67	0.93	1.81	0.83	1.96
22	1.24	1.43	1.15	1.54	1.05	1.66	0.96	1.80	0.86	1.94
23	1.26	1.44	1.17	1.54	1.08	1.66	0.99	1.79	0.90	1.92
24	1.27	1.45	1.19	1.55	1.10	1.66	1.01	1.78	0.93	1.90
25	1.29	1.45	1.21	1.55	1.12	1.66	1.04	1.77	0.95	1.89
26	1.30	1.46	1.22	1.56	1.14	1.65	1.06	1.76	0.98	1.88
27	1.32	1.47	1.24	1.56	1.16	1.65	1.08	1.76	1.01	1.86
28	1.33	1.48	1.26	1.56	1.18	1.65	1.10	1.75	1.03	1.85
29	1.34	1.48	1.27	1.56	1.20	1.65	1.12	1.74	1.05	1.84
30	1.35	1.49	1.28	1.57	1.21	1.65	1.14	1.74	1.07	1.83
31	1.36	1.50	1.30	1.57	1.23	1.65	1.16	1.74	1.09	1.83
32	1.37	1.50	1.31	1.57	1.24	1.65	1.18	1.73	1.11	1.82
33	1.38	1.51	1.32	1.58	1.26	1.65	1.19	1.73	1.13	1.81
34	1.39	1.51	1.33	1.58	1.27	1.65	1.21	1.73	1.15	1.81
35	1.40	1.52	1.34	1.58	1.28	1.65	1.22	1.73	1.16	1.80
36	1.41	1.52	1.35	1.59	1.29	1.65	1.24	1.73	1.18	1.80
37	1.42	1.53	1.36	1.59	1.31	1.66	1.25	1.72	1.19	1.80
38	1.43	1.54	1.37	1.59	1.32	1.66	1.26	1.72	1.21	1.79
39	1.43	1.54	1.38	1.60	1.33	1.66	1.27	1.72	1.22	1.79
40	1.44	1.54	1.39	1.60	1.34	1.66	1.29	1.72	1.23	1.79
45	1.48	1.57	1.43	1.62	1.38	1.67	1.34	1.72	1.29	1.78
50	1.50	1.59	1.46	1.63	1.42	1.67	1.38	1.72	1.34	1.77
55	1.53	1.60	1.49	1.64	1.45	1.68	1.41	1.72	1.38	1.77
60	1.55	1.62	1.51	1.65	1.48	1.69	1.44	1.73	1.41	1.77
65	1.57	1.63	1.54	1.66	1.50	1.70	1.47	1.73	1.44	1.77
70	1.58	1.64	1.55	1.67	1.52	1.70	1.49	1.74	1.46	1.77
75	1.60	1.65	1.57	1.68	1.54	1.71	1.51	1.74	1.49	1.77
80	1.61	1.66	1.59	1.69	1.56	1.72	1.53	1.74	1.51	1.77
85	1.62	1.67	1.60	1.70	1.57	1.72	1.55	1.75	1.52	1.77
90	1.63	1.68	1.61	1.70	1.59	1.73	1.57	1.75	1.54	1.78
95	1.64	1.69	1.62	1.71	1.60	1.73	1.58	1.75	1.56	1.78
100	1.65	1.69	1.63	1.72	1.61	1.74	1.59	1.76	1.57	1.78

Appendix C

Weights for fitting polynomial trends

The following tables are extracted by permission from Dudley J. Cowden, *Weights for fitting polynomial secular trends*, Technical Paper No. 4, School of Business Administration, University of North Carolina, 1962, Professor Cowden gives values up to N (the extent of the moving average) = 25, and also values for even N.

Except in one or two early tables, the tables give the weights required to fit at one end of the series, those for the other end being given by symmetry. For example, for fitting a straight line (a simple moving average) to nine points the weights for the first point are

$$\tfrac{1}{45} \, [17, \ 14, \ 11, \ 8, \ 5, \ 2, \ -1, \ -4, \ -7].$$

Those for the the second are

$$\tfrac{1}{180} \, [56, \ 47, \ 38, \ 29, \ 20, \ 11, \ 2, \ -7, \ -16],$$

and so on. Conversely, for the last four points, the weights are (reading the table upwards) for the last but three,

$$[8, \ 11, \ 14, \ 17, \ 20, \ 23, \ 26, \ 29, \ 32],$$

and so on.

The columns headed 0 give the weights required to extrapolate the fitting one unit beyond the end of the observed series.

The sums in the last row but one are the sums of the integral weights given in the table.

The final row in the table is the square root of the error-reducing factor, i.e. the square root of the sum of the squares of the weights.

Linear

N = 3

0	4	5	1	-1	-2
1	1	2	1	2	1
2	-2	-1	1	5	4
	3	6	3	6	3
	1.528	0.913	0.577	0.913	1.528

N = 5

0	8	3	4	1	0	-1	-4
1	5	2	3	1	1	0	-1
2	2	1	2	1	2	1	2
3	-1	0	1	1	3	2	5
4	-4	-1	0	1	4	3	8
	10	5	10	5	10	5	10
	1.049	0.775	0.548	0.447	0.548	0.775	1.049

N = 7

0	4	13	5	7	1
1	3	10	4	6	1
2	2	7	3	5	1
3	1	4	2	4	1
4	0	1	1	3	1
5	-1	-2	0	2	1
6	-2	-5	-1	1	1
	7	28	14	28	7
	0.845	0.681	0.535	0.423	0.378

N = 9

0	16	17	56	22	32	1
1	13	14	47	19	29	1
2	10	11	38	16	26	1
3	7	8	29	13	23	1
4	4	5	20	10	20	1
5	1	2	11	7	17	1
6	-2	-1	2	4	14	1
7	-5	-4	-7	1	11	1
8	-8	-7	-16	-2	8	1
	36	45	180	90	180	9
	0.726	0.615	0.511	0.422	0.357	0.333

N = 11

0	20	7	15	25	10	15	1
1	17	6	13	22	9	14	1
2	14	5	11	19	8	13	1
3	11	4	9	16	7	12	1
4	8	3	7	13	6	11	1
5	5	2	5	10	5	10	1
6	2	1	3	7	4	9	1
7	-1	0	1	4	3	8	1
8	-4	-1	-1	1	2	7	1
9	-7	-2	-3	-2	1	6	1
10	-10	-3	-5	-5	0	5	1
	55	22	55	110	55	110	11
	0.647	0.564	0.486	0.416	0.357	0.316	0.302

N = 13

0	1	2	3	4	5	6	7
8	25	44	19	32	13	20	1
7	22	39	17	29	12	19	1
6	19	34	15	26	11	18	1
5	16	29	13	23	10	17	1
4	13	24	11	20	9	16	1
3	10	19	9	17	8	15	1
2	7	14	7	14	7	14	1
1	4	9	5	11	6	13	1
0	1	4	3	8	5	12	1
-1	-2	-1	1	5	4	11	1
-2	-5	-6	-1	2	3	10	1
-3	-8	-11	-3	-1	2	9	1
-4	-11	-16	-5	-4	1	8	1
26	91	182	91	182	91	182	13
0.588	0.524	0.463	0.406	0.355	0.314	0.287	0.277

N = 15

0	1	2	3	4	5	6	7	8
28	29	91	161	35	119	49	77	1
25	26	82	146	32	110	46	74	1
22	23	73	131	29	101	43	71	1
19	20	64	116	26	92	40	68	1
16	17	55	101	23	83	37	65	1
13	14	46	86	20	74	34	62	1
10	11	37	71	17	65	31	59	1
7	8	28	56	14	56	28	56	1
4	5	19	41	11	47	25	53	1
1	2	10	26	8	38	22	50	1
-2	-1	1	11	5	29	19	47	1
-5	-4	-8	-4	2	20	16	44	1
-8	-7	-17	-19	-1	11	13	41	1
-11	-10	-26	-34	-4	2	10	38	1
-14	-13	-35	-49	-7	-7	7	35	1
105	120	420	840	210	840	420	840	15
0.543	0.492	0.442	0.395	0.352	0.314	0.285	0.265	0.258

Quadratic

N = 5

0	1	2	3	2	1	0
0	31	9	-3	-5	3	3
9	9	13	12	6	-5	-3
-4	-3	12	17	12	-3	-4
-3	-5	6	12	13	9	9
3	3	-5	-3	9	31	0
5	35	35	35	35	35	5
2.145	0.941	0.609	0.697	0.609	0.941	2.145

N = 7

0	1	2	3	2	1	0
9	32	5	1			
0	3	15	3	4	3	
-3	-1	3	4	4	1	
-4	-3	-4	2	3	4	
-3	-6	-3	1	0	1	
-1	5	0	3	1	3	
3	-1	-6	-3	-2		
7	42	14	14	14		
1.558	0.873	0.535	0.535	0.535		

N = 9

0	1	2	3	4	5
42	109	126	378	14	-21
21	63	92	441	273	14
5	27	63	464	447	39
-6	1	39	447	536	54
-12	-15	20	390	540	59
-13	-21	6	293	459	54
-9	-17	-3	156	293	39
0	-3	-7	-21	42	14
14	21	-6	-238	-294	-21
42	165	330	2 310	2 310	231
1.273	0.813	0.528	0.448	0.482	0.505

N = 11

0	1	2	3	4	5	6
135	83	270	450	55	-15	-36
81	54	199	414	87	46	9
37	30	138	373	109	92	44
3	11	87	327	121	123	69
-21	-3	46	276	123	139	84
-35	-12	15	220	115	140	89
-39	-16	-6	159	97	126	84
-33	-15	-17	93	69	97	69
-17	-9	-18	22	31	53	44
9	2	-9	-54	-17	-6	9
45	18	10	-135	-75	-80	-36
165	143	715	2 145	715	715	429
1.078	0.762	0.528	0.417	0.411	0.411	0.455

$N = 13$

0	1	2	3	4	5	6	7
99	47	33	231	121	33	− 33	− 11
66	33	25	198	132	77	33	0
38	21	18	167	138	111	86	9
15	11	12	138	139	135	126	16
− 3	3	7	111	135	149	153	21
− 16	− 3	3	86	126	153	167	24
− 24	− 7	0	63	112	147	168	25
− 27	− 9	− 2	42	93	131	156	24
− 25	− 9	− 3	23	69	105	131	21
− 18	− 7	− 3	6	40	69	93	16
− 6	− 7	− 2	− 9	6	23	42	9
11	3	0	− 22	− 33	− 33	− 22	0
33	11	3	− 33	− 77	− 99	− 99	− 11
143	91	91	1 001	1 001	1 001	1 001	143
0.979	0.719	0.524	0.408	0.373	0.386	0.408	0.418

$N = 15$

0	1	2	3	4	5	6	7	8
273	158	819	7 371	2 275	2 184	273	− 1 183	− 78
195	117	638	6 201	2 184	2 795	1 482	429	− 13
127	81	477	5 126	2 073	3 261	2 471	1 776	42
69	50	336	4 146	1 942	3 582	3 240	2 858	87
21	24	215	3 261	1 791	3 758	3 789	3 675	122
− 17	3	114	2 471	1 620	3 789	4 118	4 227	147
− 45	− 13	33	1 776	1 429	3 675	4 227	4 514	162
− 63	− 24	− 28	1 176	1 218	3 416	4 116	4 536	167
− 71	− 30	− 69	671	987	3 012	3 785	4 293	162
− 69	− 31	− 90	261	736	2 463	3 234	3 785	147
− 57	− 27	− 91	− 54	465	1 769	2 463	3 012	122
− 35	− 18	− 72	− 274	174	930	1 472	1 974	87
− 3	− 4	− 33	− 399	− 137	− 54	261	671	42
39	15	26	− 429	− 468	− 1 183	− 1 170	− 897	− 13
91	39	105	− 364	− 819	− 2 457	− 2 821	− 2 730	− 78
455	340	2 380	30 940	15 470	30 940	30 940	30 940	1 105
0.891	0.682	0.518	0.407	0.354	0.349	0.365	0.382	0.389

Cubic

N = 5

0	1	2	3	2	1	0
16	69	2	-3	2	-1	-4
-14	4	27	12	-8	4	11
-4	-6	12	17	12	-6	-4
11	4	-8	12	27	4	-14
-4	-1	2	-3	2	69	16
5	70	35	35	35	70	5
4.919	0.993	0.878	0.697	0.878	0.993	4.919

N = 7

0	1	2	3	4
16	39	8	-4	-2
-4	8	19	16	3
-8	-4	16	19	6
-3	-4	6	12	7
4	1	-4	2	6
6	4	-7	-4	3
-4	-2	4	1	-2
7	42	42	42	21
2.903	0.964	0.673	0.673	0.577

N = 9

0	1	2	3	4	5
224	85	56	-28	-56	-21
14	28	65	392	84	14
-76	-2	56	515	144	39
-81	-12	36	432	145	54
-36	-9	12	234	108	59
24	0	-9	12	54	54
64	8	-20	-143	4	39
49	8	-14	-140	-21	14
-56	-7	16	112	0	-21
126	99	198	1386	462	231
2.162	0.927	0.573	0.610	0.560	0.505

N = 11

0	1	2	3	4	5	6
96	113	48	24	-72	-51	-36
24	48	41	96	132	36	9
-16	8	32	123	232	86	44
-31	-12	22	116	251	106	69
-28	-17	12	86	212	103	84
4	-12	3	44	138	84	89
19	4	-4	1	52	56	84
24	8	-8	-32	-23	26	69
12	13	-8	-44	-64	1	44
-24	8	-3	-24	-48	-12	9
-14	-18	8	39	48	-6	-36
66	143	143	429	858	429	429
1.771	0.889	0.535	0.535	0.541	0.490	0.455

$N = 13$

0	1	2	3	4	5	6	7
176	265	33	231	−33	−407	−110	−11
66	132	25	396	132	308	33	0
−4	42	18	460	222	738	128	9
−41	−12	12	444	251	932	182	16
−52	−37	7	369	233	939	202	21
−44	−40	3	256	182	808	195	24
−24	−28	0	126	112	588	168	25
1	−8	−2	0	37	328	128	24
24	13	−3	−101	−29	77	82	21
38	28	−3	−156	−72	−116	37	16
36	30	−2	−144	−78	−202	0	9
11	12	0	−44	−33	−132	−22	0
−44	−33	3	165	77	143	−22	−11
143	364	91	2 002	1 001	4 004	1 001	143
1.526	0.853	0.524	0.479	0.501	0.484	0.441	0.418

$N = 15$

0	1	2	3	4	5	6	7	8
1 456	2 059	8 008	44 044	2 002	−19 201	−28 756	−27 846	−78
676	1 144	5 833	52 624	17 017	19 604	8 879	1 404	−13
136	484	4 048	54 709	25 762	44 294	34 244	22 599	42
−199	44	2 618	51 524	29 252	57 004	49 054	36 684	87
−364	−211	1 508	44 294	28 502	59 869	55 024	44 604	122
−394	−316	683	34 244	24 527	55 024	53 869	47 304	147
−324	−308	108	22 599	18 342	44 604	47 304	45 729	162
−189	−216	−252	10 584	10 962	30 744	37 044	40 824	167
−24	−81	−432	−576	3 402	15 579	24 804	33 534	162
136	64	−467	−9 656	−3 323	1 244	12 299	24 804	147
256	184	−392	−15 431	−8 198	−10 126	1 244	15 579	122
301	244	−242	−16 676	−10 208	−16 396	−6 646	6 804	87
236	209	−52	−12 166	−8 338	−15 431	−9 656	−576	42
26	44	143	−676	−1 573	−5 096	−6 071	−5 616	−13
−364	−286	308	19 019	11 102	16 744	5 824	−7 371	−78
1 365	3 060	21 420	278 460	139 230	278 460	278 460	278 460	1 105
1.356	0.821	0.522	0.443	0.458	0.464	0.440	0.405	0.389

Appendix D

Computer programs

This appendix provides a brief summary of the time series programs that are available in the major statistical software packages and those that were used in this book. For a broad survey of available software, see Beaumont *et al.* (1985), Mahmoud *et al.* (1986).

In the table below, the term *standard univariate* is taken to include plotting facilities, the *ACF* and *PACF*, model estimation (usually by conditional least squares, but often with backcasting available and sometimes *ML* as an option), and forecasting. The term *general forecasting* refers to methods such as *EWMA*, Holt–Winters etc., discussed in Chapter 8.

References

Beaumont, C., Mahmoud, E. and McGee, V. E. (1985). Microcomputer forecasting software: a survey. *J. Forecasting*, **4**, 305–11.

Mahmoud, E., Rice, G., McGee, V. E. and Beaumont, C. (1986). Mainframe multipurpose forecasting software: a survey. *J. Forecasting*, **5**, 127–137.

	BMDP	SAS	SPSS-X	MINITAB	AUTOBOX	RATS	STAMP
Mainframe	*	*	*	*			
PC	*	*	*	*	*	*	*
Standard univariate	*	*	*	*	*	*	*
General forecasting		*			*		
Intervention analysis	*	*			*	*	*
Transfer functions	*	*			*	*	(1)
Structural models/ state space		*				*	*
Automatic model selection		(2)			*		*
Simulation of time series					*	*	*
Multiple series		(2)			(3)	*	
Spectrum analysis univariate/cross	*	(4)				*	
Complex demodulation	*						

Notes:
(1) Models with explanatory variables included in structural framework.
(2) State-space analysis only.
(3) Additional module MTS (multiple time series).
(4) Both covariance and fast Fourier transform methods are available.
In addition to the above, the package AUTOCAST provides automatic model selection from a set of exponential smoothing models, using a minimum variance criterion.

Addresses

BMDP Statistical Software, 1964 Westwood Boulevard, Suite 202, Los Angeles, CA 90025, USA.
SAS Institute Inc, SAS Circle, Box 8000, Cary, NC 27511, USA.
SPSS Inc, Suite 3000, 444 North Michigan Avenue, Chicago IL 60611, USA.
MINITAB Inc. 3081 Enterprise Drive, State College, PA 16801, USA.
AUTOBOX Automatic Forecasting Systems Inc. PO Box 563, Hatboro. PA 19040, USA
RATS VAR Econometrics, PO Box 1818, Evanston, IL 60204, USA.
STAMP ESRC Centre in Economic Computing, London School of Economics. Houghton Street, London WC2A 2AE. England.
AUTOCAST Levenbach Associates, Morristown, NJ, USA.

References

Abraham, B. and Ledolter, J. (1986). Forecast functions implied by autoregressive integrated moving average models and other related forecast procedures. *Int. Statist. Review*, **54**, 51–66.

Akaike, H. (1974). Markovian representation of stochastic processes and its application to the analysis of autoregressive moving average processes. *Ann. Inst. Statist. Math.*, **26**, 363–387.

Ali, M. M. (1984). An approximation to the null distribution and power of the Durbin–Watson statistic. *Biometrika*, **71**, 253–261.

Almon, S. (1965). The distributed lag between capital appropriations and expenditures. *Econometrica*, **30**, 178–196.

Ameen, J. and Harrison, P. J. (1984). Discounted weighted estimation. *J. Forecasting*, **3**, 285–296.

Anderson, O. D. (1975). On a paper by Davies, Pate and Frost concerning maximum autocorrelations for moving average processes. *Austr. J. Statistics*, **17**, 87–90.

Anderson, O. D. (1980). Serial dependence properties of linear processes. *J. Operational Research Society*, **31**, 905–917.

Anderson, R. L. (1942). Distribution of the serial correlation coefficient. *Ann. Math. Statist.*, **13**, 1–13.

Anderson, T. W. (1971). *The Statistical Analysis of Time Series*. Wiley, New York.

Anselin, L. (1988). *Spatial Econometrics: Methods and Models*. Kluwer, Dordrecht.

Ansley, C. (1979). An algorithm for the exact likelihood of a mixed autoregressive-moving average process. *Biometrika*, **66**, 59–66.

Ansley, C. F. and Kohn, R. (1983). Exact likelihood of vector autoregressive moving average process with missing or aggregated data. *Biometrika*, **70**, 275–278.

Armstrong, J. S. (1985). *Long-range Forecasting from Crystal Ball to Computer*, 2nd Edition. Wiley, New York.

Armstrong, J. S. and Lusk, E. J. (1983). Research on the accuracy of alternative extrapolation models: analysis of a forecasting competition through open peer review (with discussion). *J. Forecasting*. **2**, 259–311.

Aroian, L. A. (1980). Time series in m-dimensions: definition, problems and prospects. *Commun. Statist. B.*, **9**, 453–465.

Bachelier, L. (1900). Theorie de la speculation. *Annales Science Ecole Norm. Sup.* Paris, series 3. **17**, 21–86.

Bartlett, M. S. (1946). On the theoretical specification and sampling properties of autocorrelated time series. *Suppl. J. R. Statist. Soc.*, **8**, 27–41, 85–97.

Bartlett, M. S. (1950). Periodogram analysis and continuous spectra. *Biometrika*, **37**, 1–16.

Bartlett, M. S. (1978). *An Introduction to Stochastic Process, with special reference to Methods and Applications*, 3rd edition. Cambridge University Press, London.

Bass, F. M. (1969). A new product growth model for consumer durables. *Management Science*, **15**, 215–227.

Bates, J. and Granger C. W. J. (1969). The combination of forecasts. *Operational Reserch Quarterly*, **20**, 451–468.

Beguin, J. M., Gourieroux, C. S. and Monfort, A. (1981). Identification of a mixed autoregressive moving average process: the corner method. In *Time Series* (O. D. Anderson, ed.) pp. 423–436. North Holland, Amsterdam.

Bell, W. R. (1983). A computer program for detecting outliers in time series. Proceedings, *ASA*, Business and Economic Statistics Section. 634–639.

Bell, W. R. and Hillmer, S. C. (1983a). Issues involved with the seasonal adjustment of time series (with discussion). *J. Business Econ. Statist.*, **4**, 291–349.

Bell, W. R. and Hillmer, S. C. (1983b). Modelling time series with calendar variation. *J. Amer. Statist. Ass.*, **78**, 526–534.

Bennett, R. J. (1979). *Spatial Time Series*. Pion, London.

Beveridge, W. H. (1921). Weather and harvest cycles. *Econ. J.*, **31**, 429–452.

Bhansali, R. J. (1979). A mixed spectrum analysis of the lynx data. *J. Roy. Statist. Soc. A*, **142**, 199–209.

Bhansali, R. J. (1983). A simulation study of autoregressive and window estimators of the inverse autocorrelation function. *Appl. Statist.*, **32**, 141–149.

Bhansali, R. J. (1986). A derivation of the information criteria for selecting autoregressive models. *Adv. Applied Prob.*, **18**, 360–387.

Bhansali, R. J. and Karavellas, D. (1983). *Wiener filtering (with emphasis on frequency domain approaches)*. In Brillinger, D. R. and Krishnaiah, P. R. (eds) q.v. pp 1–19.

Bhattacharya, M. N. (1982). Lydia Pinkham data remodelled. *J. Time Series Analysis*. **3**, 81–102.

Birkhoff, G. D. (1931). Proof of the ergodic theorem. *Proc. Nat. Acad. Sci.*, **17**, 650–665.

Blackman, R. B. and Tukey, J. (1958). *The Measurement of Power Spectra from the Point of View of Communication Engineering*. Dover, New York.

Bliss C. I. (1958). *Periodic Regression in Biology and Climatology*. Bulletin No. 615, Connecticut Agricultural Experiment Station, New Haven.

Box, G. E. P. and Cox, D. R. (1964). An analysis of transformations (with discussion) *J. Roy. Statist. Soc. B.*, **26**, 211–252.

Box, G. E. P. and Jenkins, G. M. (1976). *Time Series Analysis, Forecasting and Control*. Holden Day. (McGraw-Hill Book Co. New York and Maidenhead, Eng.) Revised Edition.

Box, G. E. P. and Newbold, P. (1971). Some comments on a paper of Coen, Gomme and Kendall. *J. R. Statist., Soc. A.*, **134**, 229–240.

Box, G. E. P. and Pierce, D. A. (1970). Distribution of residual autocorrelations in autoregressive-integrated moving average time series models. *J. Amer. Statist. Ass.*, **70**, 1509–26.

Box, G. E. P. and Tiao, G. C. (1975). Intervention analysis with application to economic and environmental problems., *J. Amer. Statist. Ass.*, **70**, 70–79.

Box, G. E. P. and Tiao, G. C. (1977). A canonical analysis of multiple time series. *Biometrika*, **64**, 355–366.

Brillinger, D. R. (1965). An introduction to polyspectra. *Ann. Math. Statist.*, **36**, 1351–1374.

Brillinger, D. R. (1981). *Time Series, Data Analysis and Theory.* Holt, Rinehart and Winston, New York.

Brillinger, D. R. and Krishnaiah, P. R. (1983, eds). Time series in the Frequency Domain. *Handbook of Statistics, no. 3.* North Holland, Amsterdam.

Brillinger, D. R. and Rosenblatt, M. (1967). Asymptotic theory of *k*th order spectra. In *Spectral Analysis of Time Series* (B. Harris, ed.) pp. 153–188. Wiley, New York.

Brown, R. G. (1963). *Smoothing, Forecasting and Prediction.* Prentice-Hall, Englewood Cliffs, N.J.

Brown, R. L., Durbin, J. and Evans, J. M. (1975). Techniques for testing the constancy of regression relationships over time (with discussion). *J. Roy. Statist. Soc. B.*, **37**, 149–192.

Burman, J. P. (1965). Moving seasonal adjustment of economic time series. *J. R. Statist. Soc., A.*, **128**, 534–558.

Burman, J. P. (1979). Seasonal adjustment – a survey. *Forecasting, Studies in Management Science*, **12**, 45–57. (S. Makridakis and S. C. Wheelwright, eds) Elsevier, Amsterdam.

Burman, J. P. (1980). Seasonal adjustment by signal extraction. *J. Roy. Statist. Soc. A.*, **143**, 321–337.

Burn, D. A. (1985). *The Effects of Time Series Outliers on the Sample Autocorrelation Coefficient.* Penn. State University. PhD Thesis.

Chan, H., Hayya, J. C. and Ord. J. K. (1977). A note on trend removal methods: the case of polynomial regression versus variate differencing. *Econometrica*, **45**, 737–744.

Chanda, K. C. (1962). On bounds of serial correlations. *Ann. Math. Statist.*, **33**, 1457–60.

Chatfield, C. (1978). The Holt-Winters forecasting procedure. *Appl. Statist.*, **28**, 264–279.

Chatfield, C. (1979). Inverse autocorrelations. *J. Roy. Statist. Soc. A.*, **142**, 363–377.

Chatfield, C. and Prothero, D. L. (1973). Box–Jenkins seasonal forecasting: problems in a case study (with discussion). *J. Roy. Soc. A.*, **136**, 295–336.

Cleveland, W. P. and Tiao, G. C. (1976). Decomposition of seasonal time series: a model for the Census X-II program. *J. Amer. Statist. Ass.*, **71**, 581–587.

Cleveland, W. S. (1972). The inverse autocorrelations of a time series and their applications. *Technometrics*, **14**, 277–293.

Cleveland, W. S. (1983). *Seasonal calender adjustment*. In Brillinger, D. R. and Krishnaiah, P. R. (eds) qv pp. 39–72.

Cleveland, W. S. and Devlin, S. J. (1982). Calendar effects in monthly time series: modelling and adjustment. *J. Amer. Statist. Ass.*, **77**, 520–528.

Cleveland, W. S., Devlin, S. J. *et al.* (1981). *The SABL Computer Package: Users Manual*. Computing Information Library, Bell Labs., Murray Hill, New Jersey.

Cliff, A. D. and Ord, J. K. (1981). *Spatial Processes: Models and Applications*. Pion, London.

Cochrane, D. and Orcutt, G. H. (1949). Application of least-squares regression to relationships containing autocorrelated error terms. *J. Amer. Statist. Ass.*, **44**, 32–61.

Coen, P. G., Gomme, E. D. and Kendall, M. G. (1969). Lagged relationships in economic forecasting (with discussion). *J. R. Statist. Soc. A.*, **132**, 133–163.

Cook, R. D. and Weisberg, S. (1982). *Residuals and Influence in Regression*. Chapman and Hall, London.

Cooley, J. W., Lewis, P. A. W. and Welch, P. D. (1970). Historical notes on the fast Fourier transform. *IEEE Trans. Electoacoustics*, **AU-15**, 76–79.

Cooper, J. P. and Nelson, C. R. (1975). The ex-ante prediction performance of the St Louis and FRB-MIT-Penn econometric models and some results on composite predictors. *J. Money, Credit and Banking*, **7**, 1–32.

Cowden, D. J. (1962). *Weights for fitting polynomial secular trends*. Technical paper no. 4. School of Business Administration, University of N. Carolina, Chapel Hill, N. Carolina.

Cox, D. R. and Miller, H. D. (1968). *The Theory of Stochastic Processes*. Methuen, London.

Dagum, E. (1975). Seasonal factor forecasts from ARIMA models. *Bull. Int. Statist. Inst.*, **46**, 203–216.

Dagum, E. B. (1982). The effect of asymetric filters on seasonal factor revisions. *J. Amer. Statist. Ass.*, **77**, 732–738.

Daniell, P. J. (1946). Discussion on Symposium on Autocorrelation in Time Series. *Suppl., J. R. Statist. Soc.*, **8**, 88–90.

Daniels, H. E. (1956). The approximate distribution of serial correlation coefficients. *Biometrika, 43*, 169–185.

De Jong, P. (1976). The recursive fitting of autoregressions. *Biometrika, 63*, 525–530.

Dewey, E. R. (1963). *The 18.2 year cycle in immigration, USA*. 1820–1962. Foundation for the Study of Cycles, Inc. Pittsburgh, PA.

Dickey, D. A. (1975). *Hypothesis Testing for Nonstationary Time Series*. Dept. of Statistics, Iowa State University, Ames, Iowa.

Dickey, D. A. and Fuller, W. A. (1979). Distribution of the estimators for autoregressive time series with a unit root. *J. Amer. Statist. Ass.*, **74**, 427–431.

Dickey, D. A., Hasza, D. P. and Fuller, W. A. (1984). Testing for unit roots in seasonal time series. *J. Amer. Statist. Ass.*, **79**, 355–367.

Dixon, W. J. (1944). Further contribution to the problem of serial correlation. *Ann. Math. Statist.*, **15**, 119–144.

Duncan, D. B. and Horn, S. D. (1972). Linear dynamic recursive estimation from the viewpoint of regression analysis. *J. Amer. Statist. Ass.*, **67**, 815–821.

Durbin, J. (1959). Efficient estimation of parameters in moving-average models. *Biometrika*, **46**, 306–316.

Durbin, J. (1960). Estimation of parameters in time-series regression models. *J. R. Statist. Soc. B.*, **22**, 139–153.

Durbin, J. (1963). *Trend elimination for the purpose of estimating seasonal and periodic components of time series*. In Rosenblatt, M. (ed.) q.v. pp. 3–16.

Durbin, J. (1970). Testing for serial correlation in least-squares regression when some of the regressors are lagged dependent variables. *Econometrica*, **38**, 410–421.

Durbin, J. (1980a). Approximation for densities of sufficient estimators. *Biometrika*, **67**, 311–333.

Durbin, J. (1980b). The approximate distribution of partial serial correlation coefficients calculated from residuals from regression on Fourier series. *Biometrika*, **67**, 335–349.

Durbin, J. and Murphy, M. J. (1975). Seasonal adjustment based on a mixed additive – multiplicative model. *J. Roy. Statist. Soc. A.*, **138**, 385–410.

Durbin, J. and Watson, G. S. (1950, 1951, 1971). Testing for serial correlation in least-squares regression. *Biometrika*, **37**, 409–428; **38**, 159–178; **58**, 1–19.

Efron, B. (1982). *The Jackknife, the Bootstrap and Other Resampling Plans*. Soc. Industrial and Appl. Mathematics, Philadelphia.

Elton, C. and Nicholson, M. (1942). The ten year cycle in numbers of lynx in Canada. *J. Animal Ecology*, **11**, 215–244.

Fildes, R. (1983). An evaluation of Bayesian forecasting. *J. Forecasting*, **2**, 137–150.

Findley, D. F. (1984). On some ambiguities associated with the fitting of ARIMA models to time series. *J. Time Series Analysis*, **5**, 213–225.

Findley, D. F. (1986). On bootstrap estimates of forecast mean square errors for autoregressive processes. In D. M. Allen (ed.) *Computer Science and Statistics; Proceedings of the 17th symposium on the Interface*. North Holland, Amsterdam. pp. 11–17.

Foster, F. G. and Stuart, A. (1954). Distribution-free tests in time series based on the breaking of records. *J. R. Statist. Soc. B.*, **16**, 1–22.

Friedman, M. (1937). The use of ranks to avoid the assumption of normality implicit in the analysis of variance. *J. Amer. Statist. Ass.*, **32**, 675–701.

Fuller, W. A. (1976). *Introduction to Statistical Time Series*. Wiley, New York.

Fuller, W. A. (1985). *Nonstationary autoregressive time series*. In Hannan, E. J., Krishniah, P. R. and Rao, M. M. (eds) q.v. pp. 1–23.

Fuller, W. A. and Hasza, D. P. (1980). Predictors for the first-order autoregressive process. *J. Econometrics*, **13**, 139–157.

Funkhauser, H. G. (1936). *A note on a 10th Century graph*. Osiris. 1. Bruges.

Gabr, M. M. and Subba Rao, T. (1981). The estimation and prediction of subset bilinear time series models with applications. *J. Time Series Analysis*, **2**, 155–171.

Gardner, E. S. (1985). Exponential smoothing: The state of the art (with discussion). *J. Forecasting*, **4**, 1–38.

Gardner, E. S. (1988). Simple method of computing prediction intervals for time series forecasts. *Management Science*, **34**, 541–546.

Gardner, E. S. and McKenzie, E. (1989). Model identification in exponential smoothing. *J. Operational Research Society*, **40**.

Gardner, G., Harvey, A. C. and Phillips, G. D. A. (1980). An algorithm for exact maximum likelihood estimation of autoregressive-moving average models by means of Kalman filtering. *Appl. Statist.*, **29**, 311–322.

Gleissberg, W. (1945). Eine Aufgabe der Kombinatorik in der Wahrschein-lichkeitsrechmung. *Univ. Istanbul Rev. Fac. Sci. A.*, **10**, 25–35.

Granger, C. W. J. (1963). The effect of varying month-length on the analysis of economic time series. *L'Industria*, **1**, 41–53.

Granger, C. W. J. (1981). Some properties of time series data and their use in econometric model specification. *J. Econometrics*, **16**, 121–130.

Granger, C. W. J. and Andersen, A. P. (1978). *An Introduction to Bilinear Time Series Models*. Vandenhoeck and Ruprecht, Gottingen.

Granger, C. W. J. and Hatanaka, M. (1964). *Spectral Analysis of Economic Time Series*. Princeton Univ. Press, Princeton, NJ.

Granger, C. W. J. and Hughes, A. O. (1971). A new look at some old data: the Beveridge wheat price series. *J. R. Statist. Soc. A.*, **134**, 413–428.

Granger, C. W. J. and Joyeux, R. (1980). An introduction to long-memory time series models and fractional differencing. *J. Time Series Analysis*, **1**, 15–29.

Granger, C. W. J. and Morris, M. J. (1976). Time series modelling and interpretation. *J. Roy. Statist. Soc. A.*, **139**, 246–257.

Gray, H. L., Kelley, G. D. and McIntire, D. D. (1978). A new approach to ARMA modelling. *Commun. Statist. B.*, **7**, 1–78.

Grether, D. M. and Nerlove, M. (1970). Some properties of optimal seasonal adjustment. *Econometrica*, **38**, 682–703.

Griliches, Z. (1967). Distributed lags: a survey. *Econometrica*, **35**, 16–49.

Gudmundsson, G. (1971). Time series analysis of imports, exports and other economic variables. *J. R. Statist. Soc. A.*, **134**, 383–412.

Guyon, X. (1982). Parameter estimation for a stationary process on a d-dimensional lattice. *Biometrika*, **69**, 95–105.

Hannan, E. J. (1961). Testing for a jump in the spectral function. *J. Roy. Statist. Soc. B.*, **23**, 394–404.

Hannan, E. J. (1969). The identification of vector mixed autoregressive moving average systems. *Biometrika*, **56**, 223–225.

Hannan, E. J. (1970). *Multiple Time Series*. Wiley, New York.

Hannan, E. J., Krishnaiah, P. R. and Rao, M. M. (1985 eds). Time Series in the Time Domain. *Handbook of Statistics, No. 5*, North Holland, Amsterdam.

Hannan, E. J. and Rissanen, J. (1982). Recursive estimation of mixed autoregressive moving average order. *Biometrika*, **69**, 81–94.

Harrison, P. J. (1965). Short term sales forecasting. *Appl. Statist.*, **14**, 102–139.

Harrison, P. J. and Stevens, C. F. (1976). Bayesian forecasting (with discussion) *J. Roy. Statist. Soc. B.*, **38**, 205–247.

Harvey, A. C. (1981). *Time Series Models*. P. Allan, Oxford and Halstead, New York.

Harvey, A. C. (1984). A unified view of statistical forecasting procedures. *J. Forecasting,* **3**, 245–275.

Harvey, A. C. (1985). Trends and cycles in macroeconomic time series. *J. Business & Econ. Statist.,* **3**, 216–227.

Harvey, A. C. and Durbin, J. (1986). The effects of seat belt legislation on British road casualties: a case study in structural time series modelling (with discussion). *J. Roy. Statist. Soc. A.,* **149**, 187–227.

Harvey, A. C. and Peters, S. (1984). *Estimation Procedures for Time Series Models.* LSE Econometrics Programme Discussion Paper, No. A44.

Harvey, A. C. and Phillips, G. D. A. (1979). Maximum likelihood estimation of regression models with autoregressive moving average disturbances. *Biometrika,* **66**, 49–58.

Harvey, A. C. and Pierse, R. G. (1984). Estimating missing observations in economic time series. *J. Amer. Statist. Ass.,* **79**, 125–131.

Harvey, A. C. and Todd, P. H. J. (1983). Forecasting economic time series with structural and Box–Jenkins models: A case study (with discussion). *J. Business Econ. Statist.,* **1**, 299–315.

Hasan, T. (1983). *Complex demodulation.* In Brillinger, D. R. and Krishnaiah, P. R. (eds) q.v. pp. 125–156.

Haugh, L. D. and Box, G. E. P. (1977). Identification of dynamic regression (distributed lag) models connecting two time series. *J. Amer. Statist. Ass.,* **72**, 121–130.

Heeler, R. M. and Hustad, T. P. (1980). Problems in predicting new product growth for consumer durables. *Management Science,* **26**, 1007–1020.

Hendry, D. F. (1986). Comment on 'Modelling the persistence of conditional variances'. *Econometric Reviews,* **5**, 63–69.

Hendry, D. F. and Richard, J. F. (1983). The econometric analysis of economic time series (with discussion). *Int. Statist. Review,* **51**, 111–163.

Hill, G. and Fildes, R. (1984). The accuracy of extrapolation methods: and automatic Box–Jenkins package, SIFT. *J. Forecasting,* **3**, 319–323.

Hillmer, S. C., Bell, W. R. and Tiao, G. C. (1983). Modeling considerations in the seasonal adjustment of economic time series (with discussion). In *Applied Time Series Analysis of Economic Data.* US Bureau of Census, Washington DC. pp. 74–124.

Hillmer, S. C. and Tiao, G. C. (1979). Likelihood function of stationary multiple autoregressive moving average models. *J. Amer. Statist. Ass.,* **74**, 652–660.

Holbert, D. and Son, M.-S. (1986). Bootstrapping a time series model: some empirical results. *Commun. Statist. A.,* **15**, 3669–3691.

Holt, C. C. (1957). *Forecasting seasonals and trends by exponentially weighted moving averages.* Carnegie Inst. Tech. Res. Mem. No. 52.

Hosking, J. R. M. (1981). Fractional differencing. *Biometrika,* **68**, 165–176.

Huber, P. (1981). *Robust Statistics.* Wiley, New York.

Jevons, W. S. (1879). *The Principles of Science.* London.

Johnston, J. (1984). *Econometric Methods.* (3rd edition). McGraw Hill, London.

Jones, R. H. (1985). *Time series with unequally spaced data.* In Hannan, E. J., Krishniah, P. R. and Rao, M. M. (eds) q.v. pp. 157–177.

Kalman, R. E. and Bucy, R. S. (1961). New results in linear filtering and prediction theory. *Trans. A.S.M.E. Journal of Basic Engineering. Series D.* **83**, 95–108.

Kendall, M. G. (1945). On the analysis of oscillatory time series. *J. R. Statist. Soc. A.*, **108**, 93–141.

Kendall, M. G. (1946). *Contributions to the Study of Oscillatory Time Series.* Cambridge University Press.

Kendall, M. G., Gibbons, J. D. (1990). *Rank Correlation Methods*, 5th edition. Edward Arnold, London.

Kendall, M. G. (1971). Studies in the history of probability and statistics, XXVI: The work of Ernst Abbe. *Biometrika*, **58**, 369–373.

Kendall, M. G., Stuart, A. and Ord. J. K. (1983). *The Advanced Theory of Statistics*, volume 3 (4th edition). Charles Griffin, London.

Kenny, P. B. and Durbin, J. (1982). Local trend estimation and seasonal adjustment of economic and social time series (with discussion). *J. Roy. Statist. Soc., A.*, **145**, 1–41.

Khintchine, A. Ya, (1932). Zu Birkhoffs Lösung des Ergodenproblems. *Math. Ann.*, **107**, 485–488.

King, P. D. (1956). Increased frequency of births in the morning hours. *Science*, **123**, 985–988.

Koot, R. S., Young, P. and Ord, J. K. (1992). *American and British political business cycles:* A time series approach. *J. Appl. Business Res.*, **8**, 36–41.

Ledolter, J. and Abraham, B. (1984). Some comments on the initialization of exponential smoothing. *J. Forecasting*, **3**, 79–84.

Leipnik, R. P. (1947). Distribution of the serial correlation coefficient in a circularly correlated universe. *Ann. Math. Statist.*, **18**, 80–87.

Levene, H. (1952). On the power function of tests of randomness based on runs up and down. *Ann. Math. Statist.*, **23**, 34–56.

Libert, G. (1984). The M-Competition with a fully automatic Box–Jenkins procedure. *J. Forecasting*, **3**, 325–328.

Litterman, R. B. (1986). Forecasting with Bayesin vector autoregressions: Five years of experience. *J. Business and Econ. Statist.*, **4**, 25–38.

Liu, L. M. and Hanssens, D. M. (1982). Identification of multiple-input transfer function models. *Communications in Statistics. A.*, **11**, 297–314.

Ljung, G. M. and Box, G. E. P. (1978). On a measure of lack of fit in time series models. *Biometrika*, **65**, 297–304.

Madow, W. G. (1945). Note on the distribution of the serial correlation coefficient. *Ann. Math. Statist.*, **16**, 308–310.

Makridakis, S. (1988). *A New Approach to Statistical Forecasting.* INSEAD, Fontainebleau Working Paper.

Makridakis, S., Anderson, A. *et al.* (1982). The accuracy of extrapolation (time series) methods: results of a forecasting competition. *J. Forecasting*, **1**, 111–153.

Makridakis, S., Anderson, A. *et al.* (1984). *The Forecasting Accuracy of Major Time Series Methods.* Wiley, Chichester.

Makridakis, S. and Hibon, M. (1979). Accuracy of forecasting: an empirical investigation (with discussion). *J. Roy. Statist. Soc. A.*, **142**, 97–145.

Mann, H. B. and Wald, A. (1943). On the statistical treatment of linear stochastic difference equations. *Econometrica*, **11**, 173–220.

Martin, R. D. (1983). *Robust-resistant spectral analysis*. In Brillinger D. R. and Krishnaiah, P. R. (eds) q.v. pp. 185–219.

Martin, R. D. and Yohai, V. J. (1985). *Robustness in time series and estimating ARMA models*. In Hannan E. J., Krishnaiah, P. R. and Rao, M. M. (eds) q.v. pp 119–155.

Martino, J. P. (1983). *Technological Forecasting for Decision Making*. North Holland, New York and Amsterdam.

McGee, V. (1986). The OWL: Software support for a model of argumentation. *Behaviour Research Methods, Instruments and Computers*. **18**, 108–117.

McKenzie, E. (1976). A comparison of some standard seasonal forecasting systems. *Statistician*, **25**, 3–14.

McKenzie, S. K. (1984). Concurrent seasonal adjustment with Census X-11. *J. Business & Econ. Statist.*, **2**, 235–249.

McNees, S. K. (1986). Forecasting accuracy of alternative techniques: A comparison of US macroeconomic forecasts. *J. Business & Econ. Statist.*, **4**, 5–15.

Moore, G. H. and Wallis, W. A. (1943). Time series significance tests based on signs of differences. *J. Amer. Statist. Ass.*, **38**, 153–164.

Moran, P. A. P. (1947). Some theorems on time series I. *Biometrika*, **34**, 281–291.

Moran, P. A. P. (1948). Some theorems on time series II. *Biometrika*, **35**, 255–260.

Moran, P. A. P. (1967). Testing for serial correlation with exponentially distributed variates. *Biometrika*, **54**, 395–401.

Morettin, P. A. (1984). The Levinson algorthim and its applications in time series analysis. *Int. Statist. Review*, **52**, 83–92.

Nelson, C. R. (1972). The prediction performance of the FRB-MIT-Penn model of the US economy. *American Econ. Review*, **62**, 902–917.

Nelson, C. R. and Kang, H. (1981). Spurious periodity in inappropriately detrended time series. *Econometrica*, **49**, 741–752.

Nerlove, M. (1964). Spectral analysis of seasonal adjustment procedures. *Econometrica*. 241–286.

Newbold, P. (1974). The exact likelihood function for a mixed autoregressive moving average process. *Biometrika*, **61**, 423–426.

Newbold, P. (1981). Some recent developments in time series analysis. *Int. Statist. Review*, **49**, 53–66.

Newbold, P. (1984). Recent developments in time series analysis, II. *Int. Statist. Review*, **52**, 183–192.

Newbold, P. and Granger, C. W. J. (1974). Experience with forecasting univariate time series and the combination of forecasts. *J. Roy. Statist. Soc. A.*, **137**, 131–165.

Nicholls, D. F. and Pagan, A. R. (1985). *Varying coefficient regression*. In Hannan, E. J., Krishnaiah, P. R. and Rao, M. M. (eds) q.v. pp. 413–449.

Ord. J. K. (1979). Time series and spatial patterns in ecology. In *Spatial and Temporal Analysis in Ecology*. (R. M. Cormack and J. K. Ord, eds) pp. 1–94. International Cooperative Publishing House, Fairland, Maryland.

Ord. J. K. (1983). An alternative approach to the specification of multiple time series models. In *Time Series Analysis: Theory and Practice*. (O. D. Anderson, ed.) pp. 85–103. Elsevier, Amsterdam.

Ord, J. K. (1988). Future developments in forecasting: The time series connexion. *Int. J. Forecasting*, **4**, 389–401.

Ord, J. K. and Young, P. (1989). *Model Building for Technological Forecasting*. Penn State University, Dept. of Management Science Working Paper.

Pagano, M. (1972). An algorithm for fitting autoregressive schemes. *Appl. Statist.*, **21**, 274–281.

Palda, K. S. (1964). *The Measurement of Cumulative Advertising Effects*. Prentice-Hall: Englewood Cliffs, NJ.

Parzen, E. (1961). Mathematical considerations in the estimation of spectra: Comments on the discussions of Messers Tukey and Goodman. *Technometrics*, **3**, 167–190; 232–234.

Parzen, E. (1969). Multiple time series modelling. In *Multivariate Analysis II* (P. R. Krishnaiah, ed.) pp. 389–409. Academic Press, New York and London.

Parzen, E. (1983). ARARMA models for time series analysis and forecasting. *J. Forecasting*, **1**, 67–82.

Parzen, E. (1983). *Autoregressive spectral estimation*. In Brillinger, D. R. and Krishnaiah, P. R. (eds) q.v. pp. 221–257.

Pemberton, J. and Tong, H. (1983). *Threshold autoregression and some frequency-domain characteristics*. In Brillinger, D. R. and Krishnaiah, P. R. (eds) q.v. pp. 249–273.

Pfeiffer, P. E. and Deutsch, S. J. (1980). A three-stage iterative procedure for space-time modelling. *Technometrics*, **22**, 35–47.

Phillips, P. C. B. (1978). Edgeworth and saddlepoint approximations in the first order noncircular regression. *Biometrika*, **65**, 91–98.

Pierce, D. A. (1972). Residual correlations and diagnostic checking in dynamic disturbance time series models. *J. Amer. Statist. Ass.*, **67**, 636–640.

Pierce, D. A. (1983). Seasonal adjustment of monetary aggregates. *J. Business & Econ. Statist.*, **1**, 37–42.

Plackett, R. L. (1950). Some theorems in least squares. *Biometrika*, **37**, 149–157.

Playfair, W. (1821). *A letter on our Agricultural Distress*. London.

Poirer, D. J. (1978). The effect of the first observation in regression models with first order autoregressive disturbances. *Appl. Statist.*, **27**, 67–68.

Priestley, (1962a,b). The analysis of stationary processes with mixed spectra I, II. *J. Roy. Statist. Soc. B.*, **24**, 215–233 and 511–529.

Priestley, M. B. (1965). Evolutionary spectra and non-stationary processes. *J. Roy. Statist. Soc. B.*, **27**, 204–237.

Priestley, M. B. (1981). *Spectral Analysis and Time Series*, 2 volumes. Academic Press, London.

Quenouille, M. H. (1947). A large sample test for the goodness of fit of autoregressive schemes. *J. R. Statist. Soc.*, **110**, 123–129.

Quenouille, M. H. (1956). Notes on bias in estimation. *Biometrika*, **43**, 353–360.

Quenouille, M. H. (1968). *The Analyis of Multiple Time Series*. Charles Griffin, London.

Quenouille, M. H. (1958). Discrete autoregressive schemes with varying time intervals. *Metrika*, **1**, 21–27.

Reid, D. J. (1971). *Forecasting in action: A comparison of forecasting techniques in economic time series*. Joint Conference of O. R. Society's Group on Long Range Planning and Forecasting.

Reilly, D. P. (1981). *Recent experiences with an automatic Box–Jenkins modelling algorithm*. Proceedings, ASA, Statistical Computing Section. pp. 91–99.

Reilly, D. P. (1984). *Automatic intervention detection system*. Proceedings ASA Statistical Computing Section. pp. 539–542.

Reilly, D. P. and Dooley, K. (1987). Experiences with an automatic transfer function algorithm. In R. M. Heiberger (ed.). *Computer Science & Statistics*. Proceedings of the 19th Symposium on the Interface. North Holland pp. 128–135.

Ripley, B. D. (1981). *Spatial Statistics*. Wiley, London.

Roberts, S. A. (1982). A general class of Holt–Winters type forecasting models. *Management Science*, **28**, 808–820.

Robinson, P. M. (1983). *Review of various approaches to power spectrum estimation*. In Brillinger, D. R. and Krishnaiah, P. R. (eds) q.v. pp. 343–368.

Rosenblatt, M. (1963). *Proceedings of the Symposium of Time Series Analysis* held at Brown University. 1962. Wiley, New York and Chichester, England.

Sargan, J. D. (1953). An approximate treatment of the properties of the correlogram and periodogram. *J. R. Statist. Soc. B.*, **15**, 140–152.

SAS (1984). *ASA/ETS User's Guide*, version 5 edition. SAS Institute, Cary, N. Carolina.

Schuster, A. (1906). On the periodicities of sunspots. *Phil. Trans. Roy. Soc. A.*, **206**, 69–100.

Schwartz, G. (1978). Estimating the dimensions of a model. *Ann. Statist.*, **6**, 461–464.

Shiskin, J. (1967). *The X-11 Variant of the Census Method II Seasonal Adjustment Program*. Technical Paper No. 15, US Bureau of the Census.

Slutzky, E. (1927). The summation of random causes as tne sources of cyclic processes. (Russian). English trans., 1937. *Econometrica*, **5**, 105–146.

Smith, A. F. M. and West, M. (1983). Monitoring renal transplants: an application of the multipass Kalman filter. *Biometrics*, **39**, 867–878.

Smith, R. L. and Miller, J. E. (1986). A non-Gaussian state-space model and application to prediction of records. *J. Roy. Statist. Soc. B.*, **48**, 79–88.

Spencer, J. (1904). On the graduation of the rates of sickness and mortality. *J. Inst. Act.*, **38**, 334–347.

Spliid, H. (1983). A fast estimation method for the vector autoregressive moving average model with exogenous variables. *J. Amer. Statist. Ass.*, **78**, 843–849.

Stigler, S. M. (1986). *The History of Statistics: The Measurement of Uncertainty Before 1900*. Harvard University Press, Boston.

Stuart, A. and Ord, J. K. (1991). *Kendall's Advanced Theory of Statistics*, volume 2, 4th edition, Edward Arnold, London.

Subba Rao, T. (1981). On the theory of bilinear time series models. *J. Roy. Statist. Soc. B.*, **43**, 244–255.

Subba Rao, T. (1983). *The bispectral analysis of nonlinear stationary time series with reference to bilinear time series models.* In Brillinger, D. R. and Krishnaiah, P. R. (eds) q.v. pp 293–319.

Sweet, A. L. (1985). Computing the variance of the forecasting error for the Holt–Winters seasonal models. *J. Forecasting*, **4**, 235–243.

Texter, P. A. and Ord, J. K. (1989). Automatic selection of forecasting methods for nonstationary series. *Int. J. Forecasting*, **5**, 209–215.

Texter Geriner, P. A. and Ord, J. K. (1991). Automatic forecasting using explanatory variables: a comparative study. *Int. J. Forecasting*, **7**, 127–140.

Theil, H. (1966). *Applied Economic Forecasting.* North Holland, Amsterdam.

Theil, H. (1971). *The Principles of Econometrics.* John Wiley & Sons, New York, and Chichester, England.

Tiao, G. C. and Box, G. E. P. (1981). Modelling multiple time series with applications. *J. Amer. Statist. Ass.*, **76**, 802–816.

Tiao, G. C. and Tsay, R. S. (1983). Consistency properties of least squares estimates of autoregressive parameters in ARMA models. *Ann. Statist.*, **11**, 856–871.

Tiao, G. C. and Tsay, R. S. (1989). Model misspecification in multivariate time series (with discussion). *J. Roy. Statist. Soc. B.*, **51**, 157–213.

Tong, H. (1977). Some comments on the Canadian lynx data. *J. Roy. Statist. Soc. A.*, **140**, 432–436.

Tong, H. and Lim, K. S. (1980). Threshold autoregression, limit cycles and cyclical data (with discussion). *J. Roy. Statist. Soc. B.*, **42**, 245–292.

Tsay, R. S. and Tiao, G. C. (1984). Consistent estimates of autoregressive parameters and extended sample autocorrelation function for stationary and nonstationary ARMA models. *J. Amer. Statist. Ass.*, **79**, 84–96.

Velu, R. P., Reinsel, G. C. and Wichern, D. W. (1986). Reduced rank models for multiple time series. *Biometrika*, **73**, 105–118.

Wald, A. (1955). *Selected Papers in Statistics and Probability*, McGraw-Hill, New York.

Wallis, K. F. (1972). Testing for fourth-order autocorrelation in quarterly regression equations. *Econometrica*, **40**, 617–636.

Wallis, K. F. (1974). Seasonal adjustment and relations between variables. *J. Amer. Statist. Ass.*, **69**, 18–31.

Wallis, W. A. and Moore, G. H. (1941). A significance test for time series analysis. *J. Amer. Statist. Ass.*, **36**, 401–409.

Whittle, P. (1953). The analysis of multiple stationary time series. *J. R. Statist. Soc. B.*, **15**, 125–139.

Whittle, P. (1963). On the fitting of multivariate autoregressions, and the approximate canonical factorization of a spectral density matrix. *Biometrika*, **50**, 129–134.

Wichern, D. W. (1973). The behaviour of the sample autocorrelation function for an integrated moving average process. *Biometrika*, **60**, 235–239.

Wickens, M. R. (1972). *A comparison of alternative tests for serial correlation in the disturbances of equations with lagged dependent variables.* University of Bristol Working Paper.

Winkler, R. L. and Makridakis, S. (1983). The combination of forecasts. *J. Roy. Statist. Soc. A.*, **146**, 150–157.

Winters, P. R. (1960). Forecasting sales by exponentially weighted moving averages. *Management Science*, **6**, 324–342.

Working, H. (1960). Note on the correlation of first differences of averages in a random chain. *Econometrica*, **28**, 916–918.

Wright, D. J. (1986). Forecasting data published at irregular time intervals using an extension of Holt's method. *Management Science*, **32**, 499–510.

Young, P. and Ord. J. K. (1985). The use of discounted least squares in technological forecasting. *Technological Forecasting and Social Change*, **28**, 263–274.

Yule, G. Udny, (1927). On a method of investigating periodicities in disturbed series, with special reference to Wolfer's sunspot numbers. *Phil. Trans. A.*, **226**, 267–298.

Yule, G. Udny, (1971). *Statistical papers*: Selected by Alan Stuart and M. G. Kendall. Charles Griffin, London.

Author Index

Subject Index